Brazilian Agrarian Social Movements

Contradictions between impressive levels of economic growth and the persistence of poverty and inequality are perhaps nowhere more evident than in rural Brazil. While Brazil might appear to be an example of the potential harmony between large-scale, export-oriented agribusiness and small-scale family farming, high levels of rural resistance contradict this vision. In this volume, individual contributions from a variety of researchers across the field highlight seven key characteristics of contemporary Brazilian resistance that have broader resonance in the region and beyond: the growth of international networks, the changing structure of state–society collaboration, the deepening of territorial claims, the importance of autonomy, the development of alternative economies, continued opposition to dispossession and struggles over the meaning of nature. By analyzing rural mobilization in Brazil, this collection offers a range of insights relevant to rural contention globally. Together, these contributions increase our understanding of alternative agricultural production, large-scale development projects, education, race and political parties in the contemporary agrarian context.

This book was previously published as a special issue of *The Journal of Peasant Studies*.

Rebecca Tarlau is a postdoctoral scholar in education at Stanford University, affiliated with the Lemann Center for Educational Entrepreneurship and Innovation in Brazil. She received an MA and PhD in Social and Cultural Studies from the UC-Berkeley Graduate School of Education and a BA in Anthropology and Latin American Studies from the University of Michigan. Rebecca's research focuses on the relationship between states, social movements, and educational reform. Her scholarship engages in debates in the fields of political sociology, international and comparative education, critical pedagogy, global and transnational sociology, and social theory.

Anthony Pahnke is a visiting assistant professor in political science and environmental studies at St Olaf College, Northfield, Minnesota. He spent roughly two years in Brazil, researching state and MST practices in education, agrarian reform, and agricultural production. His interests extend beyond social movements to include political economy, state theory and qualitative methods.

T0346600

Brazilian Agrarian Social Movements

Edited by
Rebecca Tarlau and Anthony Pahnke

Routledge
Taylor & Francis Group

LONDON AND NEW YORK

First published 2016 by Routledge

2 Park Square, Milton Park, Abingdon, Oxfordshire OX14 4RN
711 Third Avenue, New York, NY 10017

Routledge is an imprint of the Taylor & Francis Group, an informa business

First issued in paperback 2018

British Library Cataloguing in Publication Data
A catalogue record for this book is available from the British Library

ISBN 13: 978-1-138-66568-2 (hbk)
ISBN 13: 978-1-138-39323-3 (pbk)

Typeset in Times New Roman
by RefineCatch Limited, Bungay, Suffolk

Publisher's Note
The publisher accepts responsibility for any inconsistencies that may have
arisen during the conversion of this book from journal articles to book chapters,
namely the possible inclusion of journal terminology.

Disclaimer
Every effort has been made to contact copyright holders for their permission to
reprint material in this book. The publishers would be grateful to hear from any
copyright holder who is not here acknowledged and will undertake to rectify
any errors or omissions in future editions of this book.

Contents

CONTENTS

Citation Information

The chapters in this book were originally published in *The Journal of Peasant Studies*, volume 42, issue 6 (November 2015). When citing this material, please use the original page numbering for each article, as follows:

Chapter 1
Understanding rural resistance: contemporary mobilization in the Brazilian countryside
Anthony Pahnke, Rebecca Tarlau and Wendy Wolford
The Journal of Peasant Studies, volume 42, issue 6 (November 2015), pp. 1069–1085

Chapter 2
Institutionalizing economies of opposition: explaining and evaluating the success of the MST's cooperatives and agroecological repeasantization
Anthony Pahnke
The Journal of Peasant Studies, volume 42, issue 6 (November 2015), pp. 1087–1107

Chapter 3
Rural unions and the struggle for land in Brazil
Clifford Andrew Welch and Sérgio Sauer
The Journal of Peasant Studies, volume 42, issue 6 (November 2015), pp. 1109–1135

Chapter 4
Engaging the Brazilian state: the Belo Monte dam and the struggle for political voice
Peter Taylor Klein
The Journal of Peasant Studies, volume 42, issue 6 (November 2015), pp. 1137–1156

Chapter 5
Education of the countryside at a crossroads: rural social movements and national policy reform in Brazil
Rebecca Tarlau
The Journal of Peasant Studies, volume 42, issue 6 (November 2015), pp. 1157–1177

Chapter 6
Learning as territoriality: the political ecology of education in the Brazilian landless workers' movement
David Meek
The Journal of Peasant Studies, volume 42, issue 6 (November 2015), pp. 1179–1200

Chapter 7

The Landless invading the landless: participation, coercion, and agrarian social movements in the cacao lands of southern Bahia, Brazil
Jonathan DeVore
The Journal of Peasant Studies, volume 42, issue 6 (November 2015), pp. 1201–1223

Chapter 8

The Brazilian quilombo: 'race', community and land in space and time
Ilka Boaventura Leite
The Journal of Peasant Studies, volume 42, issue 6 (November 2015), pp. 1225–1240

Chapter 9

Can urban migration contribute to rural resistance? Indigenous mobilization in the Middle Rio Negro, Amazonas, Brazil
Thaissa Sobreiro
The Journal of Peasant Studies, volume 42, issue 6 (November 2015), pp. 1241–1261

Chapter 10

Lula's assault on rural patronage: Zero Hunger, ethnic mobilization and the deployment of pilgrimage
Aaron Ansell
The Journal of Peasant Studies, volume 42, issue 6 (November 2015), pp. 1263–1282

Chapter 11

Managing transience: Bolsa Família and its subjects in an MST landless settlement
Gregory Duff Morton
The Journal of Peasant Studies, volume 42, issue 6 (November 2015), pp. 1283–1305

For any permission-related enquiries please visit:
http://www.tandfonline.com/page/help/permissions

Notes on Contributors

Aaron Ansell is a cultural anthropologist and assistant professor of religion and culture at Virginia Tech, USA. His ethnographic work on Northeast Brazil addresses issues of democracy, patronage, state policy and capitalism from the standpoint of linguistic and symbolic anthropology. Ansell's book, *Zero Hunger: Political culture and antipoverty policy in Northeast Brazil* (UNC Press, 2014) examines the Lula administration's usage of redistributive policies to dismantle municipal patron–client hierarchies. His current research focuses on poor people's engagement with Brazil's public healthcare system in light of emerging ideals of democratic access to state services.

Ilka Boaventura Leite is a professor in the Anthropology Department at the Federal University of Santa Catarina, Brazil. She is the founder of NUER (Research Nucleus in the Studies of Identities and Interethnic Relations). She was trained as a historian (Federal University of Minas Gerais, Brazil, 1980) and an anthropologist (University of São Paulo, Brazil, 1986). Her post-doctoral work was at the University of Chicago, USA (1997) and at the New University of Lisbon, Portugal (2007). Her research is situated in the areas of anthropology, the literature on travel, interethnic relations, afro-Brazilian ethnicity, culture and black identity, quilombos, and art and ethnicity.

Jonathan DeVore is a postdoctoral associate with the Program in Agrarian Studies at Yale University, USA. He has been conducting ethnographic research with diverse land reform and squatter communities in southern Bahia's cacao lands since 2002. His research traces the historical trajectory of these contemporary land rights movements by situating them in the context of ongoing struggles to realize the promise of freedom in Brazil's post-emancipation period. DeVore holds a PhD in sociocultural anthropology from the University of Michigan, USA, and he is currently preparing a monograph based upon his research.

Peter Taylor Klein is an assistant professor of sociology and environmental and urban studies at Bard College, USA. Broadly, his work focuses on development, associational life, relations between the state and civil society and environmental conflicts in both Brazil and the United States. His current research project examines local contestations around the construction of the Belo Monte hydroelectric facility in the Brazilian Amazon, analyzing citizen demands and state responses in the context of a rapidly changing social and environmental landscape. He is also co-author of *The Civic Imagination: Making a difference in American political life* (Routledge, 2014).

David Meek is an assistant professor of anthropology at the University of Alabama, USA. He is an environmental anthropologist, critical geographer and education scholar with an area specialization in Brazil. Meek theoretically grounds his research in a synthesis of political ecology, critical pedagogy and place-based education. His interests include sustainable agriculture, social movements and environmental education. Meek's research focuses on the relationships between public policies, economic incentives and educational processes within an agrarian reform settlement in the Brazilian Amazon. In a series of recent publications he has begun advancing a theoretical framework of the political ecology of education. This perspective illuminates how the reciprocal relations between political economic forces and pedagogical opportunities – from tacit to formal learning – affect the production, dissemination and contestation of environmental knowledge at various interconnected scales.

Gregory Duff Morton is a postdoctoral fellow in international affairs at the Watson Institute, Brown University, USA. He studies changing forms of labor in the Brazilian *sertão*. His research focuses particularly on Brazil's landless movement, the MST and on *Bolsa Família*, the world's largest conditional cash transfer.

Anthony Pahnke is a visiting assistant professor in political science and environmental studies at St Olaf College, USA. He spent roughly two years in Brazil, researching state and social movement practices in education, agrarian reform and agricultural production. His interests extend beyond social movements to include political economy, state theory and qualitative methods.

Sérgio Sauer is a professor of agrarian studies at the University of Brasília (Planaltina Campus) in the post-graduate programs on environment and rural development (Mader) and on sustainability of traditional people and land (MESPT). He was a visiting professor at the International Institute of Social Studies (ISS) in The Hague, where he researched land grabbing with a grant from Brazil's Coordinator for the Improvement of Higher Education Personnel (CAPES)/Science without Borders. He also holds a prestigious PQ fellowship from Brazil's National Council for Scientific and Technological Development (CNPq).

Thaissa Sobreiro is a doctoral candidate in the School of Natural Resources and Environment and the Tropical Conservation and Development Program at the University of Florida, USA. Her dissertation investigates how households and rural communities (re)negotiate their identities and livelihood options by engaging indigenous movements and migration to identify key resources for local development in Rio Negro, Amazonas State, Brazil. Her work is based on long-term fieldwork which she has been conducting in the region since 2006. Sobreiro's research is supported by the International Foundation for Science (IFS).

Rebecca Tarlau is a postdoctoral scholar in education at Stanford University, USA. Her research analyses the relationship between social movements, the state and education, contributing to debates about state–society relations, participatory governance, international and comparative education and Freirean pedagogies.

Clifford Andrew Welch is professor of contemporary Brazilian history at the Federal University of São Paulo (UNIFESP), Brazil. He also teaches in the postgraduate program on Latin American and Caribbean Territorial Development at São Paulo State University (UNESP), where he is a researcher in the Centre for Agrarian Reform Research, Study and Projects (NERA). In 2014, he completed a senior leave postdoctorate as Humanities Research Associate in History at the University of California – Santa Cruz, made possible by a grant from Brazil's CAPES.

Wendy Wolford is the Robert A. and Ruth E. Polson professor of development sociology at Cornell University, USA. Her research includes the political economy of development, social movements and resistance, agrarian studies, political ecology, land use, land reform and critical ethnography.

Understanding rural resistance: contemporary mobilization in the Brazilian countryside

Anthony Pahnke, Rebecca Tarlau and Wendy Wolford

Contradictions between impressive levels of economic growth and the persistence of poverty and inequality are perhaps nowhere more evident than in rural Brazil. While Brazil might appear to be an example of the potential harmony between large-scale, export-oriented agribusiness and small-scale family farming, high levels of rural resistance contradict this vision. In this introductory paper, we synthesize the literature on agrarian resistance in Brazil and situate recent struggles in Brazil within the Latin American context more broadly. We highlight seven key characteristics of contemporary Latin American resistance, which include: the growth of international networks, the changing structure of state–society collaboration, the deepening of territorial claims, the importance of autonomy, the development of alternative economies, continued opposition to dispossession, and struggles over the meaning of nature. We argue that by analyzing rural mobilization in Brazil, this collection offers a range of insights relevant to rural contention globally. Each contribution in this collection increases our understanding of alternative agricultural production, large-scale development projects, education, race and political parties in the contemporary agrarian context.

Introduction

Brazil has long been recognized as a country of contrasts (Bastide 1959; Eakin 1998). Rich in natural resources from arable land to water, forests and gold, high levels of inequality have perpetuated poverty, marginalization and violence. Known for its open, welcoming culture, Brazil has been governed by a minority elite often criticized for its lack of transparency or accountability; in a country where 'all politics are personal', great emphasis is placed on 'knowing who you're talking to', in Roberto da Matta's (1991) memorable words. In recent years, as Brazil's economic growth and governmental programs have been praised for reducing poverty and hunger, mass protests throughout the country have exposed the fractures of a so-called emerging economy in which structural forms of discrimination and poverty are still evident.

Nowhere are these contrasts more evident than in the countryside, where great agricultural promise has generated boom cycles in key commodity crops while millions of rural landholders and workers live in poverty. Forty years of agricultural modernization and development in large part due to the introduction of 'Green Revolution' technologies – which have been, and remain, contested by a myriad of social actors – have made Brazil globally competitive in the production and export of major commodities, including corn, soy, cotton, rice, orange juice and livestock. At the same time, the country has enacted one of the largest agrarian reforms of the late twentieth century. Policies that support large-scale export-oriented agriculture, including subsidized credit (particularly in the early years of development in the Center–West grasslands known as the *cerrado*), deregulation and privatization, co-exist with government programs that support the redistribution of land and wealth.

Brazil is simultaneously praised as the site of a new 'economic miracle' in commodity agriculture (The Economist 2010), and a global referent for transnational peasant organizing around agro-ecological alternatives (Hardt and Negri 2004, 280; McMichael 2006). The contrasting realities in the countryside are mirrored in the creation of two separate ministries – one for agriculture, dominated by agribusiness and large farmers, and one for agrarian development, dominated by concerns for small farmers and the rural poor. As Bernardo Mançano Fernandes, Welch, and Gonçalves (2010) argue, these two ministries represent two different territorial imaginaries, one dominated by capitalist calculations of profit and extraction and the other by a peasant understanding of livelihood (see also Martins 1981).

In this collection, we analyze rural social movements as expressions of the contrasts – or contradictions – identified above. Persistent, structural rural poverty, challenged daily by peasant, small producer, indigenous and landless worker resistance, causes us to question the promise of dominant agrarian development strategies. Together, the papers argue for a nuanced understanding of the ways in which different actors negotiate new modes of political activity (and new forms of 'being' political) as they construct meaningful livelihoods through production, social reproduction and organization. We argue that these movements are not simply a *reaction* to, or rejection of, changing conditions – to neoliberalism or globalization or even to poverty and hunger. Rather, these movements produce and represent new agrarian identities that provide the symbolic, ideological and material means to make sense of – and resist – political and economic extraction in the contemporary conjuncture. These identities and meanings resonate globally even as they themselves are always context-specific, constructed in particular times and places and embedded in long histories of production and social reproduction.

All of the papers in this collection deal with social movements mobilizing in the Brazilian countryside. The focus on one country may seem narrow, but Brazil's size and diversity provides for a variety of comparisons. There are of course parallels across Latin America as a whole; all of the movements featured in this collection were formed during what scholars call the 'third wave of democracy.'[1] Regime change across the region in the late 1900s ushered in new constitutions, which in many cases explicitly sought to deepen and extend the inclusion of marginalized populations. The movements in this collection are also being shaped by a 'fourth wave' of democracy in which large-scale popular mobilizations with strong anti-neoliberal sentiment resulted in the election of left-leaning – often called populist – governments across Latin America. These governments – from Venezuela to Bolivia, Ecuador, Argentina and Brazil – engage in new ways with social movements like the ones discussed in this collection. Political actors in both state and civil society perform complicated dances between participation and inclusion on the one hand and resistance and exclusion on the other.

---

[1]The third wave of democracy is thought to begin with the Portuguese transition in 1974, and includes the Soviet-style and Authoritarian Latin American regime changes in the late 1980s. The prior two waves include the first dating back to the mid-nineteenth century, and the second, which is considered to have developed following World War II. For more discussion, see Huntington (1991).

In what follows, we outline the contemporary context in which the social movements included in this collection are operating. We discuss the relevance of Brazilian rural movements for social movement theory, and vice versa; the movements featured in this collection provide empirical examples of resistance, as well as theoretical insights into the relationship between movements and states, people and power, and nature and society. We then outline the essays in the collection, grouping them into four themes: alternative economies and development strategies; education of/in the countryside; identity and race; and party politics. We conclude the introduction by exploring some of the methodological questions and potential tensions inherent in studying and working with social movements. We ask how our position and positionality within the academy shape the questions we ask and subjects we entertain. We speak to the importance of acknowledging these tensions, insisting that the tensions are productive as long as research is responsible, accountable and transparent.

Contemporary mobilization in Brazil: a history of protesting monoculture

The economic, political and cultural histories of rural Brazil are histories dominated by large-scale agricultural production. From the first sugarcane plantations in the early 1500s to coffee, tobacco, cattle and dairy, elite land-holding families have governed in the countryside and in national politics. While much has been written about smallholders – from plantation workers with small plots of land to escaped slaves, frontier colonists and millenarian radicals – who have worked at the margins to fight for autonomy on the land, the options available to these actors have very much been conditioned by production and reproduction of the large estates. As a result, it is fair to say that collective resistance in rural Brazil has been historically and is today a rejection of landed elites and large-scale agricultural production. One key element of this rejection is certainly the size of ownership and production; Brazil has the second highest degree of concentration in land ownership in the western hemisphere. But another key element of rejection and resistance is the monocultural orientation of most large-scale plantation production. The singular focus on particular commodity crops sits in direct opposition to the diversified family farm practiced and idealized by many participants in rural social movements.

A central goal of this volume is to explore this opposition in contemporary forms of rural contention and modes of movement organization. The dominant economic model in Brazil today privileges agro-industrial exports, reducing the factors and outputs of production in rural areas to commodities. This view has worldwide reach and ambition; agribusiness elites in the United States and Europe develop the latest genetically modified crops, not in order to sustain the land or farmers or to keep students in rural schools, but so that corporations and large landowners will make a profit. This approach to farming is capital and land-intensive but usually operates with as few people as necessary. By 'farming without farmers', the countryside is reduced to a vehicle for accumulation to the benefit of people and places elsewhere. Rural movements in Brazil today need to be understood in light of this model of farming, even as different movements position themselves differently vis-à-vis the model. Presenting these movements together in this collection thus gives us a unique opportunity to understand the broader conjuncture as well as to analyze and compare mobilization efforts.

The papers in this collection examine social movements that in some way are focused on life and the land. Their campaigns shed light on the transformation of rural life; they provide us with tools for thinking through the specific ways that access to land is changing in particular places and how new modalities of land tenure shape broader changes in the regional and global economy. That these actors are at the forefront of a global 'movement of movements' (Mertes 2004), or struggles for an alternative forms of globalization, suggests that academic scholarship needs to take seriously the complexity of land–labor relationships in the global political economy.

The particular dynamics of capital accumulation today have thrown the various meanings of land – as territory, soil, livelihood, homeland, home, place, commodity, speculative asset, reservoir for future generations, and political platform – into dramatic tension. New political, social and economic imperatives and possibilities are in turn being shaped by – and re-shaping – property relations, the forces of production and new political subjects. As the authors in this collection demonstrate, the supposed death of the peasantry and the move to the industrial, modern city as part of the linear model of development has been greatly complicated by failures of that vision itself, and by everyday acts of resistance and large-scale, sustained mobilizations.

The topic of this collection is timely and important. The purpose of this collection is not to provide grand theories of agrarian change in Brazil, but rather to build an analytical toolkit for understanding contemporary struggles for alternative economies, the provision of public services, political representation and conflicts surrounding race and ethnicity, in the countryside of Brazil, one of the major agricultural powerhouses of the developing world. The key categories of analysis here are applicable to other regions. The authors suggest that in order to understand contemporary agrarian transformations, we need to draw from theories of access, accumulation and extraction in agrarian studies, including: theories of property and land, property as theft, property as accumulated labor, property as a social relationship or a bundle of rights; theories of differentiation and ongoing struggle over surplus and the means of production, primarily if not only the land; theories of moral economy and the relationship between custom, transgression and law; theories of hegemony and the tensions between consent and coercion.

Working within these theoretical frameworks, scholars in agrarian studies have shed considerable light on the presence, nature and effects of resistance. One could argue that the subaltern nature of peasants across time and place has led scholars of agrarian life to focus on contestation, whether through revolution (Moore 1993), collective mobilization (Wolf 1969; Davidson 1974; Paige 1978), or engagement in small, even hidden, acts referred to as the 'weapons of the weak' (Scott 1994), from foot-dragging to sabotaging grain supplies. Much of this literature is 'transitional', studying the transition from 'premodern' life, whether feudalism, subsistence, migratory or tributary, to market society, both capitalist and socialist. That this transition has been enacted on the backs of the peasantry through the forced re-allocation of surplus is a key insight from the field of agrarian studies. Current indigenous resistance to neoliberalism, exemplified especially by the Zapatista movement (Harvey 1998), highlights new forms of contention that challenge prior understandings of revolution by contesting identities, models of development and governmental organization. The papers in this collection provide cases that speak to these theoretical frameworks, but in grounded ways that offer a lens for connecting the specific to the abstract or general.

Existing literature and debates

The nature of contemporary rural resistance

It would be difficult to analyze agrarian politics in contemporary Brazil without engaging debates surrounding neoliberalism. For more than two decades, scholars in Brazil and around the world have investigated the rapidly shifting terrain of neoliberalism – a still-hegemonic mode of organizing politics and economy that has shaped the 'field of possibility' for actors globally. In many ways, neoliberalism has exacerbated the production principles of monoculture; the withdrawal of state support for farming in the 1990s along with the fall of tariffs and other protection for the domestic economy increased vertical integration along the agro-industrial chain. Whether conceived as a class project with specific actors ultimately responsible for creating these practices, or systems of thought generated from greater discourses, social movements such as the ones featured in this collection have organized campaigns to highlight the reductive and unequal effects of neoliberalism. At the same time, other scholars suggest that the rise of neoliberalism in the 1990s in Brazil combined with the deepening of democracy to provide a political opening for social mobilization (Dagnino 2007). Wolford (2007) argues that opposition to neoliberalism was an effective rallying cry for mobilization at a time when advances in democratic practice allowed for open contestation. Neoliberal policies also allowed for increased international investment and trade in Brazilian agriculture, while dramatic reductions in inflation under the Plano Real (1994) reduced the speculative power of land ownership.

Scholars and movements around the world have highlighted unique, innovative modes of resistance to the neoliberal logic. Indigenous movements in Latin America (Yashar 2004; Postero 2006), slum-dweller organizations in India (Appadurai 2002; Chatterjee 2004) and urban social movements in Europe and North America (Leitner, Peck, and Sheppard 2007) illustrate the diverse array of practices and groups that challenge the racial, economic and political components of neoliberal governance. The papers in this collection also highlight different kinds of resistance that address the fault lines in the contemporary neoliberal agricultural production model, specifically with respect to the ongoing privileging of monocultural production. The cases on their own have received attention in multiple places, including scholarly treatment in articles and books. However, this edition brings them together – movements composed of subjects as diverse as landless workers, small farmers, indigenous peoples and Afro-Brazilian peasants – with the intention of providing readers the opportunity to compare and assess their different modes of resistance. Our current historical moment requires an analytical – yet critical – view of *alternatives*.

A broad array of actors coheres within the transnational peasant movement. *La Via Campesina*'s seven principles of food sovereignty adhered to by its member groups – some 150 different organizations from 70 different countries – remind us that opposition to the corporate food system is global (McMichael 2005, 2006), and can incorporate a wide variety of tactics and alliances (Desmarais 2007; Borras 2008). These global movements, however, must be contextualized within their local political and economic structures (Edelman 1999, 2009). At the same time, the struggles of peasant women to achieve gender parity in national and international food sovereignty movements (Desmarais 2003) add complexity to intra-movement dynamics. Experiments in local resistance and participatory governance have also led certain groups – like the Ecuadorian-based indigenous confederation CONAIE (La Confederación de Nacionalidades Indígenas del Ecuador, The Confederation of Indigenous Nationalities of Ecuador) – to actively pursue electoral strategies (Becker 2011), while other social movements eschew formal politics and argue that the nation state has no place for them (Harvey 1998; Bob 2005). We know that rural resistance to neoliberalism continues – worldwide – yet we lack a systematized attempt to bring different modes of contention together. This collection attempts to fill that gap.

Social movements, the state and dominant economic interests

The existing literature on social movements only partially allows us to think about the relationship between rural movements, states and dominant economic elites. For example, social movement theories based in the US and Europe, including the political process model (Tarrow 1994; McAdam 1999; McAdam, Tilly, and Tarrow 2001), the resource mobilization approach (Jenkins and Perrow 1977; McCarthy and Zald 1977; Tilly 1978) and discussions of 'repertoires' of contention (Tilly 2008), are primarily focused on movement emergence and mobilization over time. In the political process perspective, social movements are defined as 'rational attempts by excluded groups to mobilize sufficient political leverage to advance collective interests through non-institutionalized means' (McAdam 1999, 37). This definition constrains our ability to analyze how rural people move in and out of social movements and can complicate the explanation of intra-movement processes, although both issues have long been the focus of various studies of agrarian movements. Prior research on revolutionary peasant organization and mobilization (Scott 1977; Popkin 1979) noted, particularly, the dynamics of recruitment and mobilization. Recently, Wolford (2010b) has argued that the appearance of a united social movement is more often a reflection of the political strategy of movement leaders, not the day-to-day realities of rural populations.

State/movement relations are also a contested issue in agrarian studies. A prior generation of scholars analyzed peasant movement organization at the periphery of state power (Wolf 1969; Hobsbwam 1973; Scott 1977), with repression and co-optation as state authority's standard rules of engagement. More current research has added nuance to our thinking of state power and agrarian contention, focusing on how rural groups may interact with their respective states to subvert status quo power relations (Das 2007). Besides focusing attention on the state and resistance to it, some of the papers in this collection explore theoretical perspectives concerning state power. In our attempt to open the 'black box' of the state, we expand the scope of what we consider social movement contestation.

Critics, and particularly scholars associated with the so-called 'new social movement' tradition, have questioned the conception of power and politics within the political process/resource mobilization approaches. These scholars point to the production of cultural meanings and practices as a form of political resistance (Alvarez, Dagnino, and Escobar 1998; Armstrong and Bernstein 2008). Drawing on Foucault's notion of power as dispersed throughout society, scholars also argue that 'collective action concerns everyday life, personal relationships, and new conceptions of space and time' (Melucci 1989, 71). Scholars of new social movements have privileged identity formation and mobilization, which has been considered central to farmer mobilization in India (Lindberg 1994). Our collection of studies embraces these various approaches to state authority and power, analyzing the relationship between contemporary rural social movements, the Brazilian government and economic practices.

One of the dominant tropes in social movement literature is that movements regularly engaging governmental institutions are destined to bureaucratize and demobilize over time. This idea can be traced back to Michels' (1915) iron rule of oligarchy, and his argument that political parties tend to become more bureaucratic and hierarchical over time, thus suppressing grassroots mobilization. Piven and Cloward (1977) took up this idea in the 1970s, arguing that once social movements get more organized, adopt formal hierarchies and begin working with the state, contentious actions become difficult and this prevents structural change from occurring. This argument is particularly difficult to sustain in Latin America, where working in and with the state has become part of the deepening of democracy and rise of leftist populism in the past 15 years (Lebon 1996; Santos 2010; Rubin and Sokoloff-Rubin 2013).

Characteristics of activism in Latin America

In what follows, we identify seven important characteristics of social movement organizing or activism in Latin America, in an attempt to synthesize some of the existing literature and debates. These characteristics include the prominence of networks; state–society collaborations; the importance of territory as an analytical and empirical unit; the claim for autonomy; dispossession as an ongoing mechanism of accumulations; alternative economies; and the importance of new ontologies that privilege nature as not only having rights but also producing alternative visions of the world.

Networks

There has been much discussion of transnational organizing through new forms of social media that allow actors in what were once isolated regions to connect their campaigns with broader networks of activists around the world. Much of this literature has focused on well-organized and visible networks such as Via Campesina and People's Global Action. These transnational movements illustrate both the deep connections between social movements across borders and also how these networks engage in knowledge production about the purpose of and need for a united struggle. In addition, many social movements, especially in Brazil, are also plugged into global networks sponsored by sympathetic organizations such as Food First, GRAIN or Focus on the Global South. Utilizing resources from social media to discourse analysis, these organizations work with movements in the Global South to help spread information, wage campaigns and mobilize resistance in key places and times. One example is the land rights network that has been able to generate global attention on large-scale land purchases in the wake of the 2007 food crisis, by naming it the Global Land Grab and producing on-the-ground research that documented the effects of dispossession, such that every major media outlet had a story on the phenomenon and multilateral institutions were forced to take action (Keulertz 2013). The creation of global networks, specifically in the form of international brigades, has been a long-term strategy of many socialist countries such as Cuba (Artaraz 2012). The extent of these connections has grown in the current era.

State–society collaborations

An important characteristic of the current moment is the election of left-leaning political parties, a transformation of the executive office that has re-shaped the landscape of political contestation. Through a combination of increased participation and the decentralization and privatization instituted during previous neoliberal regimes (Wolford 2010a), social movement actors must now navigate a more nuanced relationship with the state. From participatory budgeting (Abers 2000; Baiocchi 2005; Wampler 2007) to watershed councils (Abers and Keck 2009) and movement-state educational programs (Tarlau 2013), collaboration cannot be seen simply as cooptation, although interaction surely shapes movement and state actors alike. This volume builds on Fox (1992) and Borras' (2001) previous contributions, which have analyzed how rural movements attempt to influence and transform state actors, and in the process are themselves transformed (Borras 2001, 548). The collection also builds on the extensive literature on participatory democracy and state–society synergy, which explores 'the dynamic process of interaction across state and civil society' (Baiocchi 2005, 17) and how competing agendas within the state can produce unexpected alliances between civil society groups and state actors (Evans 1997; Ostrom 1996; Wang et al. 1999; Cornwall and Coelho 2007). The authors in this collection examine some of the most well-organized and contentious rural movements in Brazil, and how these movements are also able to work with, in and through the state to achieve their goals.

Territorial development and territorial claims

Over the past 20 years, the struggle for territory has become one of the primary political sites through which indigenous peoples have fought for recognition and the right to maintain tradition, difference and connection; in response, a counter-territorialization has developed as states use the language of territory to map their own state-sponsored identities onto polygons of productive use. The Brazilian state has turned large portions of the Amazon rain forest into territories zoned for different uses: an extractive territory, an economic development territory, conservation territories, etc. Whether territory is mapped with Cartesian technologies or with activist methodologies of *emplacing* resistance (Escobar 2010), territory has become a powerful arena for mobilization and counter-mobilization.

Autonomy

Even as movement actors are incorporated into the state, collaborating with the rules of governance to achieve political goals, communities throughout Latin America search for the space in which to define and govern themselves. The Zapatistas are the most famous case of communities struggling for – and winning – the right to self rule but there are many movements, particularly in Latin America, that include autonomy as a key political goal (and strategy). The Brazilian Landless Workers Movement (*Movimento dos Trabalhadores Rurais Sem Terra*, or MST), while sometimes mobilizing in support of political candidates who have proven their commitment to agrarian reform, claims political autonomy (or, independence) as an official stance of the movement. In addition, the struggles of these movements have led to explicitly spatialized arenas of solidarity – encampments for the MST, communities for the Zapatistas in Mexico, factories for the Recuperated Factories in Argentina – in which for all intents and purposes these movements govern (Wolford 2010b).

Dispossession

Perhaps co-constituted with the struggle for territory and autonomy is the presence of dispossession as an ongoing logic of economic and political accumulation. The so-called Global Land Grab (Borras et al. 2011) is a reflection of the many faces of dispossession – from Dharavi, an urban slum in the middle of Mumbai, to the ancestral land of the Garifuna in Honduras, land is at a premium, and social movements are fighting dispossession on the ground case by case as well as by working through their networks to influence policy at the transnational level. Movement efforts to have state authorities recognize communal areas, from indigenous populations throughout Latin America to *quilombola* territories in Brazil, challenge trends in privatization that could be argued as constitutive of dispossession by (neoliberal) enclosure. In addition to land, the movements explored in this edition highlight the coordinated resistance to other forms of dispossession, such as cultural or political dispossession.

Alternative economies, gift economies, economia solidaria, ethical economies

These are initiatives to re-ground economic interaction in the social, usually by localizing the economy and attempting to de-fetishize commodities by creating direct links between production and consumption. There has been a rise in radical efforts to re-imagine the economy, whether though popular restaurants or organic fairs and collective farms. Food sovereignty, or the demand for ecologically appropriate production and local food systems as ways to guarantee food for all peoples, has been taken up as a rallying call. Local currencies are also on the rise again, as a response to the latest economic crisis. All of these initiatives have to be understood as working with particular notions of the local, even as they are embedded in transnational flows of capital, information and activists.

Rights to nature and the right to determine what kind of nature

The rise in environmental movements of all kinds over the past decade is astonishing – movements against the privatization of water, against the sale of glaciers, against the dams on the Narmada river, against deforestation and more. In Latin America, there has been an effort to re-conceptualize human–nature relations in the state/policy arena and in a way that challenges development paradigms. In 2008, Ecuador became the first country in the world to recognize the rights of nature, in the drafting of the country's new constitution following Correa's election. In Ecuador and Bolivia, the inclusion of rights to nature is part of a larger philosophy of life and development for which indigenous groups, among others, have been fighting – the concept of *buen vivir* in Ecuador and *vivir bien* in Bolivia. The Ecuadorian constitution now states that the goal of development is *buen vivir* and that it provides a 'conceptual rupture' with previous conceptions of development over the last six decades (Escobar, quoting Acosta, 2010). While these political/legal restructurings are partial at best, they are part of the urgent process of re-articulating economic and ecological relationships.

Insights from contemporary rural resistance in Brazil

Isolating one country

Rather than select various movements from across the world, our efforts in this volume focus on innovative movement organizations and practices of resistance in Brazil. Our intention is to bring together areas, movements and conceptual discussions that sporadically and non-systematically have entered into dialogue with one another in the past. By isolating one country at one historical period, we hope to shed light on how movements in general resist economic, cultural and political authority. Again, Brazil is a particularly interesting case in which to examine rural resistance because of the simultaneous expansion of social welfare programs and the increase in monoculture production in the Brazilian countryside. Many of the movements currently resisting this economic model were founded in a very different economic era, when the countryside was dominated by rural elites who used extra-legal means to govern in the countryside, often to the detriment of Brazil's overall economic growth. In the contemporary era of a highly profitable and high-yielding industrial agriculture, these movements have had to transform their strategies, discourses and mode of organization in order to survive.

For example, the MST has endured over 30 years, through dramatic economic and political transformations, and remains an oppositional force in economic, education and agrarian reform policy. Emerging at roughly the same time, the National Movement of those

Affected by Dams (*Movimento dos Atingidos por Barragens*, or MAB) continues to contest large-scale development projects led by the Brazilian state. *Quilombos* – also known as maroon settlements – that have survived since the end of slavery at the end of the nineteenth century battle over access to land and the preservation of their culture from agribusiness expansion. Similarly, indigenous movements all across the country struggle for land rights, culturally specific education and cultural recognition. Unions – many of which were considered a tool and agent of the state during Brazil's authoritarian period (Pereira 1997; Houtzager 1998, 2001) – have emerged again as an oppositional force, challenging norms in the field of education and resisting unequal land distribution. All of these social movements have been able to sustain resistance to monoculture production and rural displacement, despite having a left-of-center government in power.

This collection

Alternative economies and development strategies

We divide this volume into four sections, organized around thematic foci. The first section, 'Alternative Economies and Development Strategies', highlights how movements challenge dominant economic actors and modes of production by implementing different strategies and alternative forms of production. In this section, we have included three papers focused on three different social movements struggling for alternative economic models and/or alternative development projects. Pahnke, writing about the MST, suggests that despite many strategic challenges, the MST *has* succeeded in regularizing cooperative productive practices that challenge neoliberal dictates and state power. He argues that MST resistance, particularly through cooperatives and agroecological production, differentiates the MST from other kinds of social movements, which can be considered either reformist (e.g. United States Civil Rights Movement) or revolutionary (Fidel Castro's July 26th Movement). Rather, the MST's alternative agricultural practices illustrate a new form of resistance, what he calls self-governmental, in which MST activists vie for the control, design and implementation of particular policies normally considered the terrain of the government.

As opposed to struggles for alternative agricultural production, Klein analyzes resistance to the Brazilian government's largest infrastructure project, the Belo Monte Dam. He examines how the semi-privatized nature of this hydroelectric project and the government's purported commitment to participatory development offer both challenges and opportunities for claims-making. On the one hand, Klein shows how these dynamics tend to lead to the fracturing of civil society. On the other hand, he shows how this context can occasionally facilitate alliances among diverse resistance groups and, in surprising ways, between these groups and the state. He examines how the dam has affected the livelihoods of local fishermen, their struggle to receive compensation for these changes and avenues for participation in the process of development planning. Klein argues that the different visions activists hold of development 'alternatives', and their relative comfort participating in this process, have direct effects on the sustainability of the long-term alliances that are necessary to actually transform the nature of Brazil's development project.

Finally, Sauer and Welch focus on the National Confederation of Agricultural Workers (CONTAG), a movement that was founded right before the start of the military dictatorship, in 1963. This paper suggests that the rural labor movement's current struggle for agrarian reform and alternative agricultural production in the countryside represents a continuity with the past, not a radical break. Through a historical analysis of rural union mobilizing during CONTAG's 50-year history, Sauer and Welch illustrate that agrarian reform has always been a central focus of this organization. However, the authors also highlight CONTAG's tendencies towards legal battles and working with the government/authorities during this period, and how the successful tactics of other movements influenced CONTAG's adaptation of more confrontational strategies. Sauer and Welch claim that the rural organizations representing the peasantry have been the principal protagonists shaping the development initiatives and peasant struggles in the Brazilian countryside over the past century.

Education of the countryside

The second section of this collection, 'Education of the Countryside', explores how social movements transform the public sphere – in this case, the Brazilian school system. The success rural social movements have had in redefining federal, state and municipal debates about public education in Brazil offers a new area of inquiry for examining contemporary relationships between the state and civil society. In this section we have included two papers examining the same social movement, the MST, but at different scales of interaction. Tarlau examines national-level educational policy and analyzes how a coalition of rural social movements successfully advocated for the incorporation of 'education of the countryside' – the notion that rural public education should be distinct from urban schooling and should encourage students to stay in the countryside – at the federal level. She argues that the concept of 'education of the countryside' emerged in the late 1990s, when MST activists decided to expand their already-well-developed educational proposal to communities beyond those living in areas of agrarian reform. However, the federal government only began to embrace these educational ideas once a strategic alliance formed between the MST and CONTAG, two historically antagonistic rural movements. Despite the incredible success the MST and CONTAG have had in transforming educational policy at the federal level, the Ministry of Education's tendencies towards hierarchy, large-scale projects and the implementation of 'best practices' have prevented these social movements from actively participating in these educational initiatives. Furthermore, Tarlau argues that the influence of agribusiness interests in the Workers' Party's governing coalition has led to a series of contradictions that has transformed the original intentions of the MST's educational proposal.

In contrast to this federal-level analysis, Meek focuses on the MST's educational initiatives in an MST settlement in the Amazonian state of Pará. Through an analysis of educational practices in a local school and federal university program – both partially governed by MST activists – Meek argues that the MST's promotion of agroecological education in the Amazon directly influences people's interactions and understandings of the land. He claims that in certain cases, depending on the form of institutionalization of the MST's educational goals, and educators' own spatial histories, this critical 'place-based pedagogy' can help activists conceptualize their local communities in relation to international, national and regional forces. Furthermore, Meek illustrates how agroecological education directly facilitates the involvement of these local communities in alternative agricultural practices that directly contest the dominant monoculture model in the region.

Identity, race and the peasantry

In the next section, 'Identity, Race and the Peasantry', scholars explore how identities – rather than constituting a residual category – are central to movements' tactical choices and self-understandings. There are three contributions in this section, focusing on issues of identity among three different sectors of the Brazilian countryside: settlers in areas of agrarian reform who *resist* identification with the MST, communities struggling for land on the basis of racial identity, and indigenous groups that organize along ethnic identification that spans the rural–urban divide. First, DeVore examines internal conflict, but this time on agrarian reform settlements in the cacao region of southern Bahia. This paper focuses on a group of settlers whom MST leaders claim as part of their movement, but who in fact reject the MST's presence in their settlements. DeVore analyzes a variety of the MST's local practices, from the brokering of government services to the movement's emphasis on collectivity, and illustrates how local settlers interpret these practices as forms of oppression and dominance. Taking the perspective of settlers who reject the good intentions of local MST leaders, DeVore sheds light on how unequal power relations are reproduced within social movements. He also examines settlers' responses to these power differentials, and argues that the MST's presence in these regions has led settlers to embrace more decentralized, 'anarchist' forms of land occupation.

Next, Leite explores how the word *Quilombo* which once referred to a historical process of slave revolt, has taken on new meanings in contemporary Brazil. This 'metaphorical process', as she refers to the changing meaning of the term, has redefined territorial rights along racial lines. Thus, while rural black communities have been traditionally one of the most exploited and marginalized groups in the countryside, these communities can now claim land on the basis of their racial self-identification. Despite these gains, Leite argues that the process of claiming land based on racial identification is extremely difficult in practice, and she describes the struggles of one rural black community in Southern Brazil to illustrate these tensions. She concludes by arguing that the *Quilombo* reform is currently limited, and she suggests continuity between past forms of racial exploitation and the currently inadequate access to territorial rights that rural black communities still face.

Finally, Sobreiro offers a case study of indigenous mobilization in the Amazon. While indigenous identity and mobilization are strongly associated with rural indigenous communities and cultures, Sobreiro illustrates how indigenous mobilization is actually strengthened by rural-to-urban migration. Rather than disrupting the territorial claims of these movements, migration facilitates urban-based organizing along ethnic and indigenous lines. Sobreiro illustrates these dynamics through a case study of one region of the Amazon where the indigenous movement first arose in an urban center, and then traveled to rural regions where territorial claims subsequently developed. This contests the traditional understanding of indigenous identity as being embedded in one particular time and space. Together, the three contributions in this section show how identity is a fluid yet powerful factor affecting rural mobilization in Brazil.

Political parties and movements

The papers in the final section, 'Political Parties and Movements', examine the consequences of two left-leaning social reforms for rural mobilizing, within the different social movements. Ansell examines the Workers' Party (PT)'s Zero Hunger initiative, and argues that the leadership of the *Quilombo* movement consciously utilized this project as a cultural engineering mission to end patronage within rural black communities. This attempt to promote more horizontal ties of solidarity among afro-Brazilians was carried out through what Ansell calls a 'pilgrimage', or a several-day trip and series of educational workshops that invoked the legacy of runaway-slave communities and collective insurrection. Despite the new lateral relationships that were formed through these experiences, this government program did not succeed in displacing alliances between the rural black community and local politicians. In fact, Ansell argues, the new horizontal ties seemed to be completely compatible with old vertical relationships, and therefore government programs might be better served by attempting to transform rather than eliminate patronage.

Finally, Morton focuses on the most famous of the PT government's social program, *Bolsa Família* – the largest cash transfer program in the world – and examines how this program has affected MST mobilization in two agrarian reform settlements in Bahía. His fine-grained analysis suggests that the common perception that *Bolsa Família* demobilizes movements by making people less desperate is incorrect. Rather, he argues, the *Bolsa Família* program is so unreliable that local communities need to remain politically active simply to ensure the continuation of their benefits. Furthermore, receiving money through *Bolsa Família* transforms the identity of local community members; they become 'managers' of a benefit. This new identity is not necessarily at odds with membership in the MST. Through his description of how people interpret and understand this far-reaching government benefit, Morton helps to explain the relationship between *Bolsa Família* and social movement mobilization in the Brazilian countryside.

Social movement researchers and researcher positionality

The guest editors have attempted to include a range of contributors to discuss these contemporary agrarian dynamics in Brazil – with different positionalities, disciplinary trainings and relationships to the rural social movements they study. For example, included in this collection are scholars with disciplinary training in geography, political science, education, history, anthropology and sociology. We also take seriously the tensions implicit in the researcher–activist relationship. As Edelman (2009) argues, local activists often feel abused by professional researchers who utilize social movements' precious energy and fail to provide timely and clear reporting of the research results. At the same time, movements often have different agendas than academics do, and the differences can result in disagreements along the way (or after publication) or they can shape the research process itself by providing a set of boundaries to the questions researchers can ask. All research involves translation and interpretation and carries with it the potential for violence, but we can rarely know ahead of time who will suffer (or who will suffer the most) from the violence, or how. The authors in this volume suggest that transparency, accountability and participatory dialogue are the only way to avoid the worst of the violence while allowing for the productive potential of research. We aspire for horizontal and collaborative forms of knowledge production, but also acknowledge the dangers in reproducing dominant movement narratives (Wolford 2007; Edelman 2009). We have included in this volume the work of several well-established agrarian scholars who have written about rural movements in Brazil for more than a decade; a younger generation of scholars from the United States who conducted fieldwork over the past five years; and several Brazilian academics who have worked directly with peasant groups for their entire careers. Regardless of their positionality, the scholars in this volume all strive to critically analyze the new modes of rural resistance in Brazil, the benefits and pitfalls of diverse forms of state–society collaborations and the consequences of social movements contesting the dominant, neoliberal model of rural development – in a left-of-center political context.

Concluding thoughts

By analyzing rural resistance around thematic categories that transcend the Brazilian context – including agricultural production, large-scale development projects, education, identity, race, and party politics – the contributions in this volume offer insights for the broader field of agrarian studies and peasant resistance. First, these papers illustrate how social movements are currently negotiating the new agricultural and development paradigms promoted by left-leaning governments. The authors show that activists are resistant rather than complacent to these initiatives, and, furthermore, are actively forming new ways to govern their own territories in response to these changes. These forms of resistance are not completely new; however, the fact that a left-leaning government is driving these developments offers a series of new dilemmas for local movement activists to negotiate.

Second, the papers highlight the interactions that are taking place between movement activists and state actors, as the state becomes more 'participatory' and inclusive. The authors illustrate the potential benefits and the risks of this participation, in *both* cases arguing that 'co-optation' does not properly characterize these developments. In other words, collaboration with state elites does not automatically entail movement demise or pacification. Rather, involvement with the government can extend movement life as well as facilitate the provision of public goods that help support counter-hegemonic practices in the countryside, while also forcing activists to expend precious energy in endless government negotiations. These state–society collaborations often represent what Evelina Dagnino (2007) has referred to as the 'perverse confluence' between neoliberal and participatory projects.

Third, this collection illustrates the continuing importance of identity, race and ethnicity in rural organizing. The authors show the various consequences of identity conflicts within movements, which can include the rejection of a movement's leadership by local communities, the establishment of new, identity-based movements or the transformation and radicalization of the movement itself to overcome conflict. The authors also illustrate the many barriers that communities face when choosing to organize along racial, cultural or ethnic grounds, and how 'racial identity' can transcend individualistic conceptions of skin color. Furthermore, strong ethnic identity is not necessarily situated in a particular place, but rather can be a powerful tool to organize across regions and urban–rural divides.

Finally, this collection has highlighted the relationship between left-leaning political parties and social movements, arguing that even self-proclaimed autonomous movements are affected by the implementation of social reforms. Whether these effects are a consequence of a direct attempt to intervene in local politics, or an indirect transformation of the subjectivity of local communities, the nature of social mobilization is currently being remade in these left-leaning contexts. In conclusion, we argue that the contemporary nature of rural mobilization is not merely an effect of neoliberal economic policies and development initiatives. Rather, this mobilization is shaping and being shaped by social reforms for the poor, racial disparities, public good provision, historical trends and current initiatives for economic and political development.

Acknowledgements

We would like to acknowledge the hard work of Jun Borras, whose support was critical for putting together this collection. We also want to thank the rest of the *JPS* editorial collective for their comments on the original proposal and their help throughout, and the authors of the papers in the collection whose patience and flexibility made this collaboration possible. We also thank the social movements whose work makes the world a better place.

Disclosure statement

No potential conflict of interest was reported by the authors.

References

Abers, R. 2000. *Inventing local democracy: Grassroots politics in Brazil.* Boulder, CO: Lynne Rienner Publishers.

Abers, R.N., and M.E. Keck. 2009. Mobilizing the state: The erratic partner in Brazil's participatory water policy. *Politics & Society* 37, no. 2: 289–314.

Acosta, A. 2010. El Buen Vivir en el camino del post-desarrollo. Una lectura desde la Constitución de Montecristi. *Policy Paper*, 9(5), 1–36.

Alvarez, S. 1990. *Engendering democracy in Brazil: Women's movements in transition politics.* Princeton, NJ: Princeton University Press.

Alvarez, S., E. Dagnino, and A. Escobar, eds. 1998. *Cultures of politics, politics of cultures.* Boulder, CO: Westview Press.

Appadurai, A. 2002. Deep democracy: Urban governmentality and the horizon of politics. *Public Culture* 14, no. 1: 21–47.

Artaraz, K. 2012. Cuba's internationalism revisited: Exporting literacy, ALBA, and a new paradigm for south-south collaboration. *Bulletin of Latin American Research* 31: 22–37.

Armstrong, E.A., and M. Bernstein. 2008. Culture, power, and institutions: A multi-institutional politics approach to social movements. *Sociological Theory* 26, no. 1: 74–99.

Baiocchi, G. 2005. *Militants and citizens: The politics of participatory democracy in Porto Alegre.* Stanford, CA: Stanford University Press.

Bastide, R. 1959. *Brasil, terra de contrastes.* São Paulo: Difusão Européis do Livro.

Becker, M. 2011. Pachakutik: Indigenous movements and electoral politics in Ecuador. Lanham, MD: Rowman & Littlefield Publishers.

Bob, C. 2005. *The marketing of rebellion: Insurgents, media, and international activism.* Cambridge: Cambridge University Press.

Borras, S.M. 2001. State-society relations in land reform implementation in the Philippines. *Development and Change* 32: 545–75.

Borras, S.M. 2008. La Vía Campesina and its global campaign for agrarian reform. *Journal of Agrarian Change* 8, no. 2–3: 258–89.

Borras, S. M., R. Hall, I. Scoones, B. White, and W. Wolford. 2011. Towards a better understanding of global land grabbing: an editorial introduction. *The Journal of Peasant Studies*, 38(2), 209–21.

Chatterjee, P. 2004. *The politics of the governed: Reflections on popular politics in most of the world.* New York: Columbia University Press.

Cornwall, A., and V.S. Coelho, eds. 2007. *Spaces for change? The politics of citizen participation in new democratic arenas.* New York: Zed Books.

Dagnino, E. 2007. Citizenship: A perverse confluence. *Development in Practice* 17, no. 4–5: 549–56.

Da Matta, R. 1991. *Carnivals, rogues and heroes: An interpretation of the Brazilian dilemma.* Notre Dame: University of Notre Dame Press.

Das, R.J. 2007. Introduction: Peasant, state and class. *The Journal of Peasant Studies* 34, no. 3–4: 351–70.

Davidson, B. 1974. African peasants and revolution. *The Journal of Peasant Studies*, 1(3), 269–290.

Desmarais, A.A. 2003. The Vía Campesina: Peasant women on the frontiers of food sovereignty. *Canadian Woman Studies* 23, no. 1: 140–5.

Desmarais, A.A. 2007. *La Via Campesina: Globalization and the power of peasants.* London: Pluto Press.

Eakin, M.C. 1998. *Brazil: The once and future country.* New York: St. Martin's Press.

Economist, the. 2010. The miracle of the Cerrado. published on August 26, 2010. http://www.economist.com/node/16886442

Edelman, M. 2009. Synergies and tensions between rural social movements and professional researchers. *Journal of Peasant Studies* 36, no. 1: 245–65.

Edelman, M. 1999. *Peasants against globalization: rural social movements in Costa Rica.* Redwood City, CA: Stanford University Press.

Escobar, A. 2010. Postconstructivist political ecologies. *The international handbook of environmental sociology*, 91–105.

Evans, P., ed. 1997. *State-society synergy: Government and social capital in development*. Berkeley, CA: University of California Press.

Fernandes, B.M., C.A. Welch, and Elena C. Gonçalves. 2010. Agrofuel policies in Brazil: Paradigmatic and territorial disputes. *Journal of Peasant Studies* 37, no. 4: 793–819.

Fox, J. 1992. *The politics of food in Mexico: State, power and social mobilization*. Ithaca, NY: Cornell University Press.

Hardt, M., and A. Negri. 2004. *Multitude: War and democracy in the age of empire*. New York: The Penguin Press.

Harvey, N. 1998. *The Chiapas rebellion: The struggle for land and democracy*. Durham, NC: Duke University Press.

Hobsbawm, E.J. 1973. Peasants and politics. *The Journal of Peasant Studies* 1, no. 1: 3–22.

Houtzager, P. 1998. State and unions in the transformation of the Brazilian countryside, 1964–1979. *Latin American Research Review*, 33, no. 2: 103–42.

Houtzager, P. 2001. Collective action and patterns of political authority: Rural workers, church, and state in Brazil. *Theory and Society* 30, no. 1: 1–45.

Huntington, S. P. 1991. Democracy's third wave. *Journal of democracy*, 2(2), 12–34.

Jenkins, J.C., and C. Perrow. 1977. Insurgency of the powerless: Farm worker movements (1946–1972). *American Sociological Review* 42, no. 2: 249–68. doi:10.2307/2094604

Keulertz, M. 2013. Land and water grabs and the green economy. In *Handbook of Land and Water Grabs in Africa: Foreign direct investment and food and water security*, ed. M. Keulertz, S. Sojamo, and J. Warner, 243–256.

Lebon, N. 1996. Professionalization of women's health groups in São Paulo: The troublesome road towards organizational diversity. *Organization* 3, no. 4: 588–609.

Leitner, H.M., J. Peck, and E. Sheppard. 2007. *Contesting neoliberalism: Urban frontiers*. New York: Guilford Press.

Lindberg, S. 1994. New farmers' movements in India as structural response and collective identity formation: The cases of the Shetkari Sanghatana and the BKU. *The Journal of Peasant Studies* 21, no. 3–4: 95–125.

Martins, J. D. S. 1981. *Os camponeses e a política no Brasil*. Petrópolis, RJ Brazil: Vozes, 61.

McAdam, Doug. 1999. *Political process and the development of the Black Insurgency, 1930–1970*. 2nd ed. Chicago, IL: University of Chicago Press.

McAdam, D., C. Tilly, and S. Tarrow. 2001. *Dynamics of contention*. Cambridge, UK: Cambridge University Press.

McCarthy, J.D., and M.N. Zald. 1977. Resource mobilization and social movements: A Partial Theory. *American Journal of Sociology* 82, no. 6: 1212–41.

McMichael, P. 2005. Global development and the corporate food regime. *Research in Rural Sociology and Development* 11: 265–99.

McMichael, P. 2006. Peasant prospects in the neoliberal age. *New Political Economy* 11, no. 3: 407–18.

Melucci, A. 1989. *Nomads of the present*. London, UK: Hutchinson Radius.

Mertes, T. 2004. *A movement of movements: Is another world possible?*. London, UK: Verso.

Michels, R. 1915. *Political parties: A sociological study of the oligarchical tendencies of modern democracy*. New York: Dover Publications.

Moore, B. 1993. *Social origins of dictatorship and democracy: Lord and peasant in the making of the modern world (Vol. 268)*. Boston, MA: Beacon Press.

Ostrom, E. 1996. Crossing the great divide: Coproduction, synergy, and development. *World Development* 24, no. 6: 1073–87.

Paige, J.M. 1978. *Agrarian revolution*. New York, NY: Simon and Schuster.

Pereira, A. 1997. *The end of the peasantry: The rural labor movement in Northeast Brazil*. Pittsburgh, PA: University of Pittsburg Press.

Piven, F.F., and R. Cloward. 1977. *Poor people's movements: Why they succeed, how they fail*. New York: Vintage.

Popkin, S.L. 1979. *The rational peasant: The political economy of rural society in Vietnam*. Oakland, CA: University of California Press.

Postero, N. 2006. *Now we are citizens: Indigenous politics in postmulticultural Bolivia*. Stanford, CA: Stanford University Press.

Rubin, J.W., & E. Sokoloff-Rubin. (2013). *Sustaining Activism: A Brazilian Women's Movement and a Father-Daughter Collaboration*. Durham, NC: Duke University Press.

Santos, C.M. 2010. Da delegacia da mulher à lei maria da penha: absorção/tradução de demandas feministas pelo estado. *Revista Crítica Da Ciencias Socias* 89: 153–70.

Scott, J.C. 1977. *The moral economy of the peasant: Rebellion and subsistence in Southeast Asia.* New Haven, CT: Yale University Press.

Scott, J.C. 1994. *Weapons of the weak: Everyday forms of peasant resistance.* New Haven, CT: Yale University Press.

Tarlau, R. 2013. Coproducing rural public schools in Brazil: Contestation, clientelism, and the landless workers' movement. *Politics & Society* 41, no. 3: 395–424.

Tarrow, S. 1994. *Power in movement.* 3rd ed. Cambridge, UK: Cambridge University Press.

Tilly, C. 1978. *From mobilization to revolution.* Reading, MA: Addison-Wesley.

Tilly, C. 2008. *Contentious performances.* Cambridge, UK: Cambridge University Press.

Wampler, B. 2007. *Participatory budgeting in Brazil: Contestation, cooperation, and accountability.* University Park, PA: Penn State Press.

Wang, X., Joel S. Migdal, Atul Kohli, Vivienne Shue, and Peter Evans 1999. Mutual empowerment of state and society: Its nature, conditions, mechanisms and limits. *Comparative Politics* 31, no. 2: 231–49.

Wolf, E.R. 1969. *Peasant wars of the twentieth century.* Norman, OK: University of Oklahoma Press.

Wolford, W. 2007. From confusion to common sense: Using political ethnography to understand social mobilization in the Brazilian Northeast. In *New Perspectives in political ethnography*, ed. Lauren Joseph, Mathew Mahler, and Javier Auyero, 14–37. New York: Springer.

Wolford, W. 2010a. Participatory democracy by default: Land reform, social movements and the state in Brazil. *Journal of Peasant Studies* 37, no. 1: 91–109.

Wolford, W. 2010b. *This land is ours now: Social mobilization and the meanings of land in Brazil.* Durham, NC: Duke University Press.

Yashar, D. 2004. *Contesting citizenship in Latin America: The rise of indigenous movements and the postliberal challenge.* Cambridge, UK: Cambridge University Press.

Institutionalizing economies of opposition: explaining and evaluating the success of the MST's cooperatives and agroecological repeasantization

Anthony Pahnke

Dominant conceptions of social movements consider their constitutive feature the *disruption* of order, not practices around *building* it. In this paper, I challenge this notion by analyzing the Landless Workers' Movement (MST)'s relatively successful efforts to *institutionalize* the practices of agricultural production developed by its members in cooperatives and agroecology. Through this analysis, I show that the movement's administration of a democratically managed form of agricultural production exemplifies a unique form of social movement resistance – namely, what I call *self-governmental* resistance. Rather than reformist or revolutionary contention, self-governmental resistance – performed by movements like the MST – *redevelops* state policies by vying for and often taking control over the design and implementation of agricultural production.

> Without cooperatives, the settlements do not survive.
>
> -Interview, MST Director, Sector of Production – Paraná State, 4 December 2011

The Movimento dos Trabalhadores Rurais Sem Terra, Landless Workers Movement or MST receives most recognition for their land occupations. Images of these actions filled the pages of Sebastião Salgado's 1997 book *Terra*, inspiring scores of ally groups (known as the Friends of the MST) throughout Europe and the United States. Besides this trademark tactic and their immense size – present in 24 of Brazil's 26 states and claiming a million and a half adherents – the MST's struggle for agrarian reform includes more than land redistribution, encompassing the implementation of an alternative educational program, development of a peasant-centered approach to health care and promotion of communal and/or peasant agricultural production techniques. In fact, much of the movement's focus – especially of late – has centered on what happens *after* families claim land. This period, known as '*a luta na terra*', or 'the struggle *on* the land', differs from '*a luta pela terra*', or 'the struggle *for* land' – namely, the occupation and encampment phases of land acquisition and development.[1] Despite this clear demarcation of phases in the

---

[1]For more, see Carter and de Carvalho (2009).

movement's struggle, scholars have yet to conduct systematic research on how the MST's efforts in *'a luta na terra'* (the period following land occupations and encampments) (1) remain contentious in ways *other than* protests and occupations and (2) succeed or fail. This paper addresses these two issues by focusing on the MST's innovative efforts to institutionalize alternative modes of agricultural production. Specifically, I focus on the movement's efforts to organize cooperatives and promote agroecological production.

This study of the MST draws our attention to two overlooked qualities of social movements in general. Concerning the first quality, I discuss how an actor – a social movement – that we typically associate with *disrupting* order can also *establish* order. I demonstrate that in the order created by the MST, we find rules and practices that constrain, as well as enable, relatively autonomous, member-led efforts to develop alternative modes of agricultural production. I see the movement's efforts to formulate rules, procedures and protocols for their own members as exemplary of *institutionalization*. I disagree with the notion – prominent in social movement studies (McCarthy and Zald 1977; Piven and Cloward 1979; Meyer and Tarrow 1998) – that social movement institutionalization *always* entails demobilization or the adoption of conventional forms of interest representation. Challenging this claim, I demonstrate how rule- and procedure-making constitute a form of social movement resistance just as disruptive as protest, but in a different way.

The second quality concerns how the MST's form of institutionalization is, in fact, exemplary of social movement resistance. This issue comes from the need to precisely identify what a movement resists and how they do so. Stated in another way, simply because 'movement' is in the MST's name does not automatically entail that they engage in resistance. For example, movement educational provision may develop because a government does not wish to, or cannot, deliver this particular service. In these situations, we do not find *collective resistance* – a defining quality of a social movement – but governmental authorities potentially ignoring social actors or delegating policy execution to them. In short, resistance has to be illustrated.

Rather than having been delegated authority by the Brazilian government, or existing in its absence, I argue that the MST's institutionalization of agricultural production remains contentious. Central to the MST's resistance, and what I discuss throughout this contribution, is how the movement encourages cooperation in its decentralized, small-group brand of agricultural production that features localized economic production as a challenge to private property and state power. MST institutionalization resists the design and implementation of 'normal' governmental agricultural production policies. The concept I develop to capture the MST's institutionalization of agricultural production as a form of movement resistance is *self-governance*.

This paper has four sections. In the first, I explore the meaning of social movement institutionalization and explain how self-governmental resistance is analytically distinct from revolutionary and reformist contention. In this section, I discuss the need for an alternative theory to explain social movement resistance, focusing on agency and, in particular, on strategy. Second, I present the Brazilian government's version of agricultural production policy that privileges state-directed, neoliberal conduct. This section identifies the MST's targets. I dedicate the third section to an extended analysis of movement cooperatives and attempts to 'repeasantize' through promoting agroecology. Exploring these features of the movement's own way of organizing agricultural production reveals precisely their oppositional nature vis-à-vis state power, private property and particular governmental policies. Lastly, I explain how the movement has institutionalized resistance, and evaluate their success. Despite achievements, the MST's efforts in production have lagged behind other efforts in education and agrarian reform. I find that more than contextual factors, the reason for limited success is strategic.

Social movement institutionalization, the state and strategy

Movement definitions (Scherer-Warren 1987; Cardoso 1992; Alvarez, Dagnino, and Escobar 1998; Aminzade 2001; McAdam and Tilly 2001; Snow, Soule, and Kriesi 2008) privilege practices that involve disrupting the status quo, formal authority of states, corporations and/or political, cultural and/or economic power. Theorists also agree on the centrality of *persistence* when conceiving of social movement resistance.[2] Within these minimal definitional constraints, Sydney Tarrow notices what he calls 'the paradox of disruptive contention': in the need to 'sustain disruption', certain movements – for example, the US Civil Rights Movement – create 'formal organizations' that 'institutionalize' through adopting moderate tactics, changing into political parties or developing interest groups (1998, 98–101).

Recognizing Tarrow's insight, yet disagreeing with how he sees social movement development, Tilly (1978) and Goldstone (1991) observe another mode of movement institutionalization in revolution. Tarrow sees movements that we can call *reformist* because they institutionalize by adopting conventional modes of opposition. In a different fashion, revolutionary movements such as the Bolsheviks or Castro's July 26th movement substitute states by executing parallel functions autonomously, e.g. by issuing currency, performing marriages, organizing legal systems, etc. In dual power, revolutionary resistance *occupies* state power in order to exercise it.[3] Revolutionary movements also, through encouraging cooperative forms of production or land redistribution, challenge private property. Reformist movements either ignore it or, as displayed by prodemocracy movements that encouraged transitions away from communist rule, promote it.

For Tarrow, and social movement theory generally, a problem arises in confusing *government* with *state*. Theories of state power have largely been abandoned to institutional analyses that isolate governmental units, such as specific administrations or bureaucracies, or that focus on vague attributes, such as repression or facilitation.[4] They use government and state interchangeably when, in fact, they are distinct. James Scott's understanding of state power as consisting of myriad attempts to homogenize culture, centralize decision-making authority and instantiate an unequal division of administrative roles between social actors and elites allows scholars to focus on power, not merely the units that exercise it (1998). The point is that states are the exercise of certain kinds of power, not equal to some fixed or specific organization. Any actor can 'perform state', for example by centralizing decision-making, homogenizing culture and creating unequal administrative roles, for example a king, a republic

---

[2]Riots, on the other hand, or even soldiers' 'undeclared desertions', as Scott reminds us (1990), are also classified as resistance. They do not constitute the mode of resistance that social movements undertake because they are not collective or sustained.

[3]I chose the term 'institutionalization' over 'organization' because the latter term entails actual buildings such as offices, general initiatives to mobilize protest or internal movement forms (e.g. compare Staggenborg 1988; Kriesi 1995; Lofland 1996; Tarrow 1998). I adopt the meaning of institutionalization as used in comparative politics (e.g. see North 1990 or Helmke and Levitsky 2004), which focuses on rules and procedures that affect action.

[4]For institutionalist analyses, see Shepsle and Weingast (1995) and Tilly, Evans, Rueschemeyer and Skocpol (1985).

or the mafia.[5] Government refers to organizations that make decisions concerning how to provide order, governing in accordance with state power. The ways that order is provided – including the design of policies and modes of implementation – is governance.[6] Because governance involves, in some sense, 'providing order', any actor, such as a movement, business, etc., may govern in accordance with state power, or against it.

Considering this discussion in light of reformist and revolutionary movements reveals how they share one striking problem: the *reproduction* of state power. Reform-style movements institutionalize resistance around demands for greater inclusion in the dominant society, for example, through lobbying, pressuring legislators and/or demanding new laws. Their resistance pressures governments to allow marginalized individuals and groups to participate in the exercise of state power. Revolutionary movements differ because of their focus on dual power. They, however, also reproduce state power by demanding and centralizing decision-making authority in some actor like a political party that arrogates power over policy, subordinating non-party actors in an unequal administrative division of labor.[7]

Reproducing state power does not mean that reform and revolutionary movements fail their members; these movements contribute to social, political and economic changes in the form of creating new policies, providing access to material goods and perhaps offering a sense of empowerment to historically marginalized populations. The problem is that governments have used state power to institutionalize policies that exclude certain actors from administrative roles and decision-making authority. By reproducing state power, reformist 'inclusion' and revolutionary 'dual power' fail to uphold principles of equality, close pathways to alternative modes of identification and allow status quo decision-making procedures to persist.

The MST's mode of institutionalization differs from reform-style and revolutionary movements. Their marches against agribusiness and promotion of communal cooperatives could be considered efforts to 'look for and create cracks' in capitalism by refusing to participate in dominant economic structures (Holloway 2010). While the 'crack' metaphor provides a general description of their actions, the movement's efforts to produce a distinct identity to represent their struggle, as well as their attempt to challenge particular governmental policies, warrant a concept that captures the complexity of their resistance. Some consider 'participatory democracy' that concept (Wolford 2010; Starr, Martínez-Torres and Rosset 2011). This draws our attention to the processes and procedures integral to internal movement affairs. It nonetheless leads us to miss how the movement's struggle is organized in specific ways against particular targets, for example state, government, policy and/or economy. It is also true that the MST's struggle, especially in agricultural production, is oriented around autonomy (Van der Ploeg 2009; Schneider and Neiderle 2010). Ploeg is especially helpful here, illustrating with the concept of 'repeasantization' how pluriactivity, reproduction, maintaining a resource base and interaction with nature together

---

[5]For more on the difference between state and government, see Nettl (1968). For more on state power see Mitchell (2006); Foucault (1991).
[6]Scholars of governance note the centrality of decision-making. Rosenau (1995) defines it as control and steering mechanisms. Schmitter (2002) discusses governance as rule-setting and decision implementation, while others include public good provision (Ronit and Schneider 1999; Pathberg 2004; Bevir 2007). Theorists of participatory democracy highlight decentralized decision-making (Avritzer 2006; Wampler 2007).
[7]While he does not make the distinction between reform and revolutionary struggles in the same way, Holloway (2002) also recognizes the same problem in social movement contention.

constitute efforts to combat deprivation and dependency on capitalist markets. Ploeg's discussion on its own, however, omits how repeasantization efforts involve and challenge governmental policy.

Focusing on its efforts to vie for control over the design and implementation of particular policies, rather than state policy overall, one can see how and why the MST defies classification as reformist or revolutionary. The movement's effort to instill and represent their struggle with their own movement-centered identity shows why their mode of resistance is not merely about governance, but a kind that *they* construct and reproduce. The MST, through which I illustrate the concept of self-governmental resistance, implements alternative rules and procedures that enable and constrain member conduct[8] – in other words, institutionalize against state power *and* private property. To challenge state power, self-governmental movements institutionalize the qualities of decentralization, heterogeneity and equality in designing and implementing *certain* policies and services. Succinctly put, reform-style movements desire *inclusion* into state power, revolutionary movements want to *occupy* it and self-governmental movements *divide* it (Table 1).

To explain the development and evaluate the success of this new mode of social movement resistance displayed by the MST, I focus on strategy. Strategy[9] – more than contextual factors like state strength, neoliberal adjustment and the presence or absence of elite allies – explains the way that resistance unfolds, because it directly connects a movement's vision of itself and its targets with how contention unfolds. *How* to achieve a goal – the tactics – is determined by *what* the plan is – the strategy.

I identify three classes of strategies that apply to social movement activity: direct-action, mediated and instrumental. Self-governmental movements, with the MST as the best case, deploy direct-action and instrumental strategies in governing particular policy areas without engaging in mediation.[10] I measure strategy in terms of strength, focusing on (1) the coherence of goals, (2) the consistent iteration of objectives and (3) the level of knowledge preparation concerning how to interact with targets. Consistently iterating objectives solidifies the movement members' commitment to goals and their identity in opposition to their targets. Rules and procedures are thus communicated with greater ease because of common themes. Also, a coherent elaboration of goals makes plans and opponents easy to understand for leaders and members, facilitating the regularization of action. High levels of preparation and knowledge acquisition assist movements to create plans on

---

[8]This could be considered a form of governmentality, borrowing Foucault's terminology. As far as I know, however, Foucault never considered governmentality as a form of resistance. Appadurai (2001) attempts to do this, in a way, yet under-specifies the dynamics of 'governmentality from below' and how it comprises a distinct form of contention.

[9]Gramsci understands the relationship of strategy and power as the difference between a 'war of position' and a 'war of maneuver', drawing our attention to the importance of dividing mobilization and preparation (1971, esp. 229–39). For game theorists, no matter the game – chicken, battle of the sexes, ultimatum – the desire to maximize payoffs drives planning and goal-directed behavior (Binmore 2007). To place content into such vague discussions of strategy, Jaspers (2006) and Maney et al. (2012) emphasize a focus on social context, time, culture, goals, emotions, dilemmas, symbols and interactions. I agree that strategy involves gain, goals and success, but not that all actors seek gain in the same way.

[10]In direct-action strategies, movements present their own demands and conduct their resistance themselves, or in a way that remains under their control. Direct action forces opponents into action outside typical, formal political channels. Mediated strategies depend on an actor other than the movement to express and implement goals. Within this general category, we find electoral, campaign or legislative strategies. In instrumental strategies, a movement requires another actor to execute an action, yet ultimate decision-making resides with the movement and not the non-movement ally.

Table 1. Kinds of movements, defining qualities, and examples.

Kind	Reform	Revolutionary	Self-governmental
Definition	Initially challenges state power, yet later reproduces it; seeks inclusion and incorporation into government; ignores or promotes private property	Initially challenges state power, yet later reproduces it; seeks parallel governments and excludes already-existing ones; demands and abolishes private property	Continuously challenges state power; governs particular services and does not practice dual power; demands and abolishes private property
Examples	United States Civil Rights Movement, LGBT Movement, Environmental Movement, Anti-Nuclear Movement	Russian Bolsheviks, early twentieth-century Fascists (e.g. Italian and German), Sendero Luminoso (Shining Path), July 26th Movement	MST, Rondas Campesinas (Peasant Leagues), Zapatistas, Piqueteros, Confederación Identidades Indigenas de Ecuador (Confederation of Indigcnous Nationalities of Ecuador) or Conaie

how to implement objectives. Plan preparation entails a movement acquiring knowledge of opponents' strengths and weaknesses, facilitating successful interaction within a movement, and with targets and allies. In the next section, I present the Brazilian government's agricultural policies to show what the MST opposes, before I apply my theory of strategy to the movement's institutionalization of self-governmental resistance.

Neoliberalism and state power in Brazilian agricultural governance

Over the last 30 years, rather than competing with state power, neoliberal governance has developed within it. We see this in two periods, with the first commonly referred to as the Washington Consensus, running from the late 1970s and early 1980s until the mid to late 1990s, and the subsequent 'Post-Washington' consensus, which continues through the present day. This division is rooted in differences between macro-economic reforms focused on changing monetary policies, privatizing state-owned enterprises and liberalizing trade (Williamson 1990), and, later, micro-level efforts grounded in local actor participation, more attention to social programs (in ways other than cutting) and greater autonomy for national governments to craft development strategies (Weber 2002; Stiglitz 2008; Birdsall and Fukuyama 2011). Peck and Tickell (2002) call these phases 'roll-back' versus 'roll-out' neoliberalism, with the shift to perceived failures a response by the practitioners who led the first round of reforms. The practices, encouraged in both, emphasize what Burchell (1996) calls 'an enterprise form of conduct' – in other words, promoting actors such as individuals, families or governments to consider themselves as individualistic, cost-benefit calculators. Furthermore, neoliberalism is 'paradoxical': in portraying itself as the 'critic' of government by calling for deregulation in favor of the 'market', it simultaneously depends on governmental authorities for implementation (Rose 1996).[11]

In Latin America, both sets of reforms came at the expense of interventionist states. Before, governments from Mexico to Argentina, as well as Brazil, crafted credit, technical

---

[11]For more on the historical context of this transition in the late 1970s, see Foucault, Senellart and Ewald (2009).

assistance and tax policies to favor large-scale agricultural production to acquire foreign currency as part of import substitution industrialization, or ISI (Haggard 1990; Frieden 1991). Large-scale agricultural production was favored usually in the form of grain production for export (Anderson and Valdés 2008). The turn to neoliberalism in the 1980s insulated macro-economic policy-making from social actors, while opening social policy to their participation (Collier and Handlin 2009). Evelina Dagnino (2007) calls this the 'perverse confluence' of neoliberalism and participatory governance, as greater social involvement is contingent on following neoliberal norms. Her insight draws our attention to co-optation, which the MST, as I shall demonstrate, avoids.

Neoliberalism's 'roll-back' phase in Brazil begins in the late 1980s and is entrenched during Collor de Mello's administration (1990–1992) with the ending of price-fixing policies and the cutting of subsidies.[12] At this time, we see a decrease in the amount of money distributed to all agricultural producers. Cuts in the late 1980s continued during the Collor administration, the following Franco era (1992–1994) and the first mandate of Fernando Henrique Cardoso (FHC; 1994–1998) (Araújo et al. 2007). In place of subsidies, large agribusiness operations began to take out loans from private banks, rather than relying on governmental support.[13]

In FHC's second mandate (1998–2002), 'roll-out' replaces 'roll-back'. Grounding this transformation – specifically for how we understand MST resistance – is how agricultural policy favors an individualistic, 'family farmer ethic' in small producers. In interviews with bureaucrats involved in agricultural credit and land distribution programs for rural actors, I was often told how the 'family farmer' is different from the 'peasant' because the family farmer is considered rational, business-oriented and superior.[14] Marking the shift towards the explicit adoption of the family farmer in the governmental discourse was the development of two ministries for agricultural policy (Fernandes, Welch and Gonçalves 2010). Special education, land reform and production policies were also created in response to massacres of landless peasants – which included MST adherents – in the states of Roraima and Pará (Ondetti 2008).

The FHC administration's 'Reforma Agraria: Compromisso de Todos', or 'Agrarian Reform: A Compromise for Everyone', details the policies used to produce the family farmer subject. In what the government called the 'o novo mundo rural', or 'the new rural world', all small producers were to receive the same treatment by 'integrating agrarian

---

[12]Prior to the neoliberal turn, the Brazilian system was characterized by price controls for specific crops, subsidized credit for technological investments (e.g. tractors) and particular governmental agencies in charge of specific crops (e.g. coffee and sugar). These programs developed in the early twentieth century, based on goals to secure rural-to-urban population shifts and providing food cheaply to the newly amassing urban populations (Barros 2008).

[13]Reliance on private banks has caused large producers to incur significant debts; in 2000, debt was 25 percent of gross agricultural production, and in 2005, it was 25 percent (Damico and Nassar 2007), leading to government debt restructuring and bailouts twice over the last 15 years (Rezende and Kreter 2007).

[14]In one interview with a Ministério de Desenvolvimento Agrário or Ministry of Agrarian Development (MDA) official who works in the Secretário de Agricultura Familiar or Secretary of Family Farming (SAF), I was told how credit policies were intended to create 'family farmers, you know, like what you have in the United States' (Interview with Representative of MDA-SAF, 11 February 2011). Others in the Instituto de Colonização e Reforma Agrária or Institute of Colonization and Agrarian Reform (INCRA) – the institution mainly involved in agrarian reform – would use 'farmer' in English, to distinguish between 'peasants' and the supposedly superior US 'farmer' (Interview with Representative of INCRA, INCRA, 2 February 2011).

reform beneficiaries into family farming' (Jungmann 2001). To 'integrate', the already-existing credit policy for agrarian reform beneficiaries, o Programa de Crédito Especial para a Reforma Agrária or Special Program for Agrarian Reform Credit (PROCERA), tripled from R$ 89 million in 1995 to R$ 250 million in 1998. Despite this increase at this brief moment, PROCERA was later eliminated when agrarian reform beneficiaries were absorbed into the already-existing credit policy for small producers known as the Programma Nacional de Fortalecimento da Agricultura Familiar, or the National Program for Strengthening Family Farming (PRONAF).

The practices constitutive of PRONAF are intended to instill neoliberal conduct. First, families are expected eventually to purchase the land and pay back credits distributed to them as a way to promote business proprietorship. Furthermore, recipients must meet certain group-based requirements, divided according to gender, age and, most importantly, income, to access credit (MDAa 2014).[15] One objective is 'expansion and improving products and their production in order to better family earnings for individuals and collectives' for 'more capital, production and profit ... creating incentives for family farmers and stimulating a sense of responsibility' (MDAb 2014). The allusion to responsibility fuses market principles and imperatives – the pursuit of profit – to the family, fashioning the latter out of the former. Such subjects, who Ploeg (2010) would call 'agricultural entrepreneurs', are intended to exist fully within the capitalist exchange economy. Despite the group-based classification system, the focus is individuals: the overwhelming percentage of recipients have been individuals (98 percent between 1999–2004), with little to no money or contracts signed with cooperatives or group associations (2 percent), or devoted to projects for infrastructure (Correa and Silva 2009, 56).

Neoliberal reforms were strengthened without fundamental changes when the center-left Lula administration came to power in 2002. To assuage business and investor anxiety, while feeling the pressure from international and domestic forces, Lula continued Cardoso's macro-economic monetary policies (Amaral et al. 2008). Additionally, the funding provided for INCRA increased, as well as credit through PRONAF for small producers, leaving unaltered the neoliberal compass in all areas of agricultural policy. Neoliberal governance in agricultural policy, particularly of the 'roll-out' phase that began during the Cardoso regime in the 1990s, became entrenched under Lula's leadership.

We find oppositional conduct during Lula's time not only by the MST, but also in credit policies implemented by INCRA. Known as 'installation' and 'promotion', and distributed solely to agrarian reform beneficiaries, these credit programs differ from PRONAF. In fact,

---

[15]The stipulations are the following: Group A (recently settled farmers), Group B (family famers not living on settlements), Group A/C (settled farmers previously in group A), agroindustry (for investments in infrastructure, crop commercialization, artisan production, rural tourism) agroecology (organic farming), Eco (sustainable development), Rainforest (sustainable development focused on reforesting areas), Semi-Arid (infrastructure development primarily in the northeastern part of the country), Woman (targeting programs for women), Youth, Defrayal and Commercialization (for farmers and their cooperatives), Quotas (for cooperatives), Microcredit and Food (for particular crops such as milk, rice, beans and wheat). PRONAF works by the government fixing interest rates on particular loans that are destined to these groups. Yearly defrayal interest rates are fixed for each PRONAF subcategory based on a farmer's income: to $5,000 Reais it is 1.5 percent; between $5,000 Reais and $10,000 Reais it is 3 percent; from $10,000 Reais to $20,000 Reais it is 4.5 percent; and from $20,000 Reais to $30,000 Reais, 5.5 percent. Investment credits are also fixed based on income levels: to $7,000 Reais, 1 percent per year; between $7,000 to $18,000 Reais, 2 percent; $18,000 Reais to $28,000 Reais, 4 percent; and from $28,000 to $36,000 Reais, 5.5 percent.

their funding target is groups, not individuals, who are to engage in construction projects collectively. They are also referred to as 'social credits' and 'lost funds', and recipients are not expected to pay them back (Interview with INCRA Representative, 31 January 2011). While the group focus and lack of repayment directly opposes neoliberalism's imperatives to produce profit-seeking, 'credit-worthy' agricultural entrepreneurs, INCRA has become forced to adopt neoliberal precepts such as cutting social programs and promoting a business ethic. The Tribunal das Contas da União, or State Audit Court (TCU), has criticized INCRA for the way it provides agrarian policy beneficiaries public support (Tribunal das Contas da Uniao 2010). TCU audits have pressured INCRA on sections of the 1988 Constitution and Law 8629 that require families to purchase land and pay back credits, which is officially known as 'emancipation'. I was told by TCU officials that INCRA's practices in land and credit policy 'promote social dependents' and 'are inefficient' (Interview with TCU lawyers, 13 September 2011). The powers granted to the TCU in the 1992, 'lei organica', or 'organic law', include oversight of all public policies accounts, finances, operations and assets, in order to 'evaluate costs', placing the institution in a position to ensure the application of a business rationality to all governmental agencies, including INCRA.

The language expressed by both INCRA and MDA officials, as well as that found in *Compromisso* and PRONAF criteria, reveals the central place of neoliberalism and state power in Brazilian agricultural policy. Policies and credit work as micro-level instruments in the state-led effort to promote neoliberal governance, placing producers on the path to become agricultural entrepreneurs. 'Experts', from officials in the MDA to INCRA, administer and deliver these credits, along with the knowledge concerning family farming, to social actors. The fusion of neoliberal governance with state power appears in how neoliberalism's 'roll-back' and 'roll-out' phases were implemented. Despite differences in terms of policy content and directives, the hierarchy established between the state agencies and producers demarcated different spaces for state and social, or economic, actors – producing a binary that subordinates a passive, rural recipient who must meet elite-determined criteria. Neoliberal forms of economic behavior encouraged and promoted are uniform and singular, attempting to homogenize a singular culture of agricultural entrepreneurship. As I will show in the next section, the MST's cooperatives and efforts to repeasantize through practicing agroecology challenge the conduct and policies encouraged by the Brazilian government.

MST cooperatives and agroecological repeasantization as oppositional production

In addition to protests against agribusiness and occupations of experimental production areas,[16] MST resistance takes the form of the systematization – what I call institutionalization – of an alternative agricultural production project that challenges state power and private property. Specifically, the movement challenges the design and implementation of particular policy areas – neither attempting to take state power overall nor simply demanding inclusion – through decentralizing decision-making, opposing the status of subordinate social actors in economic activities and proliferating local forms of production. These qualities lay the foundation for alternative ways to organize the policy area of agricultural production. This reveals how the MST divides state power, which is a central element to self-governmental resistance.

---

[16]The two main examples of the latter are the MST occupations of research centers of Aracruz cellulose in Rio Grande do Sul in 2006 and Syngenta in Paraná, 2007.

One example of the MST's efforts to institutionalize oppositional production is through its Cooperativas de Produção Agrícola or Agricultural Production Cooperatives (CPAs).[17] Since the late 1980s, the CPA has been seen by the movement as the 'superior form of cooperation' (MST 2008, 74). CPAs – totaling 49 with 2229 participating families (Carvalho 2006) – have the same guiding rules and encourage similar kinds of opposition to neoliberal agricultural policy. One CPA I visited,[18] the Cooperativa de Produção Agropecuária Nova Santa Rita or Cooperative of Agricultural Production in New Santa Rita (COOPAN), currently works with members of 30 families, decreased from the original 60 in the mid-1990s. The 30 families divide production into different areas, or sectors, rotating between pork, dairy and rice. Representatives of each sector periodically meet to discuss what they will produce and need, which they then communicate to the entirety of the cooperative general assembly for debate and approval. Participating CPA families' houses are separately arranged in a circle, known as the 'agrovila', with a communal garden, as well as a kitchen and school. Collective organization – not just of eating, but of housing – also characterizes the division of land that belongs to CPAs. In these areas, families have individual houses, but production and socialization spaces are not privately demarcated. Salaries are divided equally among members,[19] with deductions for consumption (e.g. if someone eats some pork, it is deducted from their monthly pay). Communal production and living allows some members to leave for periods of up to 2 years to contribute to the movement's political activities, while a core group maintains the operation (Field notes COOPAN 27 March 2011; MST-CONCRAB 1995; MST 2008).

In addition to supporting and promoting militants, the communal nature of property ownership resists the emphasis found in neoliberalism to create private property. Efforts to 'emancipate' – in other words, privatize – promoted by the TCU and INCRA are contested by the movement, which believes that ultimate ownership should remain with public authority while families receive usufruct rights (MST-CONCRAB 1995, MST 2008). In fact, the vast majority of families have 'titulos de concessão', or usufruct title, and not 'titulos de dominio', or full ownership. In Paraná, an INCRA representative showed me boxes of 'titulos de dominio' that were returned blank in defiance of governmental orders to privatize (Interview with INCRA Representative, PR 1 August 2011). The director of INCRA for the state of São Paulo told me that only one settlement of the state's 252 had been 'emancipated (Interview with INCRA Representative, SP 1 November 2011). Communal work and coordinated opposition to privatization are regularized forms of conduct that defy neoliberal policy and private property.

MST cooperation also resists state power through decentralizing decision-making and challenging a strict division of labor. In the democratic, group-based division of labor in the cooperative's different areas, spending and buying priorities are subject to debate. A porous

---

[17]There are two main kinds of MST-promoted cooperatives, the CPA and CPS. CPS, Cooperativas de Prestação de Servicos, or Service Cooperatives, do not collectivize production on site, but work mainly to produce technical assistance to members and facilitate production. For more, see MST (2008); ITERRA (2001).

[18]Besides COOPAN, the other CPAs I visited were Cooperativa Agropecuária Vista Alegre (COOPAVA) in São Paulo, Cooperativa de Produção Agropecuária Cascata (COOPTAR) in Rio Grande do Sul and Cooperativa de Produção Agropecuária Vitória (COPAVI) in Paraná.

[19]While this counters how typical businesses operate, it does not make the MST unique. From the cooperatives of Mondragón in Basque areas in Spain to the Kibbutz in Israel, equitable distribution of earning is a common cooperative practice. In fact, the MST crafted their form of cooperation by researching these cases (MST 2001).

division of labor characterizes each group, as nothing forces members to remain in one specific area. Such regular, systematized decision-making rules collapse divisions between 'state', 'society' and 'economy'. Neoliberal governance, through the exercise of state power, strictly separates these spheres, organizing them in hierarchies that subordinate social actors to centralized decision-making within governmental institutions that are insulated from political contestation. The way the MST institutionalizes resistance, in practice and in principle, also contests status quo agricultural production policy's individualizing efforts. Whereas the groups established in PRONAF credit programs reveal a hierarchy between the state authorities that create them and the social actors that perform them, MST production groups distribute decision-making power more widely and equally.

Another cooperative that is not confined to one area, but dispersed territorially throughout Rio Grande do Sul, is the agroecological seed cooperative Cooperativa Nacional de Terra e Vida or National Cooperative of Land and Life (CONATERRA). Better known by its brand name Bionatur, this cooperative is central to the MST's institutionalization of alternative agricultural production because of its focus on seeds. Bionatur is a 'social business', which means that losses are socialized, or assumed by all members, rather than falling on the shoulders of individuals (Interview with Bionatur Technical Team, Bage, RS 30 March 2011). The priority is not profit, but maintaining the production of diverse, agroecological seeds. Bionatur buys a certain percentage of the seeds from their members to sell to the government and private actors. As a rule, the cooperative challenges monocultural production practices by purchasing a limited amount from each producer. This encourages the producers to diversify and not become dependent on sales to the cooperative, avoiding specialization and monoculture.

Bionatur also opposes homogenizing tendencies inherent in state power and neoliberalism through localizing production practices. The groups that characterize Bionatur differ from CPAs in that they are dispersed throughout Rio Grande do Sul, rather than remaining in one settlement. Usually made up of five or six growers, each group has one lead coordinator and one MST-affiliated extension agent who communicates with all the groups in a region. Through functioning in multiple areas, different kinds of seeds can be grown, unique to specific regions (Interview with a Bionatur grower, Candiota, RS 4/2/2011). This structure allows for various kinds of seeds to be grown according to localized production methods, challenging efforts to standardize practices not just of the agroecological production of food, but of cultural modes of production more generally.

In addition to Bionatur's efforts to promote agrecology, the MST has institutionalized ways to encourage families to stop using chemicals as they diversify production. At the end of the 1990s, when many production cooperatives were in crisis economically (MST-CONCRAB 1999), movement leaders recognized the need to pursue an alternative to chemically enhanced monocrop production models and methods. This led to a period of extended debates during these cooperatives' 'transitional period' (Interview, MST Regional Director-Porto Alegre, 23 March 2011). One effect was the explicit embrace of agroecological production. Small family-centered forms of production have always sat at the center of the MST's struggle for agrarian reform. Occupying land to secure this key resource for families to reproduce themselves and sell their surplus is a key tactic in the movement's 'repeasantization' efforts. The adoption of agroecology as a movement platform in their fifth national congress in 2007 functioned to coordinate activities against government and capitalism.

Like cooperatives, repeasantization is rule-based. For example, in a visit to one site in Rio Grande do Sul, I attended a meeting of MST peasant farmers with the movement's technical assistance cooperative, Cooperativa de Serviços Técnicos or Cooperative of Technical

Services (COPTEC). Shortly before visiting one producer's site that functioned as a demonstration, COPTEC technicians led a presentation for the group of around 25 individuals about permaculture. The discussion centered on how producers were to acquire knowledge of their specific sites, including locations of water sources, nature of soil quality and layout of the terrain. They were told that 'they too, were to do research, that everyone researches, not only people with college degrees' (Fieldnotes, Settlement in the Porto Alegre Area, 3 March 2011). In Paraná state, I also attended various agricultural technical schools for movement technicians. Similar to what I observed in Rio Grande do Sul, technicians were learning how to convince member families to decommodify their production practices by learning how to do without purchased inputs and thus autonomously reproduce their means of existence (Fieldnotes, Escola Milton Santos, 6 July 2011).

MST technicians enable the proliferation of local practices through disseminating technologies and encouraging localized research. While state power favors standardized practices and centralized decision-making procedures, the movement's project contests uniformity by decentralizing decision-making power. Institutionalizing decentralization means, especially with respect to agroecological repeasantization, that the movement coordinates regular regional meetings and implements a pedagogy wherein families learn to decommodify their production practices. While occupations have always secured access to land for peasant producers, the promotion of agroecology is a continuation of the movement's repeasantization efforts. Their attempts to break patterns of dependency and organize autonomously from the capitalist economy show systematic, coordinated procedures that qualify as self-governmental resistance: the movement vies for and often takes control of agricultural policy in a way at odds with state power and private property within the context of neoliberal capitalism.

Explaining and evaluating the MST's resistance in agricultural production

The MST boasts impressive numbers concerning its institutionalization of agricultural production, claiming 49 CPAs with 2229 families participating, as well as 32 Cooperativas de Prestação de Servicio or Service Cooperatives (CPS) with 11,174 members, and seven more cooperatives that deal with credit, work and other activities (Carvalho 2006). The majority of these cases are in southern Brazil. Gonçalves estimates that around 15 percent of the families mobilized by the MST to participate in cooperatives are in the state of Paraná (2008). I was told that most cases are found in the south, with the current percentage of participants at around 5 percent (Interview, MST, Sector of Production, PR State, 11 August 2011). In the same interview, I was told that less than 5 percent of members implement agroecological production techniques. If we take the movement's own tally of membership – 1,500,000 people – this means that roughly 75,000 people participate in alternative agricultural production.

Another measure for involvement in agricultural production is the number of families accessing state policies for small-scale production and who are more likely to use agroecological techniques. From numbers available in 2009 and 2010 in the Programa de Aquisição de Alimentos or The Food Acquisition Program (PAA), the percentage of finances going to agrarian reform beneficiaries went from 8 percent to 12 percent, (R$ 28,699,236 of R$ 363,381,941 to R$ 44,643,666 of R$ 379,735,466) while participating beneficiary families went from 7 percent to 11 percent (7444 of 98,340 families to 10,440 of 94,398). Because we can deduce that roughly 50 percent of agrarian reform settlements are under MST influence (CPT 2000–2010), we can conclude that the number of movement affiliates who practice agroecology is in the thousands.

A comparison of their actions in agricultural production with other MST efforts in education and agrarian reform reveals a relatively lower level of success. In agrarian reform, since 2000 – the first year that land occupations were catalogued by movement – the MST was responsible for 51 percent of occupations and 49 percent of encampments[20] amongst the 23 active agrarian reform movements (CPT 2000–2010).[21] Given that a fraction of all settlements existed prior to the MST's origins (Leite 2004; Confederação Nacional das Associações dos Servidores do Incra (CNASI) 2011), the movement has had the most success in practicing and influencing agrarian reform policy, essentially defining it. The movement's efforts in education, while lower than in agrarian reform, show a higher level of success than in agricultural production. The movement claims to have pressured the government to construct over 2250 schools on settlements (MST 2014). However, only roughly 20 schools operate under movement control (MST 1998, 2000, 2010). One particular promising development has been movement efforts in vocational secondary education. In 2010 – the one year where we have numbers to compare movement with government – roughly 40 percent of students were affiliated with the MST (24,465 total students with 10,058 affiliated with the movement; INEP 2010; INCRA 2011). High school students went to non-settlement schools in 31 different spaces under movement control (Plummer 2008). Mobilization in agrarian reform, education and agricultural production policy varies (Table 2).

While the previous section discussed *what* the nature of MST resistance in agricultural production policy was, this section analyzes *how* this resistance takes place. In this section, I show how self-governmental institutionalization and variation in success arises from direct-action and instrumental strategic planning. The former kind of strategy outlines how the movement itself will carry out its actions, while the latter is grounded in how to use certain governmental and non-governmental actors to achieve their goals. Both involve knowledge preparation, coherently articulating objectives and consistently elaborating goals.

Concerning preparation, the MST has difficulties in creating direct-action production plans for the CPAs. The origins lie with the Catholic Church's influence on the movement. Their idea of cultivating the land in common provided few specific details concerning how to manage an agricultural operation (MST 2006). In the mid-1980s, production was incorporated into the Sistema de Cooperativas dos Assentados or Cooperative System of Cooperatives (SCA), coinciding with the movement's plan to collectivize settlements (MST-CONCRAB 1997).[22] Seen by many as an imposition, many families did not agree and, as a result, demobilized (Interview with MST State Director RS, 3 April 2011).

Those who remained in the CPAs faced problems concerning how they would actually organize production. The MST itself has noted that cooperatives face debt burdens and a lack of trained personnel, as well as poor production plans (MST-CONCRAB 1999). Visiting the cooperatives allowed me to understand these issues. At successful sites, I was told how current members started with little experience in managing agricultural production.

---

[20]While land occupations and encampments lead to settlements that are officially recognized by governmental authorities, numbers are unavailable concerning movement control of settlements.
[21]The number of families mobilized stands at a total of 494,428 in land occupations, with 311,160 of them – 63 percent – organized by the MST. The movement has also mobilized the most families in encampments, with over 58 percent of the total affiliated with the MST, or 85,205 out of a total 146,295 between 2000 and 2010. These figures also come from the CPT.
[22]This sector changes again, currently existing as the Sector of Production, Cooperation and Environment since 2006.

Table 2. Different levels of Landless Workers' Movement (MST) institutionalization.

	Agrarian reform	Education	Agricultural production
High	~50% people mobilized, thousands of encampments and settlements under control		
Medium		~40% of students in certain areas under control, thousands of schools (not under control) with > 50 sites under control	
Low			< 50 sites under control, 5–15% of potential adherents participating

This is not to claim that they had no experience in agriculture. In the First Census of Agrarian Reform, conducted in 1996 – the year when many CPAs were created – we see that the overwhelming majority of agrarian reform beneficiaries had experience in the countryside, but mostly as salaried workers, squatters or peasants – in other words, in subservient, non-decision-making positions (Schmidt et al. 1998).

Also, at this time, instead of knowledge about how to establish small-scale cooperative production, we find among workers and employers widespread aspirations for large-scale, agribusiness production. At one cooperative, I was told that beginning in the 1990s, the movement intended for cooperatives to engage in monocultural production in commodities like corn and soy (Fieldnotes COOPTAR, RS 5 April 2011). It is hard to fault the movement members for holding these objectives, given that their knowledge concerning production came from working on large-scale operations.

The problems we find in the MST's direct-action strategic planning for their cooperatives also characterize the movement's instrumental approach. For instance, prior to PRONAF, the credit program known as PROCERA incentivized groups through 'tetos', or 'ceilings', 1 and 2 (De Souza 2010). The difference between the ceilings was that to receive ceiling 2 resources, which were double the ceiling 1 amount, one had to request credit as a group. So, people formed groups during what was considered a period of 'easy money' because initial restrictions and requirements were lax (Interview with COPERLAT Director, RS 1 April 2011). Instrumentalization is seen in how movement documents on cooperatives and cooperation recognize PROCERA as a vital instrument for economic development and cooperation, showing how the MST intended to use the credit program to fuel their alternative agricultural project (MST 1995). With these intentions, members who received PROCERA resources began with limited productive means and knowledge. As a result, using PROCERA led many farmers into insolvency and, thus, to demobilize. Vague objectives also led members to focus on short-term needs like purchasing food and building supplies, not on improving productive abilities (Interview with INCRA Representative, RS 14 March 2011).

Despite the problems, cooperation and cooperatives exist. In some cases, like COOPTAT and COOPTAR, the reason was that they cut their losses and sold the larger equipment they acquired with PROCERA credit when others went into bankruptcy (Interview COOPTAT Director- RS, 12 June 2011). Also, movement leaders, after recognizing the crisis, focused on 'regional reference points', or rather, special cooperatives, for example COOPAN in Rio Grande do Sul and COPAVI in Paraná, where they trained certain members in production and financing, working with youth so that they remain on the settlements (MST 2008). Direct-action plans to organize collective production cooperatives no longer applied to all settlements, but rather became focused on select areas deemed important by the movement. I was also told that in other cooperatives, such as COOPAN, members 'learned by trial and error', acquiring know-how over time. Another adjustment is observed in how the movement recruited leaders from the Basque cooperatives of Mondragón (specifically their non-governmental organization, Mundokide) to teach members how to manage finances and production in Sergipe and Paraná (Fieldnotes, Outside of Rio Bonito, PR 23 July 2011; Interview, Director of Mundokide, 25 July 2011). Production know-how has developed over time and gradually, complemented by resources acquired through past programs and allies. Both direct-action and instrumental strategic plans explain the level – albeit low – of agricultural production institutionalization.

A low level of preparation also characterizes instrumental plans to promote agroecological repeasantization. This is where we see the movement's involvement with public policies such as Programa Nacional de Alimentação Escolar, the School Alimentation Program, or PNAE and PAA. I was told by the director of the MST's confederation of cooperatives, CONCRAB, that the plan is to foment agroecological production practices for debt-burdened, insolvent producers, having them use policies by working with leaders and members of cooperatives (Interview CONCRAB President, 1 September 2011). Concrete examples include selling vegetables to neighborhood schools via these social programs (Interview COPERLAT Director, 1 April 2011). The instrumental strategy involves using government programs for the movement's alternative production project, which requires knowledge of how to use institutions such as Ministério de Agricultura, Pecuária, e Abastecimento Ministry of Agriculture, Livestock, and Supply or MAPA. For example, accessing the PAA involves regular meetings with the 'grupo gestor', or 'management group', composed of not only movement leaders but representatives of various ministries that have little history with the movement.[23] Not only the programs but the movement's decision to interact with these institutions is new. Training producer-members how to access policies, while acquiring knowledge concerning how to interact with these more professionalized institutions, is still in the early stages.

The last complication that affects both the MST's direct-action and instrumental strategies concerns coherently and consistently articulating their relationship to a target, either agribusiness or government. While many units – individual peasant plots, collectively organized cooperatives and efforts to practice agroecology – fall within the MST's agricultural production project, they do not exist harmoniously. Historically, there have been intense and fruitful debates concerning differences, with the clearest seen at the 1985 national congress on whether or not families who receive land should remain in the movement. In agricultural production, however, tensions create problems, specifically through the competing identities of 'peasant', 'colono' and 'worker'. First, multiple movement documents note the existence of a 'peasant' consciousness, not as something to cultivate, but rather as

---

[23]These include: Ministério do Desenvolvimento Social e Combate à Fome or Ministry for Social Development (MDS); Ministério da Fazenda or Ministry of Finance; Ministério de Planejamento, Orçamento, e Gestão or Ministry of Planning, Budget and Management; the Ministério de Educação or MEC; and o Ministério da Agricultura, Pecuária e Abastecimento or Ministry of Agriculture, Livestock and Supply (MAPA).

something to be replaced by 'the worker' (MST-CONCRAB 1999, 2004). I was told how 'the mentality of the colono', a term used extensively in the south of Brazil to refer to all small producers and MST members, is 'limited and needs to be trained to produce' (Interview MST Regional Director-RS, 23 March 2011). One training document notes that the 'worker' overcomes certain 'ideological behaviors' that characterize 'emerging classes like the peasant and artisan' (MST-CONCRAB 2004, 15–16). Peasants and colonos are considered easily susceptible to manipulation, isolated and reactionary.

Coherence is compromised by the MST's explicit *privileging* of a peasant form of production in their embrace of agroecology. The annual 'Jornada de Agroecologia', or 'Journey for Agroecology', which is approaching its twelfth year in the state of Paraná in 2013, focuses on disseminating – not eliminating – peasant identities and practices. The event is by no means small: at the 10th Jornada that I attended, workshops were attended by over 4000 people, and addressed a wide array of agro-methodological problems, including everything from beekeeping to smoking meat (Fieldnotes Jornada de Agroecologia Curitiba-PR, 1 July 2011). The production practices encouraged throughout the workshops were oriented around creating peasants who independently produce. Each jornada also features a notebook of the movement's agroecological best practices and techniques, a resource in the movement's repeasantization efforts.

Contradictions in identifying who is the subject of MST agricultural production are exacerbated by a relatively low number of documents and publications that might otherwise foment the movement's oppositional project. At the time of my research in 2011, there were 10 such 'cartilhas' or notebooks from the 'jornada', as well as one 'sistematização' or 'systematization' of agroecological experiences for distribution to movement producers. CONCRAB has published 11 notebooks for instruction that usually promote the worker *over* the peasant. The movement has consistently iterated the need for production in movement bulletins and publications, specifically in *Journal dos Trabalhadores Sem Terra* and *Revista Sem Terra*, with over 4500 references since 1981.[24] Yet cooperatives alone have been discussed in publications 742 times, and agroecology, 209. In comparison, 'reforma agraria', or agrarian reform – has been mentioned 7789 times, and 'educação' or education 1845. Simply put, the movement discusses and iterates themes pertaining to agrarian reform and education more than cooperatives or agroecology. Even when discussed, the guiding identities divide attention.

Such tensions encourage different ways of organizing production, rendering planning difficult. In one case, I was told that efforts to standardize agroecological techniques outside of Paraná in all movement training centers resulted in internal movement conflicts and stalemates (Fieldnotes Escola Milton Santos, 6 July 2011). The difference between worker and producer, as well as activist identities, has generated divisions in the movement that led 50 key leaders to issue an official statement and leave the movement because they felt the MST's (and allies') emphasis on production compromises their struggle (Resignation Letter from our Organizations 2011).

This focus on strategy explains the MST's oppositional economy development better than explanations that focus on the interests and preferences of organizations, agencies and the state. A competing effort to explain their resistance might focus on the shift to neoliberalism as the main cause of self-governmental resistance. Yashar's (2005)

---

[24]I conducted the search for these terms on the electronic MST database compiled and administered by O Centro de Documentação e Memória da UNESP, the State University of São Paulo (UNESP) Center of Documentation and Memory.

study of indigenous movements points to the potential that budget cuts have on weakening governmental control, creating openings for mobilization. The turn to neoliberalism – in both its 'roll-back' and 'roll-out' phases – potentially explains that small-holders, to survive, had to organize a cooperative form of production when prior state support was withdrawn. Yet 'roll-out' neoliberalism would explain cooperatives not as resistance, but policies delegated to social actors by governmental authories. MST cooperatives and repeasantization efforts, to the contrary, contest state power and neoliberal dictates. Also, MST cooperative planning and development precedes neoliberalism's 'roll-out'. Thus, the movement's own plans – beginning in the early 1980s – provide a better explanation. Policy repeal and state withdrawal may provide the general background conditions for the need to organize alternative production, but are not the primary factor.

Other theorists would predict that the level of institutionalization would vary with the presence or absence of elite allies. According to Tarrow (1998), elites facilitate interaction with movement leaders, who then enable mobilization. The issue is that we see a growth of government/movement interaction – what we could call 'openings' in the sense of creating PAA and PNAE – and persistent, stubbornly low levels of alternative agricultural production. The development of cooperatives and agroecology shows change independent of governmental assistance or elites.

Strategic problems have led to the movement's low level of success in institutionalizing an alternative mode of agricultural production. The movement hosts events, such as the Jornada de Agroecologia, which disseminate best practices and identities. The notebooks are used to train members and cultivate forms of identification. With respect to content, we see tensions. From the examples in Paraná and recent defections, as well as the extreme stance in certain training documents concerning peasant and worker production, evidence indicates that differences contribute to low levels of resistance. Rather than cohesion, there is conflict, especially concerning who is the movement's 'base'. While direct-action and instrumental strategies cause the movement's development of self-governmental resistance in agricultural production, the movement's level of institutionalization is relatively low because of their strategic strength.

Conclusion: the theoretical importance of the MST's self-governmental struggle

This paper has two main purposes, to show: (1) that the MST's organization of cooperatives and agroecological repeasantization constitute a new mode of social movement resistance that I call self-governmental and (2) that my theory of strategy offers an explanation and a way to evaluate the success of their efforts. The MST and self-governmental resistance movements – as opposed to reform and revolutionary movements – combine direct-action and instrumental strategies. In addition to differences in strategic planning, the MST differs from these two other kinds of movements through continuously challenging state power and private property by vying for and often taking control of particular policy areas.

Describing the MST's resistance in terms of self-governance addresses a potential confusion with respect to the nature of the movement. Specifically, we are observing the practice of movement-provided order, but not towards the establishment of dual power or secession. The MST is governing – not in accord with state power and neoliberal dictates – but in ways that consistently challenge both. Defining the movement in the terms of self-governmental resistance provides positive content to the MST's form of contention that allows us to see that their actions differ from reformist and revolutionary movements.

Acknowledgements

I would like to thank Mark N. Hoffman, Rebecca Tarlau and the reviewers for the suggestions that greatly improved this contribution.

References

Secondary sources: book and articles

Alvarez, S.E., E. Dagnino, and A. Escobar. eds. 1998. *Cultures of politics/politics of cultures: Re-visioning Latin American social movements*, 293. Boulder, CO: Westview Press.

Amaral et al. 2008. *Democratic Brazil revisited*. Pittsburgh, PA: University of Pittsburgh Press.

Aminzade, R. ed. 2001. *Silence and voice in the study of contentious politics*. Cambridge University Press.

Anderson, K., and A. Valdés. eds. 2008. *Distortions to agricultural incentives in Latin America*. Washington, DC: World Bank Publications.

Appadurai, A. 2001. Deep democracy: Urban governmentality and the horizon of politics. *Environment and Urbanization* 13, no. 2: 23–43.

Araújo, P.F.C., A.L.M. de Barros, J.R.M. de Barros, and R. Shirota. 2007. Política de crédito para a agricultura brasileira. *Revista de* 27: 27–51.

Avritzer, L. 2006. New public spheres in Brazil: Local democracy and deliberative politics. *International Journal of Urban and Regional Research* 30, no. 3: 623–637.

Barros, G.S.A.C. 2008. Brazil: The challenges in becoming an agricultural superpower. In *Brazil as an economic superpower*, eds. L. Brainard and L. Martinez-Diaz, 81–109. Washington, DC: The Brookings Institute.

Bevir, M. 2007. *Encyclopedia of governance*. London: Sage Publications.

Binmore, K. 2007. *Game theory: A very short introduction*. Oxford: Oxford University Press.

Birdsall, N., and F. Fukuyama. 2011. Post-Washington consensus-development after the Crisis, The. *Foreign Aff* 90: 45–53.

Cardoso, R. 1992. *Popular Movements in the Context of the Consolidation of Democracy in Brazil*. In *The making of social movements in Latin America: Identity, strategy, and democracy*, eds. Arturo Escobar and Sonia E. Alvarez, 317. Boulder: Westview Press.

Cardoso, F.H. 1997. *Reforma agrária: compromisso de todos*. Vol. 28. Brasilia, Brasil: Presidência da República, Secretaria de Comunicação Social.

Carter, M., and H.M. de Carvalho. 2009. A luta na terra: fonte de crescimento, inovação, e desafio constante ao MST. In *Combatendo a desigualdade social: o MST e a reforma agrária no Brasil*, ed. UNESP. São Paulo.

Carvalho, H.M. 2006. In ed. B.D. Santos, *Another production is possible: Beyond the capitalist canon*. Vol. 2. Verso.

Collier, R., and S. Handlin, eds. 2009. *Reorganizing popular politics: Participation and the new interest regime in Latin America*. Unversity Park, PA: Penn State Press.

Confederação Nacional das Associações dos Servidores do Incra (CNASI). 2011. INCRA - Instrumento basico de realizacao da reforma agraria e do ordenamento da estrutura fundiara e ser mantido e fortalecido. Ministerio de Desenvolvimento Agrario, Brazil.

Correa, V., and F. Silva 2009. Perfil das liberações dos recursos do PRONAF entre 1999–2006: ocorreu alguma modificação a partir da incorporação dos grupos A e B. In *Congreso Brasileiro de economia Adminstração e sociología rural*. Vol. 47.

Dagnino, E. 2007. Citizenship: A perverse confluence. *Development in practice* 17, no. 4–5: 549–556.

De Souza, J.R. 2010. *Reforma agrária e crédito agrícola : os resultados de assentamentos rurais frente à inepta política de crédito para a reforma agrária no Brasil (PROCERA)*. Sao Paulo: UNESP.

Damico, F. S., and A. M. Nassar. 2007. Agricultural expansion and policies in Brazil. In *US Agricultural Policy and the 2007 Farm Bill*, ed. K. Arha, T. Josling, D. A. Sumner, and B. H. Thomson. Stanford University: Woods Institute for the Environment.

Fernandes, B. M., Welch, C. A., and Gonçalves, E. C. 2010. Agrofuel policies in Brazil: paradigmatic and territorial disputes. *The Journal of Peasant Studies* 37, no. 4: 793–819.

Foucault, M., G. Burchell, C. Gordon, and P. Miller, eds. 1991. *The Foucault effect: Studies in governmentality*. Chicago, IL: University of Chicago Press.

Foucault, M., M. Senellart, and F. Ewald, eds. 2009. *Security, territory, population: Lectures at the college de France 1977–1978*. Vol. 4. London: Macmillan.

Frieden, J.A. 1991. *Debt, development, and democracy: Modern political economy and Latin America, 1965–1985*. Princeton, NJ: Princeton University Press.

Goldstone, J.A. 1991. *Revolution and rebellion in the early modern world*. Berkeley, CA: University of California Press.

Gonçalves, S. 2008. *Campesinato, Resistência e Emancipação: o modelo agroecológico adotado pelo MST no Estado do Paraná*. (Doctoral dissertation, Universidade Estadual Paulista).

Gramsci, A. 1971. Selections from the Prison Notebooks, ed. and trans. Quintin Hoare and Geoffrey Nowell Smith. New York. NY: International Publishers.

Haggard, S. 1990. *Pathways from the periphery: The politics of growth in the newly industrializing countries*. Ithaca, NY: Cornell University Press.

Helmke, G., & Levitsky, S. 2004. Informal institutions and comparative politics: A research agenda. Perspectives on politics 2, no. 4: 725–740.

Holloway, John. 2002. *Changing the world without taking power*. London: Pluto Press.

Holloway, John. 2010. *Crack capitalism*. London: Pluto Press.

INCRA 2010. PRONERA results, unpublished.

INEP. 2010. Censo Escolar.

Jasper, J.M. 2006. Getting your way: Strategic dilemmas in real life.

Jungmann, R. 2001. Brasil rural na virada do milênio. In *Brasil rural na virada do milênio: a visão de pesquisadores e jornalistas*. Brasilia: Ministerio de Desenvolvimento Agrario.

Kriesi, H. ed. 1995. *New social movements in Western Europe: A comparative analysis*. Vol. 5. Minneapolis, MN: University of Minnesota Press.

Leite, S.P. 2004. *Impactos dos assentamentos: um estudo sobre o meio rural brasileiro*. Vol. 6. Livraria: UNESP.

Lofland, J. 1996. *Social movement organizations: Guide to research on insurgent realities*. Piscataway, NJ: Transaction Books.

Maney, G.M., R.V. Kutz-flamenbaum, and D.A. Rohlinger. eds. 2012. *Strategies for social change*. Vol. 37. Minneapolis, MN: University of Minnesota Press.

McAdam, D.T.S. and C. Tilly. 2001. *Dynamics of contention*. Cambridge, UK: Cambridge University Press.

McCarthy, J.D., and M.N. Zald. 1977. Resource mobilization and social movements: A partial theory. *American journal of sociology* 82, no. 6: 1212–1241.

MDAa. Ministério do Desenvolvimento Agrário (Ministry of Agrarian Development). 2014. http://portal.mda.gov.br/portal/saf/programas/pronaf/2258856 (accessed March 7, 2014).

MDAb. Ministério do Desenvolvimento Agrário (Ministry of Agrarian Development). 2014. *Credito Rural PRONAF*. http://comunidades.mda.gov.br/o/880418 (accessed March 7, 2014).

Meyer, D.S., and S.G. Tarrow. eds. 1998. *The social movement society: Contentious politics for a new century*. Lanham, MD: Rowman and Littlefield Publishers.

Mitchell, T. 2006. Society, economy and the state effect. In *The anthropology of the state: A reader*. Malden, MA: Blackwell Publishing.

Nettl, J.P. 1968. The state as a conceptual variable. *World Politics* 20, no. 4: 559–92.

North, D.C. 1990. *Institutions, institutional change and economic performance*. Cambridge: Cambridge University Press.

Ondetti, G.A. 2008. *Land, protest, and politics: The landless movement and the struggle for agrarian reform in Brazil*. University Park, PA: Penn State Press.

Peck, J., and A. Tickell. 2002. Neoliberalizing space. *Antipode* 34, no. 3: 380–404.

Piven, F.F., and R.A. Cloward. 1979. *Poor people's movements: Why they succeed, how they fail*. Vol. 697. New York, NY: Random House Digital, Inc.

Plummer, D.M. 2008. *Leadership development and formação in Brazil's Landless Workers Movement (MST)* (Doctoral dissertation, The City University of New York).

Ronit, K., and V. Schneider. 1999. Global governance through private organizations. *Governance* 12, no. 3: 243–266.

Rose, Nikolas. 1996. "Governing "advanced" liberal democracies." In *Foucault and political reason: Liberalism, neo-liberalism, and rationalities of government*, eds. A. Barry, T. Osborne, and N.S. Rose. Chicago: University of Chicago Press

Rosenau, J. N. (1995). Governance in the twenty-first century. Global governance, 13–43. Chicago, IL.

Scherer-Warren, I. 1987. Uma Revolucão no cotidiano? Os novos movimentos sociais na America Latina. São Paulo: Brasiliense.

Schmidt et al. 1998. Primeiro Censo da reforma agraria do Brasil.

Schmitter, P.C. 2002. Participation in governance arrangements: Is there any reason to expect it will achieve sustainable and innovative policies in a multi-level context. In *Participatory governance. Political and societal implications*, eds. J. Grote and B. Gbikpi, 51–69. Opladen: Leske and Budrich.

Schneider, S., and P.A. Niederle. 2010. Resistance strategies and diversification of rural livelihoods: The construction of autonomy among Brazilian family farmers. *The journal of peasant studies* 37, no. 2: 379–405.

Scott, J.C. 1990. *Domination and the arts of resistance: Hidden transcripts*. New Haven, CN: Yale University Press.

Scott, J.C. 1998. *Seeing like a state: How certain schemes to improve the human condition have failed*. New Haven, CN: Yale University Press.

Shepsle, K.A., and B.R. Weingast, eds. 1995. *Positive theories of congressional institutions*. Ann Arbor, MI: University of Michigan Press.

Snow, D.A., S.A. Soule, and H. Kriesi, eds. 2008. *The Blackwell companion to social movements*. Malden, MA: Blackwell Publishing.

Staggenborg, S. 1988. The consequences of professionalization and formalization in the pro-choice movement. *American Sociological Review* 53, no. 4: 585–605.

Starr, A., M.E. Martínez-Torres, and P. Rosset. 2011. Participatory democracy in action practices of the Zapatistas and the Movimento Sem Terra. *Latin American Perspectives* 38, no. 1: 102–119.

Stiglitz, J.E. 2008. Is there a post-Washington consensus consensus? In *The Washington Consensus reconsidered: Towards a new global governance*, ed. N. Serra and J.E. Stiglitz, 41–56. Oxford: Oxford University Press.

Tarrow, S. 1998. *Power in movement: Social movements, collective action and politics*. Cambridge: Cambridge University Press.

Tilly, C. 1978. *From mobilization to revolution*. New York: McGraw-Hill.

Tilly, C., P.B. Evans, D. Rueschemeyer, and T. Skocpol. 1985. *War making and state making as organized crime*. Cambridge: Cambridge University Press.

Tribunal das Contas da Uniao. 2010. *Relatório e Parecer Prévio sobre as Contas do Governo da República – Exercício de 2009*.

Van der Ploeg, J.D. 2009. *The new peasantries: Struggles for autonomy and sustainability in an era of empire and globalization*. London, UK: Routledge Press.

Wampler, B. 2007. *Participatory budgeting in Brazil: Contestation, cooperation, and accountability*. University Park, PA: Penn State Press.

Weber, H. 2002. The imposition of a global development architecture: The example of microcredit. *Review of International Studies* 28, no. 3: 537–555.

Williamson, J. 1990. What Washington means by policy reform. In *Latin American adjustment: How much has happened*, ed. J. Williamson, 7–20. Institute for International Economics.

Wolford, W. 2010. Participatory democracy by default: Land reform, social movements and the state in Brazil. *The Journal of Peasant Studies* 37, no. 1: 91–109.

Primary sources: Interviews, fieldnotes, documents
MST-CONCRAB. 1995. Caderno de Cooperação

MST-CONCRAB. 1999. Caderno de Cooperação

MST-CONCRAB. 2004. Caderno de Cooperação

MST. 1998. Escolas Itinerantes em Acampamentos do MST.

MST. 2000. Escola itinerante uma pratica pedagogica em acampamentos.

MST. 2001. Caderno de Cooperação.

MST. 2006. Balanço Político da Cooperação no MST
MST. 2008. Carthila de Cooperação 1.
MST. 2010. Escola itinerante do MST: historia, projeto e experiencias.
MST. 2014. *A educação do MST.* www.mst.org.br/node/8302. Accessed 7/3/2014
Resignation letter from our organizations. 2011. Available at http://passapalavra.info/2011/11/48866
Fieldnotes by Author, Settlement in the Porto Alegre Area, RS 3 March 2011
Fieldnotes by Author, COOPAN, RS 27 March 2011
Fieldnotes by Author, COOPTAR, RS 5 April 2011
Fieldnotes by Author, Jornada de Agroecologia, Curitiba-PR, 1 July 2011
Fieldnotes by Author, Escola Milton Santos, PR 6 July 2011
Fieldnotes by Author, Outside of Rio Bonito, PR 23 July 2011
Interview by Author, INCRA Representative- BR, 31 January 2011
Interview by Author, INCRA Representative- BR, 2 February 2011
Interview by Author, Representative of MDA-SAF, BR 11 February 2011
Interview by Author, INCRA Representative- RS, 14 March 2011
Interview by Author, MST Regional Director-Porto Alegre, RS 23 March 2011
Interview by Author, Bionatur Technical Team- Bage, RS 30 March 2011
Interview by Author, Bionatur grower – Candiota, RS: 2 April 2011
Interview by Author, MST State Director- RS 3 April 2011
Interview by Author, COOPTAT Director- RS, 12 June 2011
Interview by Author, MST Director, Sector of Production- PR State, 11 August 2011
Interview by Author, CONCRAB President, 1 September 2011
Interview by Author, TCU Lawyers, 13 September 2011

Rural unions and the struggle for land in Brazil

Clifford Andrew Welch and Sérgio Sauer

Studies of Brazil's agricultural labor movement have generally neglected its relationship to the struggle for land, but this is neither fair nor accurate. Analyzing the rural labor movement's historical contributions to the land struggle in Brazil, this contribution has been organized into three main periods, emphasizing social relations, institutional activism and policy changes. It argues that despite the peculiarities of different historical contexts, rural labor consistently provoked protest against policies that privileged large landholders, whose concentration of power over land and labor resources continually worsened Brazil's ranking as one of the most unequal of nations. For more than half a century, the most constant opponent of this situation among the peasantry has been the National Confederation of Workers in Agriculture (CONTAG), a corporatist organization of rural labor unions founded in 1963.

Introduction

Studies of Brazil's agricultural labor movement have generally neglected its relationship to the struggle for land, but this is neither fair nor accurate (Price 1964; Medeiros 1989; Ricci 1999; Welch 1999; Martins 2002). Since its conception in the early twentieth century, the rural worker movement has been concerned with peasant reproduction, land access and usage. Since the rural union movement pre-dates the Landless Rural Workers Movement (MST, in its Brazilian acronym) and all other contemporary land struggle organizations in Brazil, the latter movements cannot be understood without considering the former. Despite a reluctance to adopt direct action tactics in the 1980s, when land occupations – one of the most ancient forms of land access – became the signature tactic of the MST, the largest rural labor movement, represented by the National Confederation of Workers in Agriculture (CONTAG, in its Brazilian acronym), supported a significant number of these actions from the second half of the 1990s (NERA 2013).

Despite the peculiarities of different historical periods, peasant rebellions consistently provoked protest against development models that privileged large landholders, whose concentration of power over land and labor resources continued to shape Brazil's trajectory (Linhares and Silva 1999; Carvalho 2004; Fernandes, Welch, and Gonçalves 2012). From colonial times to the late nineteenth century, sugarcane planters and sugar mill owners articulated their predominance of Brazil's northeastern region with a sizable subaltern population of small producers of food and providers of skilled labor. With the decline of African slavery, most of these peasants were forced to surrender their autonomy as the planters laid claim to their farms and their lives, recreating them as dependent resident

41

workers who were only permitted to produce food for their own consumption (Palacios 2009). In nearly every corner of Brazil, peasants cleared forest and brush to create farmland, only to have it taken from them by the combined power of armed landlord henchmen and compliant government authorities. In fact, it is difficult to think of a situation in the country-side where labor issues were not directly connected to landholding, or where forms of peasant resistance did not shape the systems produced.

The beginning of sustained challenges to landlord hegemony can be dated from the anarchist, socialist and communist movements that became established in Brazil in the early twentieth century as a consequence of European immigration (Hall and Pinheiro 1985). The vast majority of immigrants came to work in agriculture, especially on coffee plantations in São Paulo state, whose plantation owners and government subsidized the voyages of thousands of mostly Italian, Spanish and Portuguese families beginning in the mid-1800s (Davatz 1980). The protests of these immigrants as well as fellow Brazilian workers, including some recently emancipated slaves of African descent, shaped a labor relations system known as the *colonato*. Important to this system was the ability to accumulate enough capital to buy one's own farm. The *colonos*, as coffee workers were called, sought to accomplish this by gaining permission from landlords to plant crops for their own sale or use in the furrows between rows of coffee trees. When these terms were not agreeable, conflicts occurred, some rising to the level of involving thousands of *colonos*, such as the coffee worker strikes of 1912 and 1913, which benefitted from the organizing efforts of anarchists and socialists. These disputes continued into the 1920s, when Brazilian radicals organized the Communist Party of Brazil (PCB, in its Brazilian acronym; Vangelista 1991; Welch 2010).

Writing for this journal in 2002, Brazilian sociologist José de Souza Martins toyed with the multiple meanings of representation when he titled his contribution 'Representing the peasantry', and yet discussed the organizations established to represent rural labor only in the last few pages. For most of the paper, Martins used the term to emphasize the variety of words other than 'peasantry' used to portray changes in the structural conditions of agriculture. These changes have depended upon varied forms of labor exploitation to extract value from the land, from enslaved to free resident workers compelled to pay rent through forced labor obligations, from dependent workers tapping rubber trees in the Amazon to immigrant families colonizing small farms in the south, from independent family farmers to today's debt-burdened contract farmers. He argued that 'the key to present agrarian struggles, and to the agrarian question itself, lies not so much in the system of landholding as in changes to the labor regime introduced by rural employers' (Martins 2002, 203). While the variety of land–labor relationships is a striking characteristic of the rural development process in a country as large and geographically diverse as Brazil (Bastos 1987), in contrast to Martins (2002), we argue that the key to present agrarian struggles lies in the organizations involved in 'representing the peasantry', including rural labor unions, this paper's main focus.

A starting point for our critique of Martins (2002) is his conclusion that 'peasantry' is not an appropriate term since its late use in Brazil is linked to political options of 'mediating agents' like the MST, and that the 'rural subject' is too complex to be 'amalgamated … into a uniform "Brazilian peasantry" with a uniform political interest' (327). First, the Portuguese word for peasant, *camponês*, can be found in colonial documents, including court proceedings in which they sought to defend their interests (Palacios 2009). In one of Martins' older studies of the Brazilian countryside (1981), he alleged that the word first appeared in the 1950s when some peasant leagues (*Ligas Camponesas*) became well known. But the Communist Party had employed the term publicly in the 1920s. Second, the uniformity alleged by Martins does not appear to be the goal of any of the organizations we discuss. In contradistinction to Martins, we opt to use the term because of its openness and flexibility, a word intended to be all-inclusive of the complex relations between those who work on the land and socioeconomic structures, as temporary or permanent wage earners, squatters or propertied small family farmers, hunters or gatherers of forest products, or rural crafts-people. Third, our emphasis on representative organizations means that our rural labor subjects are in processes of consciousness-raising in regards to their rights. Since at least the 1920s, this politically aware rural worker has frequently called him or herself a peasant.

To present the rural labor movement's historical contributions to the land struggle in Brazil, we organized the paper into three main sections. Following chronological order, each emphasizes social relations and institutional activism in particular structural contexts. In the first part, the paper analyzes the post-World War II struggle for recognition on the part of the peasantry, a period when the emerging movements operated almost entirely outside the law. In close relation to communist influences, such struggles and social mobilizations resulted in the creation of both a farm labor law in 1963 and a land law in 1964. The second part of the contribution examines attempts by the newly legalized labor movement to implement and use the land law as part of rural labor's demands in the context of the military regime that took power in a coup d'etat in April 1964. This part concludes with a discussion of the unions' struggle to shape agrarian reform law in drafting Brazil's new federal constitution in 1988, some 3 years after the return to civilian government. In the final part, the essay analyzes union efforts to modify and fulfill the promises of these new constitutional rights. As lawfully empowered entities, the unions 'needed' to work within established legal frameworks, but the strategies and tactics developed to achieve their goals shifted, depending on a variety of factors, between tendencies of negotiation with/support for authorities and opposition to/demands against the state and landlords. Throughout, we chart the tension between these two poles and note how the labor movement sought institutional recognition in the 1950s and 1960s, both challenged and collaborated with the dictatorship and added to its tool kit with redemocratization direct-action tactics such as land occupations, and yet embraced neoliberal policies as part of its historic contribution to agrarian reform struggle in Brazil.

Rural land and labor struggle for legal recognition to 1964

The organization with the longest history of action among Brazilian peasants in the twentieth century is the Communist Party.[1] In 1928, this party initiated a popular front strategy by forming the Worker-Peasant Block to institutionalize joint political mobilization of peasants and proletarians in support of the so-called national bourgeoisie and their efforts to strengthen Brazil through the nationalization of control over resources and manufacturing. Most activity revolved around electoral politics, especially the 1930 presidential election. As part of the Soviet Union's Communist International, established in 1919 to overthrow the so-called international bourgeoisie, the Communist Party was prohibited from

---

[1] Founded in 1922, the PCB was called the Communist Party of Brazil until a breakaway organization chose the same name for itself in 1962. To retain its acronym, the PCB changed its name to the Brazilian Communist Party and the new party adopted the acronym PCdoB. During the dictatorship, both gave attention to peasant mobilization, with the PCB emphasizing unionization and the PCdoB stressing armed struggle until 1974, when nearly all guerrilla groups had been repressed.

functioning openly, especially after 1930 when a rebellion brought to power a new government led by Getúlio Vargas. Nevertheless, in the 1930s, the Communist Party registered peasants as members and stimulated collective struggles in the countryside, an activity the Vargas government seemed to favor given its plans to build a corporatist state, with all economic categories organized into representative syndicates (Karepovs 2006; Welch 2009b).

Vargas made some efforts to control peasant mobilization through rural labor unionization, but the rural oligarchy resisted and ultimately blocked these plans until 1945, when a palace coup forced Vargas to leave the government (Welch 1999). The new political liberties of the post-World War II era permitted open competition by the Communist Party, which led the party to form Peasant Leagues in the countryside to defend rural labor interests and build its political base. Under liberalized voter-registration laws, the Communist leagues signed up hundreds if not thousands of new voters, who helped win the election of several Communist Party candidates to public office. Communist delegates who participated in the Constituent Assembly of 1946 influenced the content of Brazil's new federal constitution. Although many of their initiatives did not win majority support, for the next two decades constitutional articles that enabled the state to 'condition land use' to 'social welfare' standards, and specified goals of 'fixing men in the countryside' and 'justly distributing property', played important roles in determining peasant struggle (Brazil 1946; Welch 1999).

Even before May 1947, when the government canceled the Communist Party's mandate and forced members to act clandestinely, landlords and local police had repressed many leagues; others continued to function after the suppression of the Communist Party. The latest evidence of their existence can be found in police documents from 1949. In that year, the party founded a newspaper called *Nossa Terra* (*Our Country/Land*) to address the agrarian question and mobilize the peasantry to confront landlords. Until 1964, Communists used the publication under the name *Terra Livre* (*Free Country/Land*) to mobilize peasants with stories of struggles from all over the country, news of organizing efforts, legal orientation, popular poems and political journalism (Medeiros 1989; Welch 2010).

At various moments of its underground history, the Communist Party dedicated resources to defending the peasantry. In the early 1950s, it offered military training and material aid to a group of peasant families who had taken up arms to defend their possession of small coffee farms in the so-called 'Porecatu War' in the state of Paraná (Priori 2011). During the period, the party became involved in additional armed conflicts in Paraná and other southern states. In the center-west region of the country, the Communist Party also integrated itself with hundreds of peasants to defend their autonomy and influence over the micro-region of Trombas and Formoso, part of the Uruaçu municipality in the state of Goiás, where the Vargas government had experimented with an internal colonization scheme (Cunha 2007).

In September 1954, communist strategists recognized the importance of the Trombas and Formoso struggle by selecting one of its leaders, Geraldo Tibúrcio, as president of the newly formed Union of Farmers and Agricultural Workers of Brazil (ULTAB, in its Brazilian acronym). The party staged ULTAB's founding in São Paulo at a national congress of peasant delegates recruited from around the country. At the party's IV Congress in November 1954, Communist Party leaders expressed their admiration for Chinese revolutionaries and upheld the peasantry in Goiás as the nucleus of rural mobilization that would carry Brazil toward its own revolution (Welch 2010). Although ULTAB never succeeded in reproducing the Trombas and Formoso model of resistance on a broader scale, the peasants of Uruaçu were able to maintain a 'liberated region' until repressed by the military dictatorship after the 1964 coup (Cunha 2007).

Overall, the peasantry did not fare well in their confrontations with the state and large landowners, since troops and armed gunmen supported the rural oligarchy. In the Porecatu conflict, police troops and intelligence specialists from São Paulo helped defeat the resistance movement, and six peasants died. Various other conflicts exploded throughout the country, generally inspired by resistance to diverse forms of exploitation and, especially, forced expulsion from the land. As a result of these losses, the Communist Party central committee eventually concluded that revolution in Brazil would be a prolonged process, similar to the Chinese experience but with little chance of forming a popular army in the countryside. Thus, ULTAB militants followed a course of action dedicated to pressuring the state to comply with existing land and labor laws, among them the 'social welfare' test of landholding, minimum wages, paid holidays and advanced notice of contract termination. Especially important was the Communist Party's utilization of a presidential decree, passed by Vargas in 1945, which permitted the 'organization of rural life' through formal registry of small farmer 'associations' in each municipality (Welch 2010).

The Communist Party's advances in the countryside occurred in the context of an unprecedented period of electoral democracy and mass mobilization, much of it justified by an economic development discourse that called on popular participation to modernize the country. By demonstrating the ability of peasants to mobilize, the party stimulated other political actors to campaign in favor of incorporating peasants into the rural modernization project (Welch 2006a). Among these other actors were the Brazilian Labor Party, the Christian Democratic Party and the Brazilian Socialist Party (Camargo 1986). The Catholic Church also stepped up its participation. Perpetually present at the level of parishes, the church generally worked to buttress the established order, offering biblical apologies to justify rural poverty and the suffering peasants. However, increased communication and democratic politics provoked new voices and approaches, with one wing continuing to call on peasants to be patient and tolerate their exploitation in order to guarantee their place in heaven, while another wing began to practice a more activist ministry by helping the peasants understand their human and civil rights and organize to claim them (Sauer 1996).

Those more sensitive to the worldly needs of the peasantry initially attempted to form associations to join landowners and peasants in one united organization, insisting they would enhance harmony between classes rather than conflict. Certain bishops promoted these initiatives to retain peasant congregations by uniting with planters in secular associations that interpreted private property as a 'natural right', seeking to overcome the suffering of peasants through the subsidized purchase of parcels no longer useful to landlords (Welch 1999). In August 1955, they hosted the Congress for the Salvation of the Northeast in Recife, capital of Pernambuco state. Emphasizing social harmony, politicians from various states, representing various political parties, and delegates from newly formed rural associations issued resolutions calling for massive federal intervention to promote the development of the region. Such strategies would soon be criticized as 'drought industry' theater, taking advantage of peasant participation to reinforce the control of dominant groups (Almeida 1991). A segment of this same wing eventually spoke out against joint associations, supporting the need for peasants to form their own organizations (Welch 1999).

In analyzing the countryside in the mid-1950s, another group based in the northeast must be discussed. Influenced by the Pernambucan lawyer and socialist politician Francisco Julião, various peasant groups throughout the northeastern states began organizing themselves. These peasants were predominantly resident sugar cane plantation workers, subordinated by landlord prerogatives in exchange for limited land-use rights for animal husbandry and gardening (Montenegro 2004; Palacios 2009). In order to protect basic interests, such as schools for their children and the burial costs of elders, they organized themselves into mutual aid societies. The established order soon saw these associations as a threat and referred to them as peasant leagues in an attempt to link them to the old Communist Party groups and thus social disorder (Julião 2009). Julião had worked for years advocating for the civil rights of peasants using Brazil's 1906 Civil Code, which recognized 'tenure' as a key test for legitimating land rights. In January 1955, Julião legally registered the first mutual aid society – a group of 140 peasants living on the abandoned Galiléia sugar mill. The movement and Julião's political career took off in 1959 when the state government expropriated the mill and distributed its land to the peasant families who resided there (Welch 2010).

In 1961, the Communist Party central committee advised ULTAB officers to organize a peasant congress. This call came at a moment of national and international political crisis, which delayed its organization. The Cuban Revolution of 1959 marked a Cold War flare-up in the Americas, causing the United States to support a project of military and socio-economic intervention in the region called the Alliance for Progress. In August, in the Uruguayan city of Punta del Este, Brazil joined 21 other Latin American nations in signing the alliance agreement with the United States. The agreement directed each Latin American signatory to develop agrarian reform policies and plans. A few days later, Brazilian President Janio Quadros suddenly resigned, provoking a serious constitutional crisis, made all the more dramatic with Vice President João Goulart on an official visit in China. In this Cold War climate, Goulart's enemies depicted him as a communist 'fellow traveler' – an image reinforced by his presence in Communist China – and dangerous to the security of the Americas due to his close relationship to the labor movement. Centrist and leftist groups generally supported him as an able conciliator of class conflict, and defended his legal right to fill the post abandoned by Quadros. The Communist Party mobilized millions of people, including peasants, defending the constitutional transition and helping to guarantee that Goulart assumed the presidency (Ferreira 2011, Welch 2010).

Right after this crisis, ULTAB resumed the campaign for a national peasant congress, calling for delegates to participate from its small-farmer associations, Julião's Peasant Leagues and the more progressive wing of the Catholic Church. In November 1961, the peasant congress met in Belo Horizonte, with hundreds of leaders from different groups and organizations, coming from all parts of Brazil. Congress organizers wanted to unite these different organizations under the clandestine tutelage of the Communist Party in order to illuminate common peasant problems and plan a coordinated response. However, the congress's political agenda was divided between those who emphasized the rights of farmworkers and those who defended land distribution as the main cause of the peasantry.

In relation to the land question, three distinct positions characterized debate in the congress, a debate often reflective of arguments among factions within each organization. For the predominant groups within ULTAB, agrarian reform offered a means to eliminate the archaic traces of feudalism in rural areas by breaking up large, unproductive estates that served as a source of power for antidemocratic and antinationalist forces, such as traditional elites and foreign corporations. For Catholics and government representatives, agrarian reform promised to boost national economic development by lowering food costs and expanding the domestic market, thereby strengthening capitalism and the working class, as one step towards socialism. For most Peasant League delegates, agrarian reform advanced socialism by redistributing the means of production to those who worked the land, promoting social justice. While the first two groups sought to achieve change through legal means, Julião's Peasant Leagues stressed the need for a direct action approach, with the slogan 'Agrarian reform: by law or by force!' In the end, the congress declaration supported the more radical approach to agrarian reform advocated by the Peasant Leagues (Costa 1993; Stedile 2005).

President Goulart, the first lady, and the prime minister attended the congress, establishing for the first time a public connection between the social movements of the countryside and the executive branch of the federal government (Costa 1993; Welch 2010). Goulart's government, in spite of its leftist populist orientation, hesitated in seeking to fulfill the primary goal of the congress, radical agrarian reform. The 1946 constitution permitted expropriations of private properties, but the land had to be paid for in advance and in cash, making expropriations impossible as no federal funds for such payments had been appropriated. This situation resulted in promoting the secondary demand of the event – that of legally authorizing the organization of peasants, including their unionization. Thus, legally regimenting the rural labor movement would become the primary achievement of the congress (Costa 1993; Welch, 2010).

For the next 3 years, the demands for agrarian reform and rural unionization topped the list of questions debated nationally. In 1962, the minister of labor permitted the recognition of hundreds of new rural workers' unions. Later in the year, Goulart established the Superintendent of Agrarian Policies, in order to promote the formation of unions and administer the implementation of agrarian reform. In March 1963, Congress approved Brazil's first comprehensive rural labor law, the Rural Labor Statute.[2] In accordance with the statute, a vigorous campaign to organize rural labor unions ensued. The process emphasized the struggle for peasant civil rights and the improvement of working conditions through application of the law. The benefits of institutionalization and the force of the communist organizers caused many peasant leagues to transform themselves into local unions, deepening an organizational crisis already apparent in 1962. Increasingly influential was the idea that a strong unified rural labor movement had to be established to overcome the resistance of rancher, planter and landlord groups to changes in land and labor law, especially a constitutional amendment. At the end of 1963, delegates from newly formed state federations established CONTAG, the movement's maximum entity (Welch 1995; Ricci 1999).

Thus, strategists privileged expansion of the peasant union movement as fundamental to weaken the power of the rural oligarchy, creating more support for the election of candidates committed to the basic reforms deemed necessary to stimulate economic development – the major project of Goulart and his Brazilian Labor Party. The representation of rural unionization as a step toward radical agrarian reform (Pinheiro Neto 1993) provoked planter, rancher and miller groups to support unfolding plans for a military coup designed to unseat Goulart (Welch 1995). Two weeks after a public speech in March 1964, in which Goulart proclaimed agrarian reform as part of his presidential plan of action and his intention to distribute land to peasants along federal highway corridors, an armed assault, supported militarily and diplomatically by the United States, ended the Goulart government

---

[2]Established as Public Law 4,214 in March 1963, it was built on existing rights to create a specific law aimed at rural labor, establishing a corporatist system of syndical organization that was similar to those already used by some urban workers since the 1940s. It was amended and substituted by law 5,889 in 1973.

and its national reform project (Ferreira 2011). Peasant mobilization, supported by the executive branch, was a major reason for the coup (Welch 1995).

Within some 7 months of the coup, the dictatorship formulated an agrarian reform law, the Land Statute, to honor Brazil's promises to the US government, especially the Alliance for Progress accord, which established passage of the statute as a prerequisite for Brazil to receive a large, multifaceted United States Agency for International Development mission. Arguments in favor of the statute emphasized it as an instrument for securing the predominance of middle class farmers (Bruno 1995; Stedile 2005).[3] According to advocates, the statute would generate plans for eliminating the two most prevalent farm categories – the inefficient *minifundio* and unproductive *latifundio* – by stimulating the productivity of the most enterprising farmers through mechanization and chemical inputs.

In practice, the Land Statute caused the elimination of millions of smallholders and helped consolidate large landholders through favoritism. It presented a new standard for land use, replacing the term 'social welfare' with 'social function'. It specified the terms of land expropriation and structural reform in areas of conflict, but repression and demobilization initially left the peasant movement with too little force to pressure the government to implement this part of the law. To the contrary, the agricultural elite and influential investor groups pressured the dictatorship to utilize the law to eliminate 'unproductive units' by subsidizing existing 'productive units' as well as [encouraging? subsidizing?] to take over peasant land, expanding the agricultural frontier and 'modernizing' some production processes (Minc 1985; Gonçalves Neto 1997; Silva 1998). Although they eliminated the Superintendency of Agrarian Policies, many of its functions were maintained by dividing them into two new agencies: the Brazilian Agrarian Reform Institute and the National Agrarian Development Institute.[4]

In addition to the supposed paradox of the dictatorship's promulgation of Brazil's first land reform law, the military regime also maintained the recently passed farm labor law. Its corporatist nature was similarly appreciated as useful by the dictatorship. Thus, instead of eliminating CONTAG and the rural labor unions, they promoted interventions and sought to change the movement's identity to support the regime's agricultural modernization project. The military coup persecuted communists and progressive Catholics, replacing them with allies, including some military officers and, in the case of CONTAG, personnel from the conservative wing of Catholic social action groups. The regime deposed CONTAG's president, the Communist Party's Lyndolpho Silva, and replaced him with José Rotta, a conservative Catholic merchant and farmer from São Paulo state who had proved effective in organizing rural worker unions in collaboration with planters (Ricci 1999; Welch 2010).

While the ferocity of the repression cannot be denied, the literature too often emphasizes the coup as a dramatic transition in the story of the peasantry's political consciousness and the deposed government's commitment to agrarian reform. It is as if the coup leaders were correct in arguing that Brazil was on the cusp of an agrarian revolution. Evaluating the veracity of this argument depends on two very questionable 'ifs'. First, *if* Julião's peasant leagues had not been suppressed by the coup, a popular movement would have grown and

---

[3]Established by decree law 4,504 in November 1964, it had been debated for many years, but the project finally approved had been produced by a conservative think tank allied with coup conspirators. It suffered various revisions before being approved by the regime leader, Marshal Humberto de Alencar Castelo Branco (Bruno 1995).

[4]In 1970, these two institutes were merged to form the National Institute for Colonization and Agrarian Reform (INCRA, in its Brazilian acronym), which still functions today.

forced agrarian reform policies to be more radical and change the Brazilian agrarian structure. This seems unlikely given the fact that nearly 2 years before the coup, Julião and the leagues had been marginalized from public policy debates, a fact only confirmed by their total absence from the CONTAG leadership posts at the end of 1963. Second, *if* Goulart had not been overthrown, his government would have adopted agrarian policies more favorable to small farmers, including a radical agrarian reform law. Goulart's most radical proposal was a constitutional amendment that would have eliminated the need to pay cash when expropriating private land. Without that change, his proposals would have been impossible to implement, since they required peasants to purchase their plots. Moreover, plans for many mega-projects critiqued today as socially and environmentally destructive agribusiness ventures – such as the occupation of western lands by large-scale grain operations – were written during Goulart's term in office.

Finally, this period shows how CONTAG's very birth defined the 'tight spot' between tendencies to negotiate with authorities and opposition to/demands against established authority. As a public–private entity, CONTAG depended on the state and suffered interventions it could not resist. Given these examples and the regime's support for the rural labor and land statutes, there is good reason to argue that continuity rather than change should be emphasized in understanding this history.

Rural unions and the land under the statute regime (1964–1988)

The military–civilian regime that dominated Brazil from 1964 to 1984 used the rural land and labor statutes to establish and implement agricultural policies described by many as 'conservative modernization' (Moore 1966; Domingues 2002). For many peasants, however, the application of the dictatorship's policy proved more brutal than conservative – a 'painful modernization', according to agronomist José Graziano da Silva (1982) – and more traditional than modern, as it only intensified the arbitrary power of landlords.

In February 1965, the dictatorship issued an edict that altered the farm labor law to permit only one *Sindicato do Trabalhador Rural* (Rural Worker Union) per municipality – thus identifying all categories of peasants as 'rural workers'. This reflected an attempt by the regime to regiment the rural laboring classes, paving the way for the transformation of smallholders into rural proletarians by imposing a unitary formal identity on the rural poor (Grzybowski 1991).

Thereafter, the dictatorship invested heavily in agricultural modernization by capitalizing large estates through allocations of financial resources, especially agricultural credit and other fiscal incentives, creating conditions for the use of innovative inputs (from hybrid seeds to chemical fertilizers) and mechanized agricultural implements (from chainsaws to harvesters). On the frontiers of agricultural expansion, the regime granted financial privileges to urban entrepreneurs to stimulate the purchase of large tracts of undeveloped land (Oliveira 2010), and provided them with research and technical assistance that reinforced historic trends of planting monocultures for export (Gonçalves Neto 1997). 'Modern' quality and uniformity standards also weighed heavily on small-scale growers of cacao, coffee and other crops, for whom the dictatorship provided no subsidies or technical support. These processes deepened land concentration and forced the displacement of millions of peasants, causing them to either move to urban areas or participate in projects designed to colonize less desirable parts of Brazil's vast hinterland. In fact, these policies expelled more than 25 million peasants from their homes in the 1960s and 1970s (Martine 1987).

Many of these new migrants sought work in the newly forming agro-industrial centers in the central-south region, especially as cane cutters on the sugar plantation and mill complexes of São Paulo state. But the situation of these peasants, forcibly uprooted and transformed into rural proletarians, had become so sorry that a new expression was invented to describe them as a 'cold lunch' (*bóia-fria*), reflecting working conditions that caused them to leave home before dawn, carrying with them a pot of hot food that turned cold in the fields before the mid-morning lunch break. More than 100,000 *boias-frias* lost labor law protections in 1965 when the São Paulo-based industry determined that the law applied only to industrial workers in their sugar mills. It took nearly 10 years to legally reverse the cane-cutter exclusion (ESP 1973, 16), and by that time the industry had developed new ways of exploiting these workers. Instead of contracting the cutters directly, the industry stimulated the creation of a third-party system of labor cooperatives and labor contractors, called 'cats' (*gatos*) for their sly recruitment strategies. These techniques merely served to disguise the employment relation and allow the industry to avoid the added legal and financial obligations of employers under rural labor law. These tactics also denied workers the possibility of joining unions to defend themselves, turning the *boia-fria* into a new form of modern slavery (D'Incao 1975).

In the political sphere, the dictatorship produced a new constitution in 1967, which suspended national elections, and in 1968, it suspended Congress and intensified persecution of opposition groups, especially left-wing and popular leaders, violently smashing any form of dissent (Alves 1984). According to political scientist José Murilo de Carvalho (2004), besides the repression, the dictatorship also adopted a strategy of rural populism, looking for support among planters and peasants alike. The strategy was two-pronged, with modernization depicted as a nationalist economic development project and rural unionization represented as a means to advance the social rights of peasants. As Carvalho comments (2004, 172): 'The initial repression exercised against these unions, in addition to the assistance tasks now assigned to them, contributed to reduce their political combativeness and offered political dividends to the military governors', including electoral support. These strategies provoked internal conflict within the rural labor movement, but the repression did not cause the unions to break away from the corporatist dilemma. While these strategies provoked dependency on state benefits, the institutional relationship seemed important to maintain the unity needed to resist repressive times (Grzybowski 1991).

On the other hand, to avoid alienating the landlords who resisted both the labor and land statutes, the regime removed 'agrarian reform' from its first national development plan, which oriented economic policy from 1972 to 1974 (Carvalho 2004; Gonçalves Neto 1997). The plan crowned the achievements of President Medici, who ran the country during the bloodiest years of repression (1969–1974). Paradoxically, these 'heavy years' of dictatorship saw dramatic growth in rural unionization (Maybury-Lewis 1994; Houtzager 2004). In 1968, a rival faction of militant labor unionists took control of CONTAG, led by a skillful young sugarcane cutter from Pernambuco, José Francisco da Silva. Silva first became a union official by working as one of a team of 'field' representatives on a sugar plantation near Nazaré da Mata. Under threat of intervention, the union local strived to express its commitment to bread-and-butter questions while decentralizing power to protect its autonomy by promoting greater grassroots agility in resolving labor disputes in the field. From field delegate to local union president, Silva quickly rose to lead the state federation. The dictatorship, with its intervention in the confederation, had centralized power in the director. Silva's allies used the centralized structure to win majority support for his presidential bid from the small group of delegates permitted to vote. In the next few years, Silva and allies sought to revive CONTAG from the grass roots with an aggressive national union organizing campaign (Maybury-Lewis 1994; Ricci 1999; Medeiros 2012). In 1971, the dictatorship tried to co-opt the initiative with the Rural Fund (FUNRURAL, in its Brazilian acronym), its project for making the unions administrators of this rural social welfare program. Euclides Nascimento, a historic Pernambuco union leader, called the fund a

> problem ... dropped inside the union movement. It anesthetized the movement. Many local rural unions fell into this trap, losing their line, the line of combativeness, fighting for land and for rights, and went over to attending the affairs of doctors, dentists, retirement. (Maybury-Lewis 1994, 76)

All the same, the number of rural labor unions in Brazil more than doubled, from 938 in 1967 to 2068 in 1975. The number of workers represented thereby grew from around 1 million to more than 4 million, and continued to grow at a somewhat slower rate throughout the remainder of the dictatorship. In 1986, CONTAG's 2856 affiliated unions reported a membership total of nearly 10 million peasants (Maybury-Lewis 1994). This reinvigorated rural labor movement had to work to keep the dream of land reform alive in Brazil. According to rural sociologist Leonilde Medeiros (1993), CONTAG reinvented agrarian reform as a common goal to unify diverse rural struggles and demands. Anthropologist Moacir Palmeira and economist Sérgio Leite (1998, 132) similarly wrote that 'the workers were maturing into their own project of agrarian reform that counter-posed the policies elaborated by the military government'. Despite CONTAG's coordination of various conflicts under this unifying banner, Medeiros notes that 'the expropriation of land for social interests did not unfold as suggested by the Land Statute, except in a few specific settings where collective actions helped guarantee the permanence of workers on the land' (2012, 706).

CONTAG based its agrarian reform line on demanding recognition for the rights expressed in the land statute, holding the regime to the legal standards it represented for itself (Minc 1985; Martins 1988; Pereira 2005). In the context of the political persecution then suffered by activists of all stripes, union leaders considered the defense of the regime's own laws as the most sensible form of militancy (Palmeira and Leite 1998). While union leaders viewed the 'politics of the possible' option as a cost-effective approach to opposing regime policies (Maybury-Lewis 1994), the rural fund and land statute could be used to address only a small portion of the enormous challenges peasants faced.[5] To help resolve problems the unions seemed unwilling or incapable of addressing, the peasantry sought help from other institutions, especially the Roman Catholic Church, which responded in 1975 by joining with the Lutheran Church to found the Pastoral Land Commission (CPT, in its Brazilian acronym).

The CPT embraced the peasant cause as its mission and eventually became involved in the movement to revitalize rural unions, making it highly relevant to a discussion of rural labor unions and land struggle (Almeida 1991). According to CPT insider and sociologist Ivo Poletto, the priests and pastors involved created the land commission based on the 'necessity of overcoming isolation, of improving the knowledge of reality, and making the Amazonian pastoral work more dynamic' (1985, 38). Like CONTAG, the land commission initially raised the agrarian reform flag 'to support the demands of rural workers for a

---

[5]Ricci (1999) argues that the rural union movement lost its way in the 1970s, when it privileged land reform despite the fast growth of farm labor membership more interested in wages and conditions. Ricci claims this failure to represent *the* rural labor voice made CONTAG ill-prepared to head off the formation of rural social movements in the 1980s.

land reform within the letter and spirit' of the land statute (CPT 1983, 76), but did so criticizing the rural labor unions as being too passive in pursuing similar goals.

Despite the use of Marxist texts and ideas like class struggle at the grassroots level, the Catholic Church hierarchy continued to oppose communists in positions of authority. The church challenged established CONTAG rural union leaders not merely to replace supposedly co-opted union officers, who accommodated the dictatorship, but also to challenge officials with links to the Communist Party. The presence of a band of some 60 communist guerrillas in the Amazonia diocese of Bishop Pedro Casaldaglia had helped him convince the church to support founding the CPT, especially after military operations began to repress the incipient movement, causing the displacement, internment and death of an untold number of peasants. Thus, the Catholic Church's confrontation with communism added fuel to the fire as the CPT sought to strengthen radical grassroots leadership among peasants. The development of these leaders was accomplished gradually through varied forms of meetings, advocacy, classes and popular education techniques. Those identified as potential leaders were encouraged to form opposition slates to take control of their unions (Maybury-Lewis 1994; Balduíno 2004; Canuto 2006; Welch 2006b).

In the 1970s, one alleged Communist Party militant in the rural labor union structure was Roberto Toshio Horiguti, president of the São Paulo state federation of rural unions and a vice president of CONTAG (Aly Junior 2013). In 1979, he participated in a conference on the land statute sponsored by the Brazilian Agrarian Reform Association (ABRA, in its Brazilian acronym).[6] The ABRA conferees concluded that the land law had been implemented 'only in ways that interested the owners ... in benefits that stimulated ranchers and enriched the large landowners' (Horiguti 1979). Horiguti noted the misuse and subversion of the law, asserting that the union movement was 'a constant organ of pressure' for implanting 'agrarian justice' by 'proclaiming the reign of law' (Horiguti 1979, 23). For activists in the unions and the CPT, fulfillment of the land statute actually meant the opposite of what happened in practice. For the dictatorship, the law served to justify agricultural 'modernization', but the unions and CPT emphasized the need to break up large estates and distribute land to small-scale farmers (Silva 1971; Minc 1985).

In the late 1970s, the frequency and intensity of strikes in the cities and countryside grew, with metalworkers setting the pace in industrial centers and sugarcane cutters leading the way in agriculture with their demands for better working conditions and pay (Siguad 1980; Alves 1984). In 1979, an amnesty law allowed the return of exiled political leaders and activists and stimulated the re-formation of leftist political parties. The union movement played a central role, with CONTAG supporting the legalization of the Communist Party and the CPT participating in founding the Workers Party (PT). In the context of a political 'opening' promoted by the dictatorship, the Communist Party and PT represented two different nodes in the network of the democratic left. With its position supporting the autonomy of working-class organizations, the CPT contributed to forming the new union movement, as well as other popular and political organizations (Poletto 1990). In 1981, it played a key role in resisting Communist Party attempts to restore 'union unity' under communist political leadership at the crucial National Working Class Congress (CONCLAT, in its Brazilian acronym; Thomaz Junior 2002; Medeiros 2012).[7] In 1983, the CPT also helped organize the Unified Workers' Central (CUT, in its Brazilian

---

[6]Founded in 1967, ABRA gradually took the form of a 'think tank' dedicated to agrarian reform advocacy, and was led by José Gomes da Silva, who carefully constructed an alliance with the rural labor movement, including CONTAG officers, CPT figures like Poletto and other rural leaders.

acronym) and, in 1984, the MST, thereby escalating the ongoing tension between the CPT and CONTAG (Sauer 1996).

In the process of rebuilding old political parties and building new popular organizations, the broad sources of rural mobilization kept agrarian issues and demands for land on the political agenda. Despite their tensions, CUT, CONTAG and CPT were united in supporting agrarian reform, creating in 1982 the National Agrarian Reform Campaign, with the participation of ABRA and the Brazilian Social and Economic Analysis Institute (Palmeira and Leite 1998). Popular mobilization for political freedom 'raised expectations regarding the possibility of agrarian reform' as part of the political process of democratization that helped provoke the military to step down from power at the end of 1984 (Deere and Medeiros 2008, 83).

In January 1985, José Sarney – the first president democratically selected since 1961 – took office, and land reform advocates such as CONTAG lobbied for dramatic changes through the creation of a national agrarian reform plan, a policy orientation instrument originally called for in the land statute. Despite sugarmill owner and planter protests, Sarney appointed as president of INCRA ABRA's coordinator, José Gomes da Silva, who embraced the task of preparing the plan (Minc 1985). Silva (1989, 14) believed that the laws were in place to allow the executive branch 'to initiate a change in the agrarian structure in Brazil', requiring 'only an indispensable decision, will and political action' to put such a plan in place. He believed that the break-up of unproductive estates and the establishment of small farms thereon would serve to benefit not only the peasantry but the entire political economy, with gains to be realized in the marketplace and public square, due to the enhanced economic and political participation of peasants (Silva 1971, 1987).

Launched in May 1985, at CONTAG's fourth national congress, the agrarian plan promised to set aside land and settle 1.4 million families in 4 years (Sauer 2002). As soon as Sarney signed the plan in October 1985, Silva – whose health was not good – left INCRA triumphant. But landlord groups worked tirelessly to undermine the plan, and Sarney lacked the important political will to challenge them sufficiently (Silva 1987). The agrarian plan failed to achieve its goals and in 5 years, the Sarney government settled only 125,000 families, not even 10 percent of the projected total.

In the midst of the struggle to implement the agrarian plan, CUT created a leadership post on its board that would form the central's National Rural Unionization Department (DNTR, in its Brazilian acronym). From 1986, the union central invested in a grassroots rural unionization campaign meant to gain control of CONTAG (Ganzer 1997; Medeiros 2012). The struggle for control of peasant unions focused on forming competing slates to dispute official elections (Iokoi 1986; Medeiros 2012). Only by controlling a majority of rural worker unions in each state could CUT take control of the state federations, which then would allow it to gain control over the confederation. Ironically, this meant that CUT had organized itself to dominate a system it had initially opposed, the corporatist structure that privileged unified representation over union autonomy (Ganzer 1997; Ribeiro 2013). Theoretically, unity suppressed rank-and-file demands, whereas autonomy stimulated more aggressive grassroots activism, promising to more thoroughly represent the peasantry. CONTAG was not a still-life, however, so the confederation stepped up its union organizing activities while, starting with the CONCLAT in 1981, it became involved in

---

[7]According to Poletto (1990, 19), the final document of the 1986 general assembly overemphasized the CPT's support to the organization's 'conditioning the "value" of the struggle for land to the discovery of the political importance of the unions' and party's organizations'.

founding another 'centralizing' labor institution, reviving the Communist Party's old General Workers Command, where CONTAG's President Silva was made vice president.

In the midst of these union struggles, the inadequate implementation of the agrarian plan led the movements to shift their mobilizations to the fight over the composition of the country's new constitution, in 1987 and 1988. The importance of union politics for Brazil's land reform came into sharp relief during these debates and mobilizations, which deeply influenced the constitution that was promulgated in 1988 and remains in effect today (Sauer 2002). The backdrop for the 1987–1988 constitutional fight includes the membership of CONTAG's president on a commission appointed by Sarney in 1985 to draft a proposal for consideration by the elected delegates of the constituent assembly, which was established in February 1987.

Silva brought to the discussion two important CONTAG documents generated in the context of the confederation's 1985 congress: a list of 10 directives for agricultural policies and a manifesto called *The rural workers and the constituent* (Silva 1989, 32–34). They demanded laws to facilitate the expropriation of large farms, limit the maximum size of agricultural properties, restore public lands surreptitiously stolen from the state and automatically restrict the property rights of large landlords (defined as those holding more than three 'fiscal modules',[8] where more than 50 percent of the land was left fallow, constituting a failure to fulfill the 'social function' standard of land, as demanded by the 1964 land statute; CONTAG 1985). These labor movement proposals represented an explicit break from 20 years of defending the land statute and insisting on its strict application, as each of these demands expressed a determination to expand its parameters and CONTAG's commitment to defining agrarian reform as a path to constructing a democratic society (Duarte 1998).

By supporting activist opposition slates and land occupations, the CUT's rural department, the landless movement and CPT challenged CONTAG's emphasis on institutional struggle, but found common cause with CONTAG and ABRA in mobilizing to influence the constituent assembly to incorporate agrarian reform. All of the social movements, including the just-born MST, contributed to the popular mobilizations (Fernandes 1996; Sauer 2012; Medeiros 2012). They organized several mass rallies in Brasilia and sponsored national circulation of a 'people's amendment' demanding agrarian reform's inclusion in the constitution. The amendment attracted 1.2 million signatures of support and contributed to ensuring a place for agrarian policy and reform in the 1988 Constitution (Sauer 2012). An article specified the criteria necessary for land to fulfill its 'social function', conditioning the privilege of retaining agricultural land on its productive use as well as respect for labor and environmental laws. The federal government was charged with expropriating for agrarian reform purposes the lands of those landowners who violated these conditions (Brazil 1988; Silva 1989; Sauer 2010).

The rural social and labor movements were not alone in their attempts to mobilize pressure to influence and participate in the constitutional process. Landlords also got organized. The rural employers' union counterpart to CONTAG as well as both traditional and new groups of planters and landlords, including many cattle ranchers who then established the Rural Democratic Union (UDR, in its Brazilian acronym), formed the so-called Broad Agriculture Front to more aggressively defend landlord interests in the constituent assembly

---

[8]A 'fiscal module' (*modulo fiscal*) is a Brazilian legal term defined in the 1964 land statute as an area of land sufficient to support a family. The calculation of its size varies from municipality to municipality, depending on agricultural conditions.

(Silva 1989; Simon 1998). In the intense dispute over each word of the constitution, land-owners imposed their interests through an additional article that excluded 'productive property' from the expropriation process. Moreover, they succeeded in blocking the inclusion of popular demands, like limits on the size of rural properties and punitive, uncompensated expropriation. As experience later demonstrated, the implementation of the constitution's agrarian reform policy chapter privileged productivity as the main criterion taken into account to evaluate if landed property meets social function criteria, becoming one of the greatest impediments to the fulfillment of the agrarian reform goals established by those fighting to represent the peasantry in Brazil (Fernandes, Welch, and Gonçalves 2012; Sauer 2013).

Rural unions and land struggle under Constitutional law (1988 to 2013)

At the end of the 1980s, after the constitutional debates, CONTAG continued to make agrarian reform a priority. However, several changes can be charted in its form of mobilization and struggle for land. A variety of factors affected change in the confederation's approach to this goal. Highlighted below are four key processes: (1) the influence of occupations as a primary form of struggle; (2) the involvement of the union base (locals and state federations) in direct-action land struggle, including the organization of land occupations in some states; (3) CONTAG's 1995 affiliation to CUT and that movement's emphasis on direct action; and (4) the government's expropriation of land, especially after social movements found the leverage to force a response from President Fernando Henrique Cardoso in the mid-1990s. Even while CONTAG moved closer to the MST in some ways, it defined a distinct approach that guaranteed its negotiating credentials with authorities.

Into the 1990s, rural unions continued to be 'a necessary reference and an effective support system' for almost all of the struggles in the countryside (Grzybowski 1991, 62). Challenged by a growing landless movement, various unions and some state federations in the Northeast, North and Central-West regions of Brazil began to organize peasant families to occupy land in order to demand its expropriation for agrarian reform settlements (Sauer 2002). This direct union involvement in such grassroots actions gradually altered CONTAG's internal understandings and deliberations, leading the peasant labor movement to heighten its call for agrarian reform. Given the nature of occupations as encampments involving entire families, CONTAG also gained experience mobilizing women and children, not only predominantly male workers or heads of families, adding further complexity to its identity and practices (Medeiros 2012).

During the 1990s, the effects of these developments on political change within CONTAG – including official recognition of occupations as legitimate and decisions to encourage the participation of locals in mobilizing families – can be detected in the way its organizational documents changed from one congress to another. In 1991, for example, CONTAG's fifth congress reaffirmed that agrarian reform should be put in place through 'radical land redistribution and the formation of settlements of rural workers on land that is economically viable' (CONTAG 1991, 64). The congress documents further stressed that agrarian reform policy should have as its objectives: (1) increasing employment in the countryside, (2) reducing rural exodus, (3) increasing food production for the internal market, (4) increasing rural worker salaries and (5) decreasing regional inequality (CONTAG 1991). Theoretical and political understandings developed in prior decades strongly influenced these objectives, including notions of the 'functionality' of small-scale production, which provided the majority of cheap food for the urban population (Sauer 2002).

However, until the mid-1990s, the rural union movement's direct land actions had generated few concrete results, with some state federations promoting land mobilizations in Goiás, Minas Gerais and Pará (Medeiros 1993). Direct-action advocates discussed manifestos, promoted new leadership and organized land struggle support mechanisms (CONTAG 1991). However, the union movement faced many difficulties putting these political decisions in practice and mobilizing families to occupy land.

The final document of the next CONTAG congress clearly reflected these barriers (CONTAG 1995; Medeiros 2012). Internal and official evaluations recognized the mismatch between political decisions and the practices of the union movement in the struggle for land. Officially, it admitted that agrarian reform objectives had been 'placed as a priority in union documents and in union encounters, however, these goals were only put into practice by a small portion of local unions, as these unions were confronted with other struggles' (CONTAG 1995, 43). Indeed, the majority of union work involved dealing with the problems of salaried farmworkers and small farmers. In terms of agrarian reform, the principal novelties of this period included the recognition of occupations as legitimate political actions 'to guarantee access to land and production', and the leadership's willingness to critically evaluate the organizational challenges the unions faced to participate in this movement (CONTAG 1995, 43).

From 1995, when confederation delegates voted to affiliate with CUT, leaders essentially endorsed the land struggle then practiced by some union locals and state federations. Given the direct-action roots of the unionists who participated in CUT's rural department, the final merger between CONTAG and CUT promised to reduce internal disputes and increase the confederation's investment in achieving objectives like agrarian reform (Medeiros 2012).[9] This reflected the growing participation of the local and state federations that supported direct action and reasserted the importance of struggling for agrarian reform through the organization and mobilization of landless rural workers (CONTAG 1995). These actions represented a significant break with the legacy imposed by the dictatorship of confining union practice to the legally proscribed parameters that long limited the unions to representing salaried agricultural workers and small farmers, despite outreach to *boias-frias* and other underrepresented segments of the peasantry.

From the mid-1990s, CONTAG prioritized land occupations as a means of achieving agrarian reform goals, especially with the attention the national media gave this form of mobilization and struggle once the MST had established itself in São Paulo's Pontal do Paranapanema region. As Brazil's richest state, a leading commodity producer, headquarters not only to many landlord organizations but also to the banking industry, the seeming contradiction between advanced capitalism and peasant struggle fascinated the press. Initially, heightened press coverage built public sympathy for the landless and increased pressure on the labor movement to express itself regarding the issue. The press incited the public to expect the rural labor movement to be involved in the land struggle (Coletti 2002; Sauer 2002; Welch 2009a). Although statistics on the involvement of diverse

---

[9]Two internal disputes continued to influence the peasant labor movement. These were the 1989 founding of the Federation of Rural Employee Unions (FERAESP, in its Brazilian acronym) in the state of São Paulo, and that of the Federation of Family Farm Workers (FETRAF, in its Brazilian acronym), both affiliated to CUT. FERAESP militated mostly among seasonal cane cutters and fruit pickers, increasingly supporting them in direct-action land struggle. Founded in 2001, FETRAF-Sul remained a regional organization in the southern state of Santa Catarina until 2005, when it expanded activities into other states, stimulating its transformation into a national organization.

organizations only began to be gathered in 2000, they consistently placed CONTAG as second only to the MST in terms of the entities identified with the land struggle. From 2000 to 2012, CONTAG led 519 land struggles involving more than 53,000 families, in comparison to the 2701 settlements involving more than 450,000 families led by the MST (NERA 2013, 30).

This defense of agrarian reform, and the increasing involvement of CONTAG-affiliated unions in the struggle for land, led the confederation to participate in the National Forum for Agrarian Reform and Justice in the Countryside (FNRA, in its Brazilian acronym). Established in 1995, this forum eventually congregated some 40 entities and movements, such as the CPT, MST, CUT and other grassroots, labor and non-governmental organizations. Initially, its agenda was consensual, reflecting shared criticisms of the land-tenure policies of President Fernando Henrique Cardoso. Under this pressure, the Cardoso government deftly shifted strategy, committing itself to settle hundreds of thousands of landless peasants and establish a separate ministry to handle the challenge, while also working to disintegrate the unity of the forum for agrarian reform by embracing CONTAG and isolating the MST. Consistent with the administration's acceptance of neoliberal Washington Consensus policies, President Cardoso aligned the government with World Bank initiatives to implant market-based agrarian reform policies. By gaining CONTAG's endorsement and participation in the implementation of these policies, the Cardoso administration significantly fractured the cohesion of the forum (Sauer 2012; Medeiros 2012).

In this period, CONTAG adopted the 'family farmer' concept, which had already been developed by international institutions like the Food and Agriculture Organization of the United Nations (FAO; Lamarche 1993; FAO/INCRA 1994).[10] For CONTAG, the term was quickly used as a substitute for 'small production' or 'small farmer' (CONTAG 1991) – and even the slightly older generic term 'rural worker' (Palmeiras 1985). The change in terminology indicated changes in the labor movement's subsequent theoretical formulations and negotiation agendas, especially CONTAG's approach to agrarian reform. In 1995, the confederation first applied the term to identify the subject of CONTAG's new sustainable rural development policies (CONTAG 1995, 51). From this basis, CONTAG broadened its demands and formulated model agricultural policies for 'family farmers' (CONTAG 1997; Sauer 2002).

The reorientation became apparent when CONTAG launched a series of protests initially called National Days of Struggle, which in May 1995 gained the name 'Brazil's Cry of the Land' (*Grito da Terra Brasil*; Grito 1995). Noteworthy is the fact that at this time the MST lent its support to policies designed to favor 'family farmers', as it was one of the top-listed signers of the original manifesto.[11] In August 1995, the government responded to the demands of *Grito* organizations by establishing the National Program for Strengthening Family Farmers (PRONAF, in its Brazilian acronym; Schneider, Mattei, and Cazella 2004). CONTAG considered PRONAF a major political victory, despite the program's restricted credit lines and its initial use only by family farmers in the South of Brazil. For CONTAG, this was an important victory because it was the first government program created exclusively for 'family agriculture', recognizing and

---

[10]For some authors, the family farmer idea came to be seen as an important means for defusing a growing international anti-globalization peasant resistance movement, especially the growing alliance of rural social movements in the Via Campesina (Fernandes 2002; Borras, Edelman, and Kay 2008).
[11]The manifesto of 1995 Grito was signed by several agrarian, indigenous and fishing movements and organizations, including CUT and MST, and stressed the need for the government to design a 'Differentiated Agricultural Policy for Family Farming' (Grito 1995, 4).

'making official' the existence of this 'social category' (Schneider, Mattei, and Cazella 2004, 1).

On the other hand, despite the appearance of victory, some critics of CONTAG's 1995 theoretical formulations questioned the ways in which the emphasis on family farming began to dislodge agrarian reform as a top priority (thus reducing the importance of land struggle). In many documents, the government used terms such as 'expansion', 'extension' and 'strengthening' of family agriculture as euphemisms for increasing the productivity of existing farms, rather than for increasing the number of farms and farmers. Agrarian reform meant breaking up large estates and distributing these lands as well as public lands to the landless, thereby producing more farmers (FNRA 2005). Excessive preoccupation with 'family farmers' risked diluting long-established dichotomies that cast the world in black-and-white hues of either/or counter-positions, especially between the struggle for land and the growth of family farming (Carneiro 1997; Sauer 2002; Paulino 2014).

An increase in the number and intensity of rural conflicts also influenced CONTAG to revise its theories and tactics. Two massacres in the mid-1990s received the most attention. On 9 August 1995, gunmen killed 12 landless campers in Corumbiara in Rondônia state and, on 17 April 1996, police massacred 19 MST protesters and permanently crippled 50 more in Eldorado dos Carajás in Pará state. The latter had the greatest national and international repercussions, influencing rural land and labor disputes and policies into the twenty-first century (Sauer 2010; Medeiros 2012). These violent episodes, together with the long MST march to the nation's capital in 1997, and a burst of land conflicts all over the country, kept the theme of agrarian reform on the national agenda, including CONTAG's agenda. These developments also forced the administration of President Cardoso to formulate an agrarian reform program and create, in 1996, a cabinet post for land policy that served as the basis for creating a separate ministry in 1999, today's Agrarian Development Ministry (MDA, in its Brazilian acronym). These entities became responsible for administering INCRA and PRONAF, and developed a variety of initiatives related to agrarian issues, including a project to settle 400,000 landless families in 4 years (Deere and Medeiros 2008, Sauer 2010).

In 1998, CONTAG hosted its seventh national congress and consolidated its proposal for an 'alternative project of rural sustainable development' based on strengthening family farms. Consequently, the defense of agrarian reform lost additional political space in the union movement because it was not 'taken as an end in itself', but rather as 'an essential part of constructing a strategy of development, centered on expanding and strengthening family agriculture' (CONTAG 1998, 72). Its leaders did not see any contradiction in simultaneously adopting policies promoting 'the direct participation of union entities in the coordination of occupations and the accompaniment of these processes, administratively and judicially' (1998, 72) and calling for affiliated union locals to focus their energy primarily on already expropriated areas – in other words, the development of the settlements as part of the consolidation of family agriculture (CONTAG 1998; Sauer 2002).

Reading between the lines of these documents, one can reflect on the pre-coup tensions between Communist Party militants and Catholic priests. The militants provoked strikes and the priests privileged institution building. In 1998, a CONTAG increasingly controlled by delegates representing CPT and CUT opposition slates hewed more toward the Catholic legacy. Rather than stimulating land occupations, CONTAG directed union officers to preoccupy themselves more with probing the 'multiple possibilities for and barriers to forms of organization and production' and the 'economic benefits' of settlements (CONTAG 1998, 76). These concerns had two principal foundations: the new family farm theoretical reference and settler demands. In fact, the number of families settled by the agrarian reform initiatives of the Cardoso administration had grown greatly, demanding greater attention to their specific sustainability problems and the need to enhance public policies supporting settlements.

For CONTAG, as well as the government, the family farmer concept provided a useful focus for exploring policy options for settlers. Settlement families linked to the labor movement began to discuss their demands for credit, technical assistance and the organization of production, commercialization, education, health and such through the decision-making apparatus of the unions. They generated new perspectives such as the affirmation that the 'implementation of public policies for settlements should not be restricted to the mere distribution and ownership of land' (CONTAG 1998, 75). Rather, the agrarian reform settlement projects should be governed by a set of public policies leading to family farmers' agricultural productivity and profit.

In this period, CONTAG made several critiques of the Cardoso government's policies, especially its adherence to various neo-liberal theories, including the minimal, selling off state resources and decreased spending on agricultural programs. In addition, CONTAG joined many other social movements and entities in criticizing Cardoso for a series of anti-land occupation orders that his government began to issue in 1997.[12] These decrees sought to curtail the actions of rural social movements, as they transformed land struggles into criminal acts and prohibited INCRA from evaluating the agrarian reform expropriation suitability of land that had been occupied. Arguably, these measures ended 'once and for all the possibility that occupations would lead to expropriation' (Deere and Medeiros 2008, 88) by attempting eliminate the political pressure caused by land occupations. While occupation had long served as the landless movement's most effective tool for advancing the cause of agrarian reform, the labor movement came late to the tactic and, despite its discourse against criminalization, its actions in favor of the government's family farm policies moved CONTAG further away from concern over the issue. In spite of these changes, some CONTAG-affiliated unions remained involved in land struggles from 2000 to 2012, organizing more than 1000 families in 14 occupations, in comparison to the nearly 14,000 families the MST led in 108 conflicts during this recent period (NERA 2013, 31).

As shown, the adoption of the family farm concept as both a theoretical orientation and a political identity required for its justification economic results within the agrarian reform settlements. The need for success in the settlements – measured especially by gains in productivity and profits – became fundamental to CONTAG's ability to demonstrate 'the viability of the development of Brazilian family agriculture through the democratization of access to land' (CONTAG 1998, 77). By 2001, it argued that the settlement projects themselves served as 'an effective way to multiply family agriculture' (CONTAG 2001, 34). This justified the demands for public policies concerning credit, investment and technical assistance, and especially the increase in lines of credit from PRONAF (CONTAG 2001). It stressed the protagonism of the productive family farmer as the guarantor of democratization through land access, not agrarian reform in and of itself (CONTAG 2001).

From a political point of view, however, CONTAG continued to affirm agrarian reform as 'a strategic instrument in which to transform Brazilian society' due to its ability 'to destroy the unjust concentration of land, of income and of power by Brazil's political

---

[12]Executive Order 1.577/1997, reissued under the number 2.027-38 on 4 May 2000, and definitively replaced by Executive Order 2.183-56 on 24 August 2001, prohibited 'rural property to be the object of judicial dispossession if subjected to invasion motivated by agrarian conflict or collective land conflict' for two years or more following the end of the dispute (Brazil 2001, Article 4, Section 6).

rural elite' (CONTAG 2001, 12).[13] It also reaffirmed the importance of occupations as 'essential in guaranteeing the expropriation of rural property and consequently the settlements of rural workers' (CONTAG 2001, 39). Its political formulations at the start of the twenty-first century did not alter the confederation's attachment to the family farmer concept, which its practices and demands aimed to advance by gaining government support for public policies to implement its sustainable rural development project (CONTAG 2001; Sauer 2002).

As mentioned above, to the chagrin of most other FNRA member organizations, CONTAG supported the Cardoso government's 'market-based agrarian reform' initiative (CONTAG 2001). In contrast, the majority of agrarian forum members vehemently opposed these policies. Initiated in 1996, the government's program of buying and selling land, called 'land credit for combating rural poverty', was a 'market-driven agrarian reform' mechanism (Deere and Medeiros 2008; Pereira and Sauer 2011). Despite systematic opposition by social movements, the Cardoso administration continued to make more credit available for buying and selling land, through World Bank programs. Given the MST's vigorous opposition, only CONTAG's support made possible the approval of a second World Bank loan in 2000. This generated more internal conflict within the agrarian reform forum, especially as the loans saddled many families with serious debt problems as they accessed land through these credits (Deere and Medeiros 2008; Pereira and Sauer 2011).

These developments eventually provoked the dissolution of the precariously established forum unity (Wanderley 2003). Geographer Bernardo Mançano Fernandes (2002) argued that the 'family farmer' concept was the root cause of this dissension. According to his interpretation of the concept, the family farmer was a peasant 'metamorphosized' into a small businessman through the exploitation of family and friends. Peasants would have no future unless they embraced the market, either as agribusiness operators or workers. Thus, the elimination of the peasantry was implied by family farm policy like PRONAF. Capitalist experiences based on these ideas, such as the agricultural history of the US, showed how family farmers were aggressively squeezed out of business and eliminated by fewer and fewer families (turned corporations, like Cargill) operating at larger and larger scales (Welch 2005). Thus, the concept represented a threat to the landless and small farmers that groups associated with the Via Campesina sought to represent. They sought to grow and fortify, not reduce and eliminate the peasantry. Thus, at the start of the twenty-first century, the MST and other rural social movements, such as the Small Farmer Movement and Peasant Women's Movement, broke with CONTAG and distanced themselves from the concept partially by contributing more and more to the idea that Brazil's rural poor should be identified as peasants. This is one more reason for opting to recuperate the term 'peasant' from the confusion created by Martins in his 2002 contribution.

As Cardoso's second and last term of office came to a close in 2002, opposition presidential candidate Luis Inacio Lula da Silva (Lula) received overwhelming support from both the rural social and labor movements due to his long personal and political support for radical agrarian reform. To achieve victory in his fourth presidential campaign, however, Lula and his advisors had moved closer to what the Communist Party used to

---

[13] According to the 2001 congress document, 'the democratization of the use and possession of land is an essential means for altering the set of social, economic, environmental, and political conditions necessary to the process of development in the country' (CONTAG 2001, 30).

call the 'national bourgeoisie', including agribusiness firms. Despite the decreasing importance of agrarian reform in Lula's electoral platform, once elected, the Workers Party leader fulfilled his promise to launch Brazil's second National Agrarian Reform Plan at the end of 2003. Following a period of decline in the number of settlements established and families served, Lula's government responded to the pressures of old comrades by significantly increasing the number of settlements created and families settled, especially in 2005 and 2006 (NERA 2013, 19). Like his predecessors, however, Lula no longer had the will to fulfill the ambitious initial goals of his agrarian reform plan. His government added numbers by expanding support for market-based agrarian reform, previously criticized by the Workers Party. In 2006, Lula also signed a law for family farmers, defining them more broadly, including other social groups like the indigenous peoples and maroon communities (*quilombolas*), who thereby gained access to governmental programs like PRONAF.[14] He added more families and hectares by 'regularizing' land occupied but not titled by both peasants and large-scale land grabbers, especially in the Amazon region, where the actions meant little in terms of the redistribution of land (FNRA 2008; Oliveira 2010; Welch 2011).

While CONTAG continued to share with other agrarian movements the demand that Lula increase his commitment to agrarian reform, it tended to maintain political support for Lula and his party. With the continuance of PT rule that came with electoral victory in 2010 for Lula's chosen replacement as president, his former chief of staff Dilma Rousseff, this support translated into a convergence between the government's rural development strategy and some of the policy positions CONTAG had projected in its 1998 congress, especially those emphasizing the consolidation of agrarian reform settlement projects as healthy, productive and profitable for family farmers. But the convergence did not translate into unquestioned support for Dilma, as her near abandonment of agrarian reform provoked factions within the labor movement to increase their criticisms of the government.[15] CONTAG had in fact become much more complex by the twenty-first century. CUT was no longer the only labor central to which member federations were affiliated, and some of these other centrals, such as Força Sindical, were openly aligned with the PT's leading rivals, especially Cardoso's Brazilian Social Democracy Party (PSDB, in its Brazilian acronym).[16] In 2012, CONTAG officially welcomed the outreach of rural social movements that sought to include the confederation in a broad alliance of the rural poor to condemn Dilma's policies. In August 2012, CONTAG participated with the MST and other Via Campesina organizations in organizing the so-called Second National

---

[14]Public law 11,326 was approved on 24 July 2006, to establish the concept and directives for the development of national policy for family farmers and their businesses (Brazil 2006).

[15]By the end of 2012, President Dilma had created one of the lowest numbers of settlements and settled one of the lowest numbers of families per year of any other president since 1988. As her reelection campaign started to gear up toward the end of 2013, she suddenly approved 100 agrarian reform land expropriations, still a very low number.

[16]To explain the contradiction between support and negotiation at this point in history, it is essential to clarify that CONTAG member federations and local unions are affiliated with at least four major labor movement confederations. Each has a distinct relationship with important political parties. While the CUT predominates as an ally of the PT that seeks to provide unconditional support for Dilma, the General Confederation of Workers is aligned with the Brazilian Socialist Party, a friendly rival to the PT, as is the Male and Female Workers Central of Brazil, which follows the line of the PCdoB, which is also a tendentious player in the PT governing coalition. As a PSDB ally, the Força Sindical affiliated members are the most critical of Dilma and supporting the PSDB's candidate for presidency in 2014.

Peasant Congress.[17] Joint activities continued in 2013, as these same groups pressed for the recognition of the violent repression suffered by the overall peasant movement since World War II. It seemed a new chapter had begun in the collaborative struggle among social and labor movements.

Conclusion

While agrarian reform remained a labor movement priority, both CONTAG's definition of reform policy (its theoretical approach to the agrarian question) and especially its actions (the practice of rural unionism) changed considerably over the years. This paper argues that, through various socioeconomic and political changes, including runaway land concentration and transitions in and out of authoritarian rule, each of which produced a different land governance regime, the union movement remained committed to an agrarian reform that privileged a concept of land usage defined by individual property rights, family labor and production. This conclusion challenges the one presented in 2002 by Martins, in which he alleged that CONTAG was historically committed to 'class struggle against capitalists and landlords' and that the CPT and MST saw 'no need for radical systemic transformation and adhere[d] ... to a communitarian vision in which capitalists and workers enjoy[ed] a tension-free parity of esteem' (327). To the contrary, as we have shown, it was CONTAG that saw little need for transforming the capitalist system and strived to harmonize peasant interests with those who supported market-oriented family farming.

Martins (2002, 324) correctly highlights CONTAG as the 'far older grassroots organization ... in the Brazilian countryside', which 'has been an authentic "voice from below" ... representing millions of rural workers'. But as Brazil embraced Green Revolution policies and practices, millions of these same rural workers participated in CPT and MST actions in the 1980s and 1990s (Alves 1991; Coletti 2002). With its roots in the Communist Party, class struggle is an important criterion for evaluating CONTAG, but the criticisms of the Communist Party and CONTAG are legendary for downplaying class struggle to build alliances with nationalist ruling groups in order to construct capitalism in Brazil, under the theory that it is a necessary stage to advance toward socialism (Welch 1999). As evidenced above, CONTAG now urges implementation of a rural sustainable development project that it deems 'alternative', without explicit reference to either socialism or structural change.

Historically, the rural labor movement in Brazil emphasized the necessity of implementing an agrarian reform in combination with its defense of expanded labor rights of farm workers. This was CONTAG's main political banner in the years of repression and dictatorship, especially in light of the possibilities provided by the dictatorship's administration of the Land Statute and FUNRURAL. However, in this context, the policy emphasis rarely turned into effective land struggle actions. The confederation only came to officially recognize the legitimacy of land occupations and support their organization in the mid-1990s. Among the many reasons analyzed for this mismatch, we stressed differentiation at the

---

[17]The formal name of this event was the United Meeting of Male and Female Rural Workers and Peoples of the Countryside, Waters and Forests. Organizers reflected on its similarities with the First Peasant Congress organized by the Communist Party in 1961 (see above), but the gathering rejected being called a "congress" as participants did not aspire to deliberate over specific political decisions.

grassroots level in the union constituencies between agricultural employees, small farmers and the landless, and also the organizational structure of CONTAG itself, in which national decisions rarely flow unencumbered to the entire trade union movement and local unions face unanticipated challenges, including legal limitations on their geographical and organizational boundaries.

Despite the conflicts and disputes for leadership in the countryside, the emphasis the union movement has given to agrarian reform since the 1920s has been fundamental to the issue's longevity in Brazil. Especially in the 1990s, CONTAG's continued commitment to agrarian reform, combined with the actions of the MST, ensured its place as a priority concern in Brazil's transition to free market democracy. The labor movement's defense of agrarian reform gave consistency to the political struggle for land, including changes to the federal constitution, land expropriations and the settlement of more than a million families since 1995 (NERA 2013).

The union movement maintained agrarian reform on the national political agenda. Even though it has lost momentum in recent years, due largely to political calculations and policies supported by large-scale agribusiness and fed by dependency on income from agricultural exports, the rural labor movement's combination of protest and negotiation has contributed greatly to institutionalizing not only land reform but also programs of financial and technical support for small-scale family farmers. Despite the ideological debate, CONTAG's family farmers are essentially the same sort of people the MST calls peasants. While criticisms of the labor movement's willingness to collaborate with the government have consistently been raised by supposedly more radical groups, like the Peasant Leagues and the MST, each of these has also sought to win government concessions to help peasants and support their continuity as organizations. Given the union movement's long history of ups and downs, what most stands out is its consistent support for agrarian reform. For more than half a century, CONTAG has been the most constant carrier of that banner, and those who diminish its role ignore the diverse and important contributions the rural labor movement has made to contemporary Brazilian history.

Acknowledgements

The authors thank issue organizers Rebecca Tarlau and Anthony Pahnke, as well as Editor Jun Borras for soliciting this contribution and having enough patience to let us 'finish' it. We are also grateful to Rebecca for translating several parts of the manuscript. Two anonymous reviewers offered many helpful comments and questions to improve the piece. We are also grateful for diverse grants from Brazilian funding agencies such as Brazil's National Science and Technology Research Council (CNPq) and CAPES, that have been important sources for the scholarship, research materials, travel financing and time needed to complete the paper.

References

Almeida, A.W.B. 1991. *Aggiornamento agônico*. Rio de Janeiro. Unpub. Mss.
Alves, F.J.C. 1991. Modernização da agricultura e sindicalismo: lutas dos trabalhadores assalariados rurais da região canavieira de Ribeirão Preto. Thesis (Ph.D). Universidade de Campinas.
Alves, M.H.M. 1984. *Estado e oposição no Brasil (1964–1984)*. Petrópolis: Editora Vozes.
Aly Junior, O. 2013. Interview by C. Welch. Tarija, Bolivia.
Balduíno, T. 2004. Interview by C. Welch. Ribeirão Preto, SP.
Bastos, E.R. 1987 [1985]. Sindicalismo no campo no Brasil: direitos trabalhistas e conflitos de terra. In *Processo e relações do trabalho no Brasil*, 2nd ed. eds. M.T.L Fleury and R.M. Fischer, 122–31. São Paulo: Atlas.
Borras, S.M., M. Edelman, and C. Kay, eds. 2008. *Transnational Agrarian Movements: Confronting Globalization*. Oxford: Blackwell Publishing.

Brazil. 1988. Constituição da República Federativa do Brasil de 1988. http://www.planalto.gov.br/ ccivil_03/constituicao/constituicao.htm (accessed August 2, 2014).

Brazil. 2001. Medida Provisória n° 2.183-56, de 24 de agosto. Brasília, Presidência da República. http://www.planalto.gov.br/ccivil_03/mpv/2183-56.htm (accessed December 20, 2014).

Brazil. 2006. Lei n° 11.326, de 24 de julho. Brasília, Presidência da República. http://www.planalto. gov.br/ccivil_03/_ato2004-2006/2006/lei/l11326.htm (accessed September 14, 2014).

Brazil, Government of. 1946. Constituição dos Estados Unidos do Brasil (de 18 de setembro de 1946). http://www.planalto.gov.br/ccivil_03/constituicao/constituicao46.htm (accessed August 2, 2014).

Bruno, R. 1995. O Estatuto da Terra: entre a conciliação e o confronto. *Estudos sociedade e agricultura* 5(Nov): 5–31.

Camargo, A.deA. 1986. A questão agrária (1930–1964). In *História geral de civilização brasileira. O Brasil republicano: sociedade e política*, ed. B. Fausto, 121–224. São Paulo: Difel.

Canuto, A. 2006. Interviewed by C. Welch. Goiânia, GO.

Carneiro, M.J. 1997. Política pública e agricultura familiar: uma leitura do Pronaf. *Revista Estudos - Sociedade e Agricultura* 8: 70–82.

Carvalho, J.M. 2004. *Cidadania no Brasil: o longo caminho.* Rio de Janeiro: Editora Civilização Brasileira.

Coletti, C. 2002. Ascensão e refluxo do MST da luta pela terra na década neoliberal. *Ideias* 9, no. 1: 49–104.

Costa, L.F.C. 1993. *O Congresso Nacional Camponês: trabalhador rural no processo politico brasileiro.* Rio de Janeiro: Universidade Rural.

CONTAG. 1985. *Anais do IV Congresso Nacional dos Trabalhadores Rurais: sincialismo forte e reforma agrária já.* Brasília: DF.

CONTAG. 1991. *Anais do 5°. Congresso Nacional de Trabalhadores Rurais.* Brasília: DF.

CONTAG. 1995. *Anais do 6°. Congresso Nacional de Trabalhadores Rurais: nem fome, nem miséria; o campo é uma solução.* Brasília: DF.

CONTAG. 1997. *Desenvolvimento local sustentável baseado na agricultura familiar: construindo um projeto alternativo.* Brasília: DF.

CONTAG. 1998. *Anais do VII Congresso Nacional de Trabalhadores e Trabalhadoras Rurais: rumo a um projeto alternativo de desenvolvimento rural sustentável.* Brasília: DF.

CONTAG. 2001. *Documento-base do 8° Congresso Nacional de Trabalhadores Rurais.* Brasília: DF.

CPT – Comissão Pastoral da Terra. 1983. *CPT: pastoral e compromisso.* Goiânia/Petrópolis: CPT/Ed. Vozes.

Cunha, P.R.da. 2007. *Aconteceu longe demais: a luta pela terra dos posseiros em Formoso e Trombas e a revolução brasileira (1950–1964).* São Paulo: Edunesp.

Davatz, T. 1980 [1850]. *Memórias de um colono no Brasil (1850).* São Paulo/Belo Horizonte: Editora da Universidade de São Paulo: 105–73.

Deere, C. D., and L. S. Medeiros. 2008. Agrarian reform and poverty reduction: lessons from Brazil. In *Land, poverty and livelihoods in an era of globalization: perspectives from developing and transition countries*, eds. A.H Akram-Lodhi, S.M. Borras Jr., and C. Kay, 80–118. London/ New York: Routledge.

D'Incao, M.C. 1975. *O 'bóia-fria': acumulação e miséria.* Petrópolis: Editora Vozes.

Domingues, J.M. 2002. A dialética da modernização conservadora e a nova história do Brasil. *Dados* 45: 459–82.

Duarte, E.G. 1998. Do mutirão à ocupação de terras: manifestações camponesas contemporâneas em Goiás. Thesis (PhD.), Universidade de São Paulo.

ESP. 1973. Industriários de cana fazem dia 19 a primeira reunião. *O Estado de S. Paulo* 18 May: 16.

FAO/INCRA. 1994. *Diretrizes de política agrária e desenvolvimento sustentável para a pequena produção familiar.* Brasília, DF: FAO/INCRA.

Fernandes, B.M. 1996. *MST: formação e territorialização.* São Paulo: Editora Hucitec.

Fernandes, B.M. 2002. Agricultura camponesa e/ou agricultura familiar. http://www.geografia.fflch.usp. br/graduacao/apoio/Apoio/Apoio_Valeria/flg0563/2s2012/FERNANDES.pdf (accessed June 6, 2013).

Fernandes, B.M., C.A. Welch, and E.C. Gonçalves. 2012. *Land governance in Brazil: a geo-historical review of land governance in Brazil.* Roma: International Land Coalition. http://www. landcoalition.org/publications/land-governance-brazil (accessed November 1, 2012).

Ferreira, J. 2011. *João Goulart. uma biografia.* Rio de Janeiro: Civilização Brasileira.

FNRA – Forum Nacional pela Reforma Agrária e Justiça no Campo. 2005. *Análise conjuntural - reforma agrária*. Brasília: D.F. http://terradedireitos.org.br/en/2005/12/19/analise-conjuntural-reforma-agraria/ (accessed July 29, 2014).

FNRA – Forum Nacional pela Reforma Agrária e Justiça no Campo. 2008. Nota do Fórum Nacional pela Reforma Agrária e Justiça no Campo sobre um novo marco legal e institucional para a Amazônia. Brasília: D.F. http://www.cnbb.org.br/comissoes-episcopais-1/caridade-justica-e-paz/495-nota-do-forum-nacional-pela-reforma-agraria-e-justica-no-campo-sobre-um-novo-marco-legal-e-institucional-para-a-amazonia (accessed July 29, 2014).

Ganzer, A. 1997. Interviewed by C. Welch. Taguatinga: DF.

Gonçalves Neto, W. 1997. *Estado e agricultura no Brasil: política agrícola e modernização econômica brasileira, 1960–1980*. São Paulo: Editora Hucitec.

Grito da Terra Brasil. 1995. *Pauta nacional de reivindicações*. Brasilia: DF. 24 May.

Grzybowski, C. 1991. *Caminhos e descaminhos dos movimentos sociais no campo*. Petrópolis/Rio de Janeiro: Vozes/Fase.

Hall, M.M. and P.S. Pinheiro. 1985. Alargando a história da classe operária: organização, lutas e controle. *Remate de Males* 5: 95–119.

Horiguti, Roberto Toshio. 1979. O Estatuto da Terra e a posição do trabalhador rural. *Boletim da Associação Brasileira de Reforma Agrária* 9, no. 6: 21–28.

Houtzager, P.P. 2004. *Os últimos cidadãos: conflito e modernização no Brasil rural*. São Paulo: Editora Globo.

Iokoi, Z.M.G. 1996. *Igreja e camponeses: teologia da libertação e movimentos sociais no campo Brasil e Peru*. São Paulo: Editora Hucitec.

Julião, F. 2009 [1962]. Que são as Ligas Camponesas?. In *Camponeses brasileiros: leituras e interpretações clássicas*, eds. C.A. Welch, E. Malagodi, J.S.B. Cavalcanti, and M.N.B. Wanderley, 271–97. São Paulo: Editora da Unesp.

Karepovs, Dainis. 2006. *A classe operária vai ao parlamento. o bloco operário e camponês do Brasil (1924–1930)*. São Paulo: Alameda Casa Editorial.

Lamarche, H., ed. 1993. *A agricultura familiar, comparação internacional*. Campinas: Editora da Unicamp.

Linhares, M.Y. and F.C.T. Silva. 1999. *Terra prometida: uma história da questão agrária no Brasil*. Rio de Janeiro: Editora Campus.

Martine, G. 1987. Êxodo rural, concentração urbana e fronteira agrícola. In *Os impactos sociais da modernização agrícola*, eds. G. Martine and R.C. Garcia, 59–79. São Paulo: Editora Caetes.

Martins, J.S. 1981. *Os camponeses e a política no Brasil*. Petrópolis: Editora Vozes.

Martins, J.S. 1988. *Não há terra para plantar neste verão: o cerco das terras indígenas e das terras de trabalho no renascimento político do campo*. Petrópolis: Ed. Vozes.

Martins, J.S. 2002. Representing the peasantry?. Struggles for/about Land in Brazil. *Journal of Peasant Studies* 29, no. 3–4: 300–35.

Maybury-Lewis, B. 1994. *The politics of the possible: The Brazilian rural workers' trade union movement, 1964–1985*. Philadelphia: Temple University Press.

Medeiros, L.S. de. 1989. *História dos movimentos sociais no campo*. Rio de Janeiro: FASE.

Medeiros, L.S. de. 1993. *Reforma agrária: concepções, controvérsias e questões*. Rio de Janeiro, Unpub. mss. http://www.cefetsp.br/edu/eso/reformaagrariaquestoes.html (accessed December 20, 2014).

Medeiros, L.S. de. 2012. Sindicalismo rural. In *Dicionário de educação do campo*, eds. R. Caldart et al., 706–13. São Paulo: Expressão Popular/Fiocruz.

Minc, C. 1985. *A reconquista da terra: Estatuto da Terra, lutas no campo e reforma agrária*. Rio de Janeiro: Zahar.

Montenegro, A. T. 2004. As Ligas Camponesas às vésperas do golpe de 1964. *Projeto História*, v.29, Tomo 2, p. 391–416.

Moore, B. 1966. *Social origins of democracy and dictatorship: lord and peasant in the making of the modern world*. Boston: Beacon Press.

NERA – Núcleo de Estudos, Pesquisas e Projetos de Reforma Agrária. 2013. *DATALUTA Banco de Dados da Luta pela Terra: Relatório 2012*, eds. E.P. Girardi and B.M. Fernandes, Presidente Prudente, São Paulo: UNESP.

Oliveira, A.U. 2010. A questão da aquisição de terras por estrangeiros no Brasil - um retorno aos dossiês. *Revista Agrária* 12: 3–113. http://www.revistas.usp.br/agra/article/view/702/711 (accessed September 17, 2012).

Palacios, G. 2009 [1987]. Campesinato e escravidão: uma proposta de periodização para a história dos cultivadores pobres livres no Nordeste oriental do Brasil (1700–1875). In *Camponeses brasileiros*, eds. Welch et al., 145–78.

Palmeira, M. 1985. A diversidade da luta no campo: luta camponesa e diferenciação do campesinato. In *A Igreja e questão agrária*, ed. V. Paiva, 43–51. São Paulo: Edições Loyola.

Palmeira, M., and S. Leite. 1998. Debates econômicos, processos sociais e lutas políticas. In *Política e reforma agrária*, eds. L.F.C Costa and R. Santos, 92–169. Rio de Janeiro: Editora Mauad.

Paulino, E. T. 2014. The agricultural, environmental and socio-political repercussions of Brazil's land governance system. *Land Use Policy* 36: 134–44.

Pereira, A.W. 2005. *Political injustice: authoritarianism and the rule of law in Brazil, Chile and Argentina*. Pittsburgh: Pittsburgh University Press.

Pereira, J.M., and S. Sauer. 2011. A 'reforma agrária assistida pelo mercado' do Banco Mundial no Brasil: dimensões políticas. *implantação e resultados. Sociedade e Estado* 26, no. 3: 587–612.

Pinheiro Neto, J. 1993. *Jango: um depoimento pessoal*. Rio de Janeiro: Editora Record.

Poletto, I. 1985. A CPT, a Igreja e os camponeses. In *Conquistar a terra, reconstruir a vida*, eds. D.P. Casaldáliga, et al., 29–66. Petrópolis: Vozes.

Poletto, I. 1990. A pastoral da terra e a construção da democracia. *Cadernos de Estudos da CPT* no. 1: 1–24.

Price, R.E. 1964. *Rural unionization in Brazil*. Land Tenure Center Report, 14, Madison: University of Wisconsin.

Priori, A. 2011. *O levante dos posseiros: a revolta camponesa de Porecatu e a ação do Partido Comunista Brasileiro no campo*. Maringá, PR: Editora da Universidade Estadual de Maringá.

Ribeiro, A.S. 2013. Entrevista pelo Memórias DNTR-CUT. http://memoriasdntrcut.wordpress.com/2013/09/03/trampolim-foi-uma-semente-que-a-gente-plantou/#more-132 (accessed on November 14, 2013).

Ricci, R. 1999. *Terra de ninguém: representação sindical rural no Brasil*. Campinas: Editora da Unicamp.

Sauer, S. 1996. The land issue as a theological problem: the Roman Catholic and Lutheran Churches' social and political commitment to the struggle for land in Brazil. Thesis (MA), Stavanger School of Mission and Theology.

Sauer, S. 2002. Terra e modernidade: a dimensão do espaço na aventura da luta pela terra. Thesis (PhD), Universidade de Brasília.

Sauer, S. 2010. *Terra e modernidade: a reinvenção do campo brasileiro*. São Paulo: Expressão Popular.

Sauer, S. 2012. Articulações em defesa da reforma agrária. In *Dicionário de educação do campo*, eds. R. Caldart, et al., 103. São Paulo: Expressão Popular/ Fiocruz.

Sauer, S. 2013. O Governo Lula no campo: compromissos e embates nas políticas agrárias e agrícolas. In *Trajetória e dilemas da reforma agrária no Brasil*, ed. L. Mattei, Florianópolis. (forthcoming).

Schneider, S., L. Mattei, and A.A. Cazella. 2004. Histórico, caracterização e dinâmica recente do PRONAF. In *Políticas públicas e participação social no Brasil rural*, eds. S. Scheider, M.K. Silva, and P.E.M. Marques, 21–50. Porto Alegre: Editora da UFRGS.

Siguad, L. 1980. *Greve nos engenhos*. Rio de Janeiro: Editora Paz e Terra.

Silva, J.G. da. 1971. *A reforma agrária no Brasil: frustração camponesa ou instrumento de desenvolvimento?*. Rio de Janeiro: Zahar Editores.

Silva, J.G. da. 1982. *A modernização dolorosa: estrutura agrária, fronteira agrícola e trabalhadores rurais no Brasil*. Rio de Janeiro: Zahar Editores.

Silva, J.G. da. 1987. *Caindo por terra: crises da reforma agrária na Nova República*. São Paulo: Busca Vida.

Silva, J. Gomes da. 1989. *Buraco negro: a reforma agrária na constituente de 1987/88*. São Paulo: Editora Paz e Terra.

Silva, M.A.M. 1998. *Errantes do fim do século*. São Paulo: Editora da Unesp.

Simon, C.G.B. 1998. *Os campos dos senhores: UDR e elite rural - 1985/1988*. Londrina: Editora UEL.

Stedile, J.P., ed. 2005. *A questão agrária no Brasil: programas de reforma agrária: 1946–2003*. São Paulo: Expressão Popular.

Thomaz, Jr., A. 2002. *Por trás dos canaviais, os 'nós' da cana: a relação capital x trabalho e o movimento sindical dos trabalhadores na agroindústria canavieira paulista*. São Paulo: Annablume Editora e Comunicação.

Vangelista, C. 1991. *Os braços da lavoura: imigrantes e 'caipiras' na formação do mercado de trabalho paulista (1850–1930)*. São Paulo: Editora Hucitec.

Wanderley, M.N.B. 2003. Agricultura familiar e campesinato: rupturas e continuidade. *Estudos Sociedade e Agricultura* 21 (Oct): 42–61.

Welch, C.A. 1995. Rivalry and unification: mobilising rural workers in São Paulo on the eve of the Brazilian golpe of 1964. *Journal of Latin American Studies* 30, no. 2: 61–89.

Welch, C.A. 1999. *The seed was planted: the São Paulo roots of Brazil's rural trade union movement*. State College, PA: Penn State Press.

Welch, C.A. 2005. Agribusiness: Uma breve história do modelo norteamericano. In *Anais do X Encontro de Geógrafos da América Latina*, 16467–505. São Paulo: Universidade de São Paulo.

Welch, C.A. 2006a. Movimentos sociais no campo: a literatura sobre as lutas e resistências dos trabalhadores rurais do século XX. *Revista Lutas e Resistências (Londrina, PR)*, 1, 60–75.

Welch, C.A. 2006b. Keeping communism down on the farm: the Brazilian rural labor movement during the cold war. *Latin American Perspectives* 33, no. 3: 28–50.

Welch, C.A. 2009a. Camponeses: Brazil's peasant movement in historical perspective (1946–2004). *Latin American Perspectives* 36, no. 4: 126–55.

Welch, C.A. 2009b. Os camponeses entram em cena: a iniciação da participação política do campesinato paulista. In *Formas de resistência camponesa: visibilidade e diversidade de conflitos ao longo da história. v. 2: Concepções de justiça e resistência nas repúblicas do passado (1930–1960). Coleção História Social do Campesinato no Brasil*, eds. M. Motta and P. Zarth, 29–51. São Paulo: Editora da UNESP.

Welch, C.A. 2010. *A semente foi plantada: as raízes paulistas do movimento sindical camponês no Brasil (1924–1984)*. São Paulo: Expressão Popular.

Welch, C.A. 2011. Lula and the meaning of agrarian reform. *NACLA: Report on the Americas* 44, no. 2: 27–30.

Engaging the Brazilian state: the Belo Monte dam and the struggle for political voice

Peter Taylor Klein

This contribution uses the case of Brazil's largest infrastructure project, the Belo Monte hydroelectric facility, to examine the challenges and opportunities for resistance and claims-making in the face of contemporary development projects. It shows that the confluence of the privatized nature of hydroelectric projects and the government's purported commitment to democratic, participatory development has impacts. I argue that this context, on the one hand, contributes to the fracturing of civil society. On the other hand, it presents opportunities for the creation of surprising alliances among diverse resistance groups and the state. I further argue that direct acts of resistance in this context can encourage the state to work for the public good.

In June of 2010, just 12 months before the start of construction of Belo Monte, the world's third largest hydroelectric facility, Brazilian President Luiz Inácio Lula da Silva visited the Amazonian city of Altamira, Pará, the city closest to and most affected by the dam, to express his support for the facility. He addressed the project's critics and anti-dam activists directly:

> I know that what many well-intentional people … don't want is to repeat the errors committed in this country throughout hydroelectric dam construction. We never again want a hydroelectric dam that commits the crime of insanity that was Balbina, in the state of Amazonas. We don't want to repeat Tucuruí. We want to do something new. So, let me give some advice to the comrades who are against [the dam]: instead of being against it, propose alternative to use the R$ 4 billion that we are making available in the process to take care of the social and environmental question. Let's discuss how it is that we will use this 4 billion, to better the life of the ribeirinho people, to better the life of the Indians, to better the life of farmers. There are R$ 4 billion, money that Pará has never seen, to take care of social questions. (da Silva 2010)[1]

In this speech, President Lula (as he is commonly called) captured the contentious nature of Belo Monte and the government discourse that it will be constructed differently than large dams of the past. The project has been debated for over three decades, with proponents arguing that the dam will produce necessary and sustainable energy for Brazil while also bringing development to the region. It has become the country's signature development project under a national plan to accelerate economic and infrastructure growth, and the

---

[1] Author's translation.

government has repeatedly stressed that affected communities can take part in decision-making processes to ensure the negative effects are mitigated. Opponents, however, argue that it will lead to significant and irreparable damage to the region's environmental and social landscape. The first 2 years of construction did, indeed, bring changes to the region, disproportionately affecting marginalized communities, such as riverine and agricultural communities forced to leave their homes, and economically poor residents of Altamira who faced rising prices and the inability of government services and basic infrastructure to keep pace with population growth.

This contribution uses the Belo Monte case to investigate the challenges and opportunities faced by local resistance efforts as they make claims in the face of Brazil's contemporary development projects. I suggest that the combination of the semi-privatized nature of infrastructure projects and the Brazilian government's stated commitment to developing a deeper, more participatory form of democratic development both enables and constrains resistance and claims-making. First, these dynamics lead to the constantly shifting strategic alliances within and between resistance groups. Second, the landscape of participatory processes is quite different in this context, particularly with the introduction of the private sector as a primary actor in decision-making. The government has shown that it is committed to citizen participation, but the logic of the private sector conflicts with that of democratic development. This can actually create alliances between resistance efforts and government agencies, and lead to the 'activation' of the state (Abers and Keck 2009) in the creation and implementation of participatory spaces.

Democratic development, privatization and dams in contemporary Brazil

The contemporary Brazilian state under the left-leaning *Partido dos Trabalhadores* (PT – Worker's Party) couples the goals of a more participatory and just democracy with economic development that is driven by the construction of large, semi-privatized infrastructure projects such as hydroelectric facilities. These large development projects and their associated negative consequences on local communities present challenges in upholding the discourse of the government that purportedly seeks to serve the needs of the poor through social programs and opportunities for citizen participation in state decision-making processes.

In 2002, when the people of Brazil first elected Lula, the PT candidate, to the presidency, social movements and engaged citizens across the country celebrated. They hoped that Lula, who aligned himself with historically marginalized social movements (Wright and Wolford 2003; Welch 2006), would finally deepen democracy, increase citizen participation in decision-making processes and reduce social inequities. President Lula and his PT successor, President Dilma Rouseff, did introduce more formalized and diverse citizen participation initiatives (Secretaria-Geral da Presidência da República 2011) and social programs such as *Bolsa Família*, a conditional direct cash transfer program that provides financial assistance to the poor.

However, President Lula's election also led to a renewed interest in an aggressive developmental strategy (Hochstetler and Montero 2013), which promoted state intervention in economic and industrial growth by building linkages between the state and private enterprises (Johnson 1982). While developmentalist ideas – prevalent in Brazil and Latin America from the 1930s to the 1980s (Schneider 1999) – were never completely replaced by the neoliberal ideology that took hold in Brazil from 1995 to 2002 under President Fernando Henrique Cardoso, they were reinvigorated by the Lula administration and adapted for a more market-oriented and globalized economy (Hochstetler and Montero 2013).

President Lula's flagship development program, the *Programa de Aceleração do Crescimento* (PAC – Growth Acceleration Program), sought to bolster Brazil's infrastructure system while creating jobs. From 2007 to 2010, the program invested approximately USD 250 billion to promote 'the resumption of the planning and implantation of large social, urban, logistics, and energy infrastructure projects in the country, contributing to its accelerated and sustainable development' (PACa 2013).[2] Hydroelectric projects were a central feature of PAC and have remained that way in PAC 2, President Rouseff's continuation of the program. Belo Monte, the signature PAC program with a USD 15 billion price tag, is the largest and most costly infrastructure project in the country.

Although these dams are government-driven projects, they are not completely government run. After the late 1980s and early 1990s, when state-led construction of large dams was put on hold due to significant anti-dam organizing and the state's inability to receive development loans, the energy sector was partially privatized under President Cardoso, prompting proposals for hydroelectric facilities from private consortia (Rothman 2001; Kingstone 2004). The PAC hydroelectric projects have continued in this vein. Each is constructed and operated by a consortium of public and private companies brought together to oversee that particular facility. In nearly all cases, private companies hold more than 50 percent of the stake in the consortium, which is then considered a private enterprise (PACa). *Norte Energia*, a consortium of companies in which the private sector holds 50.01 percent, oversees the Belo Monte facility.

The impact of this privatization on dam resistance efforts is not well understood. Much of the literature on the privatization of Brazil's electric sector tells of the significant mobilization against privatization in the 1990s (Hochstetler 2011; Kingstone 2004), the difficulties in state regulation of privatized projects (Kingstone 2004) and how the state has ultimately maintained control of the energy sector despite privatization (de Oliveira 2007; Hochstetler 2011). Hochstetler (2011, 353) suggests that the private sector has always been a part of an 'energy-building enabling coalition', a mix of state and market actors whose shared goal is to build energy projects, which confronts a 'blocking coalition' made up of state and civil society actors opposed to construction. Less is known about how a privately run enterprise impacts the ability of locally based citizens and social movements to make demands during the installation of hydroelectric facilities.

There are dozens of planned, in process and completed PAC hydroelectric facilities around the country (PACb), many of which are surrounded with controversy. Seen as a clean and sustainable form of energy by its proponents, hydroelectricity makes up nearly 30 percent of Brazil's energy supply and more than 80 percent of the country's electricity (Empresa de Pesquisa Energética 2013). In 2011, the Ministry of Mining and Energy announced plans to invest over USD 100 billion in new hydroelectric projects, mostly in the Amazon region (Ministério de Minas e Energia 2011). Advocates of hydropower argue that dams such as Belo Monte can and will be built in ways that respect the local environmental and social communities. However, large dams have historically left devastating social and environmental impacts in their wake, disproportionately affecting poor and disadvantaged communities (McCully 2001; Khagram 2004). They flood significant amounts of land, often displacing local communities and emitting greenhouse gas emissions (Fearnside 2006), cause the loss of biodiversity and, due to the increased availability of work, rising population rapidly increases the cost of living near the construction site

---

[2] Author's translation.

(Fearnside 2006). These contrasting perspectives of dams have led to ongoing and intense debates over the viability and necessity of projects such as Belo Monte.

Political participation, social movements and accountability in the contemporary moment

The controversies over large dams point to the contradictions of democratic development under a leftist government that has sought to align with historically marginalized social movements and open possibilities for citizens to participate in government decision-making. How does a state simultaneously encourage economic development, reduce inequality and poverty and support the democratic involvement of its citizens? Further-more, how are the state and other powerful actors held accountable to these stated, yet sometimes conflicting, goals? These have become central concerns of scholars interested in development, participatory democracy and the politics of accountability.

Scholars have argued that a balanced form of development can emerge when there is a synergistic relationship between the state and society, in which a capacity-filled govern-ment and society work together to mutually reinforce one another (Evans 1996; Heller 1999). Fox (2007) argues that this synergy is important in efforts to hold the state accoun-table, as pro-accountability actors in both the state and civil society form coalitions to gain more power and voice. State-initiated participatory spaces are avenues through which this state–society synergy can form, and scholars discuss the need for a 'deepening' of demo-cratic institutions in which the state opens up participatory spaces that allow decisions to be made through deliberation by everyday citizens, civil society actors and government repre-sentatives (Fung and Wright 2003).

The first institutionalized participatory processes emerged in Brazil with PT-initiated participatory budgeting in Porto Alegre in the 1990s (Abers 2000; Baiocchi 2005; Wampler 2007). These experiments have since expanded to other municipalities around Brazil (Baiocchi, Heller, and Silva 2011), and President Lula later formalized participation at the federal level. According to a report released by the *Secretaria-Geral da Presidência da República*, 'Starting in 2003, the country adopted a new form of government, based on the ongoing and qualified dialogue with diverse segments of society, creating a co-respon-sibility between the State and the population' (2011, 6).[3]

With the celebration and rise of institutionalized participation, social movements have had to make choices about how to engage with the state to make claims. As Dagnino (2002) shows, resistance groups, social movements and civil society, in general, are often ambiva-lent about participatory spaces created by the state. Some feel deceived by state-led parti-cipatory efforts that utilize the discourse of citizenship to mean something completely different, others feel torn regarding how they will participate and yet others reject the idea of working with state agencies altogether. This ambivalence comes from what she calls a 'perverse confluence' (2003) of the neoliberal ideology in the 1990s under Cardoso and the increase of democratic participation ushered in by the PT government, albeit with less than successful results at the national level (Goldfrank 2011). I suggest that this confluence led to further opportunities and challenges for resistance efforts. In par-ticular, the privatization of formerly state enterprises that accompanied neoliberalism has remained a distinguishing feature under the PT administrations, as has the drive for more participatory, socially conscious ways of implementing development initiatives.

---

[3]Author's translation.

This context has not meant that social movements have necessarily had to choose between participation and contentious action. Instead, some of the most active organizations in participatory institutions have also been the most likely to engage in public protests and make demands through multiple channels (Lavalle, Acharya, and Houtzager 2005; Tarlau 2013). Social movements such as the Landless Workers' Movement have managed, in some cases, to maintain their autonomy from the state, allowing them to use disruptive politics against the state while also participating in decision-making spaces with that same state (Tarlau 2013). Social movements' concurrent participation in institutionalized spaces of participation and continued collective acts of resistance can build social networks and have an overall positive impact on the ability of civil society actors to make demands (Carlos 2011). In the case of Belo Monte, I show that as many longstanding social movements come to terms with their relationship with the PT, they choose one approach or the other. Others, however, utilize direct acts of resistance to call attention to their situation and open spaces of participation.

Social movements and affected communities have participated in influential ways during the planning stages of hydroelectric facilities in Brazil. Participation during the licensing process has allowed anti-dam groups to occasionally stop dams from being built (Rothman 2001). Even when dams are built, this citizen participation slows down the planning process and usually leads to less-damaging dams with fewer conflicts (Hochstetler 2011). McCormick (2007) argues that new collaborations between lay people and experts offer avenues for affected populations to make claims and impact the policy process. This literature, however, has been largely limited to the opportunities and advantages that participatory spaces offer during the pre-construction phases. I build on these accounts by examining participation during the installation of a large development project when affected communities seek avenues to make claims as their lives change.

The literature on participatory democracy most often pays attention to the formal spaces created by the state to incorporate more citizens and civil society groups in decision-making processes, but participation between government and citizens takes many forms beyond the formalized, planned and policy-prescribed spaces to which scholars have up until now paid attention. Wendy Wolford (2010) insightfully describes how, in the absence of a strong and capable state, social movements demanding land reform were able to force the development of participatory processes by transgressing acceptable and legal boundaries of action. In this paper, I show a similar process of the use of direct action in opening a participatory space, but argue that the presence of a power-yielding private company, rather than a weak state, can provide the setting for the creation of participatory forums of decision-making. This also builds upon Rebecca Abers's and Margaret Keck's (2009) work on the mobilization of the state; they argue that a weak state can be 'activated' by civil society through participatory spaces, thus mobilizing the state to act on behalf of the public interest. The case of Belo Monte shows a similar process at work, but suggests that it is also possible for direct acts of resistance to mobilize the state to *open* a participatory space, particularly when a private company is overseeing the development project. The *possibility* of participation, rather than a predetermined and policy-prescribed participatory space, seems to be the requisite condition for activating the state to work for the public good.

Through this analysis, we can also see that the debate over dam construction cannot be reduced to two coalitions of dam proponents versus opponents, as much of the scholarship has depicted. Baviskar (1999, 234), in her analysis of the conflict over the Narmada Valley dams in India, illustrates how the anti-dam movement is not held together by a unified ideological position or alternative vision of the future, but rather by diverse experiences that 'placed them in opposition to the developmental state'. In so doing, she shows that resistance efforts are much more complex than usually depicted by scholars. Nevertheless, she still describes a two-sided conflict between community members, with all their disparate interests, and the developmental state. Similarly, but with focus on the diverse actors that drive dam construction, Ribeiro (1994, xxi) details the 'economic and political processes through which [several powerful capitalist agencies] come together' to build the Yacyretá dam in Argentina and Paraguay. Although he disentangles the network of powerful actors at local, national and transnational levels, he also describes an ultimately two-sided scenario in which those affected by the dam have little ability to stop the push towards economic expansion. Hochstetler (2011) begins to break down this dichotomy by showing the many linkages between the state, civil society and market actors in the debate over Belo Monte; however, in her analysis, these linkages occur within either the 'enabling coalition' or the 'blocking coalition', as described above. In contrast, this paper moves beyond the dichotomy by showing the complex and ever-changing set of relations within and between the state and civil society that arise as the effects of the dam are felt.

Finally, this essay contributes to the literature on participation and accountability politics by offering an examination of the role of the private sector in spaces of engagement. One concern of the devolution of state decision-making to participatory spaces has been the potential for private interests to appropriate public interests (Abers and Keck 2009), but little is known about the dynamics of participation and citizen claim-making when private companies are actually in control of the development initiatives in which the space is situated. Similarly, scholarship on 'accountability politics' has focused on holding national governments and international organizations, such as the World Bank, accountable (Clark, Fox, and Treakle 2003). As this contribution shows, the ability to do so is further complicated in a context in which the state has turned over much of its control to a private entity.

Setting and methods

The Belo Monte hydroelectric facility has a long history, and its changes over time represent how it has come to symbolize the contemporary moment of infrastructure development in Brazil. The idea of damming the Xingu, the river on which Belo Monte is now being constructed, was first introduced inside the military government in the mid 1970s. Initial plans concluded in 1979 and detailed six dams along the Xingu and the connected Iriri River (Moya, Franco, and Rezende 2007). Kararão, the name for the part of the complex that is now Belo Monte, would have flooded an estimated 1225 square kilometers of land, including the villages of indigenous groups (Fearnside 2006; Moya, Franco, and Rezende 2007). In the late 1980s, after the official plans were released, local and international social movement groups, environmental advocacy organizations and local citizens, including indigenous groups, protested against the dam. Around the same time, the World Bank pulled its funding from the project, and the dam was temporarily halted. By the time new plans were released over 10 years later, the dam was renamed and significantly redesigned in an effort to reduce the environmental and social impacts. The five dams upstream of Belo Monte were no longer considered, and the new plans decreased the flooding by two-thirds and avoided flooding any indigenous lands.

Belo Monte now typifies the most recent wave of infrastructure projects being built in the name of Brazilian development. Proponents have argued that it is and will continue to be carried out in participatory, sustainable and responsible ways that support local communities while providing the energy that Brazil's growing economy needs. Opponents have long contested that the negative effects are too great to warrant damming the Xingu River. After nearly three decades of debate, construction began in 2011, and the effects on the local community were immediately noticeable. The population of Altamira grew quickly due to the increased availability of work and business opportunities; local researchers and officials reported that Altamira's population had increased from less than 100,000 in 2010 to over 140,000 by 2012, and some expected a doubling of the population by 2015. This population growth strained the local infrastructure and day-to-day living. Within the first year of construction, residents of Altamira said that they had already noticed sudden and significant rises in living expenses – rents had more than tripled in much of the city over the course of the year and many food items had doubled in cost – as well as increased violence and traffic problems. The already-insufficient infrastructure in Altamira could not handle the increasing demands, and rising river levels would displace at least 20,000 individuals from their homes (Ribeiro and Redondo 2011).

The challenges faced by social movements and marginalized populations in making claims on the state amidst these changes can be difficult to measure without a deeper under-standing of their ever-changing day-to-day lives. Thus, in order to understand what anima-tes claims-making in the context of Belo Monte, I carried out a political ethnography that paid close attention to the discourses and practices of various groups engaging in myriad ways (Schatz 2009). I lived in Altamira for a total of 1 year between July 2011 and June 2013, the first 2 years of Belo Monte's construction, where I participated in activities of and conducted interviews with social movement organizers and participants, fishermen and riverine people who were making demands on the state, the state-level public defender, the federal level public defender, government officials and representatives from Norte Energia.[4] Table 1 provides a chart of the relevant groups discussed throughout this contri-bution and their relation to Belo Monte.

The fracturing of civil society and the shifting of alliances under the leftist government

The PT government under Presidents Lula and Dilma has stressed time and again that the Belo Monte dam will be built better than dams in this past. As mentioned in President Lula's speech, Tucuruí and Balbina led to significant social and environmental impacts and produced far less electricity than expected (Fearnside 2001; Khagram 2004). Belo Monte, on the other hand, has been credited by proponents as a sustainable and responsible development project that will be accompanied by participatory processes through which affected populations will be well informed and even have opportunities to participate in decision-making. The federal government has taken steps in this regard. Nearly USD 2 billion is being invested in programs to reduce the negative social and environmental impacts of its construction and construct schools, hospitals and other projects to develop the infrastructure of the region. Approximately USD 250 million will be allocated by a par-ticipatory decision-making space – the Steering Committee of the Regional Sustainable Development Plan of the Xingu (CGDEX) – comprised of representatives from civil

---

[4]I recorded nearly 1000 pages of fieldnotes from over 150 events and interviews. I recorded some of the meetings I attended and some interviews when appropriate. In many cases, however, recording was not permitted or possible. In those cases, I took handwritten notes and typed longer field notes later. I also collected hundreds of documents, flyers and other printed materials from the groups I studied.

Table 1. Summary of groups related to Belo Monte.

Name (Portuguese)	Name (English)	Acronym or abbreviation	Description and relation to Belo Monte
Casa de Governo	Government House	CG	Local representative of the federal government that works with citizens, civil society and Norte Energia to find solutions to demands made by the population
Colônia dos Pescadores	Fishermen's Union	Colônia	Government-recognized union that represents fishermen; the union has intermittently partnered with other organizations, smaller subsets of fishermen, and government officials to make demands
Comitê Gestor do Desenvolvimento Regional Sustentável do Xingu	Steering Committee of the Regional Sustainable Development Plan of the Xingu	CGDEX	Presidentially mandated committee made up of representatives from civil society and government; tasked with allocating USD 250 million for projects that will bring sustainable development to the region and reduce the dam's negative impacts
Fundação Viver Produzir e Preservar	Living, Producing, and Preserving Foundation	VPP	Non-governmental organization that supports small-scale farmers who live and work along the Transamazon highway; they have long opposed damming the Xingu but have focused on how to support the rural population in the face of dam construction
Instituto Brasileiro do Meio Ambiente e dos Recursos Naturais Renováveis	Brazilian Institute of Environment and Renewable Natural Resources	IBAMA	Federal agency responsible for environment regulation and management of natural resources; IBAMA assesses and issues the licenses required to construct and operate Belo Monte
Ministério da Pesca e Aquicultura	Ministry of Fish and Aquaculture	MPA	Federal ministry that attends to the needs of fishermen and fish farmers while developing the fishing sector; they started to play an integral role in supporting the fishermen facing the effects of Belo Monte's construction

(*Continued*)

Table 1. Continued.

Name (Portuguese)	Name (English)	Acronym or abbreviation	Description and relation to Belo Monte
Movimento dos Atingidos por Barragens	Movement of People Affected by Dams	MAB	National social movement that organizes and aims to support people affected by dams; MAB has been active in the national debate over Belo Monte for many years and established permanent representatives in Altamira in 2009
Movimento Xingu Vivo Para Sempre	Xingu Forever Alive Movement	XV	The locally based social movement opposed to the construction of Belo Monte
Norte Energia			The consortium of public and private companies that owns the Belo Monte project
Partido dos Trabalhadores	Workers' Party	PT	Left-wing political party that grew from the social movements in the 1980s and has controlled the executive branch since 2003

society of 11 dam-affected municipalities and officials from federal, state and municipal government. This presidentially mandated space is responsible for approving projects that will mitigate negative impacts of the dam and increase development in the region.

Despite these changes, Belo Monte is still bringing negative effects. Engaged citizens in Altamira – including long-time activists who have for years been unified in their fight against Belo Monte, participants of national and international social movements who have moved to Altamira and residents who had not been engaged but who began feeling the effects of the dam's construction – have had to determine the best approach to making claims in this context. This section shows how long-time regional and national activists have had to come to terms with changes to the political party, the PT, that they have long supported but that is implementing a large infrastructure project that they have long opposed. I argue that this process has fractured civil society and resistance efforts.

Long-time leftist activists coming to terms with the contemporary PT

President Lula's 2010 speech in support of Belo Monte was a significant moment for civil society in Atlamira. Many leaders of social movements and non-governmental organizations (NGOs) said that this was a watershed moment for civil society that fractured resistance efforts and divided a once unified community. For decades, many people had come together to protest against the dam. They had rallied in the streets and pressured the government and international actors to stop construction. They had mobilized support for the PT and celebrated President Lula's victory less than 8 years previously. Now, the government they had elected with hope for a different path forward was committed to constructing Belo

Monte in the name of democratic and economic development. With fewer and fewer barriers to construction, these anti-dam activists had to make decisions: would they continue
protesting against the dam, would they engage the government in efforts to mitigate the
negative effects or could they find middle ground?

Andreas,[5] the head of an NGO with a base in Altamira, explained the dynamics within
civil society in 2008–2009, prior to the now-infamous presidential visit:

> Everyone was against the dam, even the politicians … . There was such a strong history of
> social movement activity in the region that many of the people who had worked with or in
> the social movements were now in positions of power. Many had worked through the PT,
> the party that was born out of social movements, and so they could use this as a vehicle to con-
> tinue their activism around their work, but through the political sphere and their position of
> relative power.

He said that this unity changed when President Lula came to Altamira in 2010: 'Everybody
knew the president was going to speak in favor of the dam, so it became a confusing time for
people'. Some decided to stay home, others went to the stadium for the event and even helped
with the logistics, and yet others protested in the streets. 'It was quite apparent to people who
was on what side … at least in their perspective'. He was hesitant to paint it as a black-and-
white moment of people going to one 'side' or the other, but he was suggesting that a significant split occurred among resistance efforts because many people saw it this way.

One of the principal groups that chose to protest that day, the Xingu Forever Alive
Movement (XV), stands out for their vocal and staunch anti-dam position. The leaders of
the organization, which was founded in 2008, have been active in struggles for citizenship
rights since the opening of the Transamazon highway in the 1970s and have fought against
the idea of damming the Xingu River since plans were first announced in the late 1980s.
While other groups have chosen to negotiate with government agencies or participate in
decision-making processes, XV has refused to do so, utilizing traditional social movement
techniques, such as street protests and rallies, to raise awareness regarding the harms that
the Belo Monte dam may cause. In an interview with Carla, a leader of XV, she talked
about the divisions amongst the social movements and NGOs in Altamira, confirming
that the split came when President Lula came to speak:

> There were people there who used to be with XV who had buttons on that had some sort of
> government logo. They stopped [those of us from XV] from entering … . They had said
> they were against the dam, but …

She trailed off, clearly frustrated by the memory. She also talked about the PT, the changes
since President Lula was elected, and the reasons she left the party in 2009:

> The parties on the left, Lula, are supposed to be of the people … . They are said to be populist
> parties. As [Lula] entered into power it weakened the fight against the government as it changed
> to support the government. In all parties there exists the desire to have power to dominate … I
> won't affiliate with a party anymore. I believe in the popular organization of the people. I
> believe in going to the streets (luta da rua), fighting for people's rights, for quality education,
> for quality health. For the rest of my life I will fight for these things. As it says in our consti-
> tution, we have these rights … . Any party that gets into power will try to keep their power.
> They will dominate. They want the power of domination.

---

[5]Personal names, with the exception of public figures, are pseudonyms.

Carla continued by talking about the power of people uniting in order to make demands so that that large projects like Belo Monte do not just benefit big capital. 'These projects are disrespecting the people who live here. People don't have a way to live with dignity'. She often spoke of how Belo Monte is a 'crime against humanity committed by the government'. Carla, along with hundreds of other activists, had partied in the river when President Lula was first elected, and as his presidency went on she, like many others, experienced a great deal of disappointment with their government. In a speech about Belo Monte, a lawyer who often works with XV expressed similar disappointment with the government: 'I was overjoyed in 2002 when Lula was elected and I have never felt so deceived and let down by my government'. As Carla and others like her searched for ways to reconcile their past hopes with the PT and the great disenchantment they felt as Belo Monte became a reality, they refused to negotiate or participate in any way with the PT project.

In contrast, others who had worked very closely with Carla and those fighting against the construction of the dam chose a different path once dam construction seemed imminent. Simone, a leader from the Living, Producing, and Preserving Foundation (FVPP), a local organization that supports rural farmers, stressed in an interview with me that FVPP was 'always part of the fight against the dam'. She explained more about their change:

> But with the building of the dam, the movement has a different posture … . It's a fight for the rights of the people. The other side is a discussion with the government. Every big project has resources for mitigation of the effects … . It's more comfortable to stay against [the dam], stay the same line and continue being against. This is really strong, but comfortable, too. But thinking about people 10–15 years from now. That's not easy.

I also spoke with the Claudio, another leader of FVPP, who ran for vice-mayor as a PT candidate while I was living in Altamira. He shared similar perspectives to Simone, explaining, 'We [the people fighting against the dam] were beaten because our biggest goal [of stopping the dam's construction] did not happen. The political force of the government was bigger than our force of the people … We are small in comparison'. But he stressed that that it is important to recognize that the 15 years of civil society work against Belo Monte was not done in vain. He listed off the changes in plans to the dam that reduced flooding, the list of conditions that must be met by Norte Energia in order to receive the license to operate the facility, and the increased opportunities for participation with government officials. He went on to talk about the divisions within civil society: 'This question of "who is for and who is against" is a strategy to divide society and transform the debate over this big project into a competition as though it were a soccer game'. He wanted civil society to come together and work to ensure the rights of the people were met. In order to do so, he remained an active member of the community and engaged in discussions with the government and Norte Energia, serving as the primary civil society representative in CGDEX, the participatory space for sustainable development. While he was not happy that the dam was being built, he had embraced the opportunities to participate directly with government, and had showed through these activities and his candidacy for vice-mayor that he remained allegiant to PT.

Despite Claudio's and Carla's shared belief in the power of popular organization, their mutual desire to bring unity to their struggles for the people in the region and their long history of working together, the connections between their organizations were completely severed after President Lula spoke in Altamira. I often heard XV express their disappointment in those, like Claudio, who had remained loyal to the PT and who were engaging in any kind of direct participation and negotiation with government actors or Norte Energia.

Relationships between groups are, of course, constantly shifting, and the fissures between groups are not always as long lasting as those between XV and FVPP. This was the case between XV and the national-level Movement of People Affected by Dams (MAB), who first arrived in Altamira in 2009. MAB was founded over 20 years ago and has been successful in influencing national-level policy and organizing dozens of communities in anti-dam movements, even preventing some dams from being built (McCormick 2006). While they are opposed to the construction of dams, they work with communities who will be affected by dams in order to support and teach residents as they make demands on their local government. David, one of the long-time MAB organizers who began working in Altamira in 2012, explained their philosophical problems with the drive for dam construction:

> Brazil is governed by capital. Decisions are made in favor of capital, so the issues go well beyond Belo Monte. And the issues here are more than just the actual dam. It's about energy, the actual dam, and transmission of the energy … . The problem isn't necessarily the particular government but the government has a responsibility to better conditions for the people … . MAB is against the dam because it is an environmental and social crime, but they fight for the people to live a good life. So, they don't just stop at being against the dam. The problem is the model being used by the government, a model based on capital. Capital is the problem.

One would think Carla and the others at XV would agree with David's perspective. After all, they both used the same language of Belo Monte as a 'crime' and are both fighting for the people affected by the dam. However, in my interactions with these groups, it was clear that while they occasionally worked together, they did not see eye to eye on everything. The reasons for the strained relationship became clearer in my interview with Carla, when she talked about MAB:

> MAB says they are fighting big capital and that to do so the country needs to move towards a more socialist society. But here, in these cases, big capital has names and addresses. There are names of big capital: Lula, Dilma, the courts, and the company, Norte Energia. MAB defends PT – they say it's not the fault of Dilma and Lula but that it's just big capital. But it's not. I've been saying this for a long time.

For XV, they wanted nothing to do with the government or PT. MAB, on the other hand, found it possible to engage in both contentious activities and more direct participatory processes with government officials. David, like Claudio, told me that labeling people as either in favor of or against the dam disrupts the ability to organize because it creates enemies and makes it difficult to have a more complex view. He also stressed that MAB is not political, meaning not associated with a particular political party, but he shared his personal opinion: 'The PT was born of social movements but the PT program has since been abandoned. That said, I think PT is the best option for now'.

This case highlights how a range of engaged citizens have come to terms with their former expectations of the PT in very different ways. Most expected and hoped that President Lula's election meant a new kind of government, and most recognize that the PT they supported years ago had not transformed the federal administration in the ways that they had hoped (Goldfrank 2011). The PT they rallied behind in the past would not have advocated for the construction of the Belo Monte dam. Once it was apparent that the dam would be constructed, we see how citizens dealt with broken promises in different ways, and the fracturing this caused within civil society. Activists could no longer rally around the discourse of 'the left' to unify with similar goals, nor could they rally around an anti-Belo

Monte position. Instead, we see how some groups chose the path of engaging with the state through participatory spaces created by their PT government or calling on that government to uphold its promises. Others wanted nothing to do with the PT, felt as though they or others would be coopted if participating with the state. These actors were all dealing in different ways with the 'perverse confluence' (Dagnino 2003) of neoliberalism and democratic participation while trying to hold the government accountable in different ways.

The challenges faced by social movements and activists in the context of large development projects are undoubtedly great and go far beyond the difficulties in creating a unified resistance effort. The introduction of a large infrastructure project like Belo Monte leads to impacts felt by people not traditionally in resistance struggles. In what follows, I detail a fishermen's protest to highlight the surprising opportunities that arise for claims-making efforts in this difficult context, highlighting how resistance groups can still find moments of unity and showing how they can mobilize the state (Abers and Keck 2009) to open up participatory decision-making spaces.

Calling on the state in the presence of a private entity

By August of 2012, just over a year after construction of Belo Monte began, fishermen and riverine communities were noticing changes to the river on which their livelihoods depended, and they began searching for ways to make their claims heard. They blamed the *ensecadeira*, the cofferdam that dries the river behind it in order to build the dam that will generate power, for the drastic reduction of fish and the pollution of the water where they live and work. On a trip with Paulo, the president of the fishermen's union, to an island just past the *ensecadeira*, I talked with residents who complained about the changes.[6] 'Last year at this time you could see fish jumping out of the water right here', a shop owner explained as he pointed to the water just a few steps away. A fisherman who stood in the doorway added, 'I recently went fishing near here and caught six or seven kilos of fish. Last year at the same time it was over 20!' These residents, like many others who lived and worked along the river, were becoming increasingly concerned with their situation and dissatisfied by the support that they expected but were not receiving from Norte Energia and government officials. In addition to receiving no compensation for the effects the dam was having on their lives, they also felt as though they were receiving little information about what was happening.

These concerns led to a bubbling of support for direct action. In mid-September, a small group of about a dozen fishermen, supported by the union and the staunch anti-dam movement XV, gathered on the banks of the river in Altamira to begin a protest with the goal of calling attention to their situation. Paulo explained to the local press, 'We are not against the dam, but we are fighting for our rights, for our livelihoods, and the respect we deserve'. The group of fishermen decided to fish in prohibited zones near one of the three major

---

[6]The structure of fishermen unions in Brazil is similar to that of Brazilian rural workers' unions. Local fishermen unions, *colônias dos pescadores*, operate at the municipal level. They respond to the state-level organization, the *federação*, which is overseen by a country-wide organization, the *Confederação Nacional de Pescadores e Aquicultores* (CNPA). This system of unions, which is designed to defend the rights and interests of fishermen, has been federally regulated since the 1970s and was officially sanctioned in 2008. The founder of Altamira's colônia explained that in the early 1990s, he began noticing that fishermen in other municipalities 'had some power because they had a colônia'. Thus, he founded the colônia in 1997 in order to 'fight for the dignity that the fishermen deserve'.

construction sites and camp on an island in the same area. Paulo explained to me that he hoped the protest would raise awareness of their situation among the government and the population, and as they met with officials from Norte Energia.

Rodrigo, an outspoken fisherman and member of the union, shared similar sentiments when asked by the local press about the situation for fishermen, considering that IBAMA had granted the license for Norte Energia to close the river:

> We need this area [of the river] … . We now see a company, Norte Energia, preventing us from working in this area. I've been a professional in this area for 28 years, and really we aren't against this enterprise [Belo Monte]. We want them to understand that we need our rights as fishermen of the Xingu River … . We are using lawyers that are representing the union to request our rightful compensation that we professional fishermen have. And until this moment Norte Energia hasn't given us anything. They have abandoned us and they are trying to do what was done in Tucuruí … . We don't accept this … . Norte Energia will make a mountain of money off of this dam and the fishermen will be left with what? We need the directors of Norte Energia to actually discuss with us the situation of the fishermen!

As the interviews with the press concluded, the fishermen fashioned small Brazilian flags to the front of their boats and set off for the 4-hour journey in their small, motorized boats. The view was striking – about 12 small, wooden boats, each with a Brazilian flag fluttering in the wind as the city slowly disappeared behind us. With these flags, they seemed to be making a statement that this was their country, too, and while the government and others in power were intent on building this dam, they would have a say in how this was done. In an enthusiastic speech on the first day of the protest at the island where they camped, Paulo told the fishermen, 'Every one of us here has rights. Let's fight for our rights! It is only like this that we will show that we are Brazilians and that our rights have to be secured!'

The protest endured for a month, drawing more than 200 participants at its peak and serving as a place of unity for a number of social movements, including XV and MAB, to work together in their goals of supporting people affected by the dam. It concluded with a 10-day occupation of the construction site that halted work. The fishermen, boat pilots, riverine dwellers and farmers who participated drafted a list of demands that were negotiated at a 'conciliatory hearing' with the federal prosecutor, the state level public defender, officials from a variety of government agencies and high-ranking employees from Norte Energia. The fishermen's principal demand was compensation for the disruption the dam was having on their livelihoods. In the end, they settled for an agreement to participate in further studies to investigate the changing river conditions, which would then determine whether they could be financially compensated. While their demands were not met, the protesters celebrated. They had engaged in direct negotiations with Norte Energia and many agreed that they would protest again if their other demands were not met.

Perhaps the most significant outcome, however, was the attention they were able to draw to their situation from the federal government. After just a week of protesting, a high-level representative from the Ministry of Fish and Aquaculture (MPA), Fernando, met with over a hundred fishermen at the headquarters of the union, explaining how they would work with the union and other local groups representing the community in order to ensure that the rights of the fishermen would be met and that their demands would be heard by Norte Energia. Fernando immediately arranged a second meeting with representatives from each of the groups who attended the first meeting, including the union, two other associations that represent different sectors of the fisheries, MAB and officials from *Casa de Governo* (CG), the local representative of the federal government whose objectives include to accompany the demands of the region and find solutions together with federal and municipal organs and together with the company – Norte Energia –, guaranteeing the presence of the State locally [in Altamira]. (Ministério do Planejamento 2013)

In the beginning of this second meeting, he explained,

I haven't yet requested a meeting with [various federal-level governmental organs] and Norte Energia to bring the demands of the fishermen forward. I first wanted to hear from you ... make a quick evaluation to understand the process ... and determine how we can improve the deliberation process that was originally created by the fishermen.

Throughout the meeting, it was clear that he and the federal representatives from CG were interested in discussing the concerns of the local community and putting pressure on Norte Energia in the name of the fishermen. As the participants explained their numerous concerns with the changing river, the threats to their livelihoods, the difficulties with their equipment and the failure of Norte Energia to respond or offer compensation for those being affected, Fernando and the other federal representatives asked clarification questions and provided opinions on what could be realistically expected. As the group developed ideas of how to improve the situation, many stressed the importance of Norte Energia's presence at meetings and the mitigation role that the federal government could and should play. By the end of the meeting, they had drafted a document that summarized their ideas for moving forward, primary among them being the creation of a steering committee run by the federal government that would be dedicated especially for the fishermen.

It was clear that the local community was utilizing the moment to put pressure on parts of the federal government that they felt could actually force Norte Energia to act. They did not expect compensation immediately, but they wanted to be able to participate in direct talks that could (1) ensure that the company understood their situation and position, and (2) open avenues to make demands directly on the both the state and Norte Energia. The federal government, with its stated focus on democratic and participatory decision-making, could provide this opportunity. The direct act of resistance seemed to be opening the door to direct participation between civil society organizations, government representatives and Norte Energia.

Activating the state for and through participation

The fishermen's protest had long-lasting consequences. The months following the protest were marked by direct negotiations between the fishermen's union, government officials and Norte Energia. They did not use direct action again, although the fishermen who joined the protest initially said they would be willing to go back to the dam site to make more demands. One municipal-level official who works closely with the fishermen explained the change: 'Now is a moment for dialogue. The fishermen are trying to organize better now than in the past ... they have been making connections with people within the [municipal and federal] government to try and get what they want'. It was true; they were invited to participate in one of the official, federally-mandated 'social accompaniment forums', a space coordinated by Norte Energia that was designed to provide an avenue through which the community could hear about and discuss the progress of Belo Monte and its effects on the region. Norte Energia restricted access to these meetings to the municipal governments and a limited number of officially sanctioned civil society groups, so securing an invitation was a real success.

At the first forum meeting to which the fishermen groups were invited, the MPA initiated the Commission on Fish and Aquaculture, the steering committee that Fernando discussed during the protest, to address specific concerns related to fisheries. In addition to MPA, the commission is comprised of federal agencies that represent environmental and indigenous concerns, state level secretary of fisheries, municipal governments from the region, and the fishing sector, including the union and other civil society associations that represent the interests of various types of fishermen. Over time, the principal concern for the Altamira-based civil society groups turned to the housing situation for fishermen, and the commission became the principal avenue through which these concerns could be voiced. The rise in the water level due to the dam would flood many of the fishermen's homes and nearly all of the commercial businesses that support their work, and the commission allowed for direct negotiations between state agencies, Norte Energia and the fishermen in developing a plan of action.

Nearly a year after the beginning of the protest, I asked Julia, a representative of the MPA who participated in nearly all of the meetings in Altamira and who worked closely with many of the groups related to the region's fisheries, about the impacts of the protest on the creation of the special Commission of Fish. I had heard from Paulo and others that the protest led to their opportunities for direct negotiations. She confirmed: 'If it were not for the protest, the commission would not have happened. That [protest] put the pressure on Norte Energia. The fishermen called for the Ministry, so Norte Energia had to respond'. She added, 'This was important for the Ministry, as well, because [other sectors of the] government started to recognize the MPA'.

Julia's statements are telling for two reasons. First, as is evident throughout the story, the act of direct resistance led to multiple opportunities for fishermen to participate in negotiations with various sectors of the government and Norte Energia. The union wanted the support of the federal government and saw the opportunity for partnerships, and the protest opened avenues for these partnerships to build. Second, her follow-up statement reflected some of the frustration felt within the ministry over not being included in other processes related to Belo Monte. The protest brought attention to the fishermen's situation, and when the union called on the MPA, it forced the government to recognize the work of the ministry.

The union and other fishermen found an alliance with the MPA in making demands on Norte Energia. Julia explained that the MPA helps to 'orient Norte Energia', holding meetings with technical teams from IBAMA, Norte Energia and MPA. In these meetings, they talk about what the fishermen want and what Norte Energia is obligated to provide. She indicated that, in communicating to Norte Energia what the fishermen are requesting, she uses language such as 'According to the law ... ' in order to pressure Norte Eneriga.

During the Commission meetings, it was clear she played a similar supporting role. As the group discussed a particularly contentious housing settlement for the fishermen whose houses and businesses would be flooded, members of the union and other fishermen expressed their concerns. One fish shop owner conveyed her frustration:

> Norte Energia does the best for itself, picks the cheapest option. The population of Altamira did not ask for Belo Monte. We are happy in our houses even though Norte Energia says they're poor quality ... . You, [Norte Energia], need to think more and respect the population ... . I'm not against Belo Monte. I'm against the pain you have brought to us. You don't listen to us, but this will not go forward because we will go to the streets and get what we deserve.

Others shared similar thoughts, pleaded with the company to understand their situation and threatened to use direct action. Julia then stood to speak, bringing the conversation back to concrete requests they had made and that, if fulfilled, could allow the fishermen to play a more significant role in the deliberation process:

> We are trying to facilitate an agreement and solution. The ministry has asked for the potential sites to be visited with the fishermen, and until now I do not know if that has happened. Second, we asked that the documents [regarding the housing options] be given to the fishermen [by Norte Energia] …. Neither I nor the fishermen have had a say in the selection of a site … and the lack of technical information from Norte Energia is causing huge problems. The fishermen have been very clear about their concerns and desires, and Norte Energia has not addressed them.

The direct act of resistance by the fishermen's union was a way for them to call on government officials to uphold the discourse of participatory, responsible development as the dam was being built and affecting their lives. By encouraging the government to hear their demands, they were also building state capacity. This was possible not because of a weak state, as other accounts have suggested (Wolford 2010; Abers and Keck 2009), but in part due to the presence of a power-yielding private company, Norte Energia.

Conclusion

The stories of resistance efforts in the face of the Belo Monte hydroelectric facility highlight the challenges and opportunities in the ongoing struggles for political voice in the context of contentious development projects in Brazil. The Brazilian government is committed to a kind of development that will grow the economy and serve its energy and infrastructure needs, while also purporting to do so in participatory and responsible ways. In order to hold the state accountable for these promises during the construction of Belo Monte, locally based residents and activists had to navigate their relationships with two new sets of actors: the federal PT government and Norte Energia, the private company in charge of building the facility. The federal-level PT government, which had largely been absent in the region prior to Belo Monte, was suddenly present in people's lives. Not only was the government-touted project a symbolic presence, but it also brought a host of government officials who were physically present in the region. For long-time activists, they had to come to terms with their failed expectations of how the PT government could transform society. Instead, they were faced with a PT infrastructure project that they had for many years fought against. Some, like XV, chose to turn away from the PT and instead fight against the dam and the government that they helped elect. Others, such as FVPP, embraced the participatory opportunities that made them proud to be aligned with the PT and sought to utilize the resources made available by the government that they felt was at least building this dam in a different way from those of the past. Yet others, in this case MAB, tried to find a middle ground. They were not completely satisfied with the PT and thus not willing to give up their contentious resistance, but they also saw the importance of working with local and federal government officials in order to support the needs of dam-affected people. I argue that the diverse ways that activists dealt with this disjuncture led to a fracturing of social movements and resistance struggles.

For residents joining the resistance efforts in order to make demands, they also had to navigate the opportunities afforded them by the PT government, but they were not, necessarily, reconciling their past PT relationships with the current situation. Instead, they – the fishermen, in this case – were leveraging the government that offered participatory opportunities and that claimed infrastructure development would be done differently. They were able to mobilize the state (Abers and Keck 2009; Wolford 2010) to open a participatory space of debate and negotiation with the private company, Norte Energia. Thus, their direct act of resistance established an alliance with the government that supported their struggles against the company.

This story suggests that the presence of the private company can actually provide motivation for and an avenue through which the state can act on behalf of the public good, but this is not to be celebrated outright. The outcomes of the participatory spaces that the fishermen were able to open are yet to be seen; the fishermen have an undoubtedly difficult task, even with the support of federal officials, to achieve their material wishes. The challenges associated with dealing with a private company are significant. In the meetings of the special Commission, Norte Energia officials consistently claimed they would have more information in the future, that they were doing what was mandated by law and that they would continue to be open to dialogue. None of these statements, however, meant they would actually compensate the fishermen or shift their plans to accommodate the principal concerns.

These difficulties reflect an aspect of what Dagnino (2003) calls the 'perverse confluence' of neoliberalism and democratic development. In addition to the reduced, minimal state that she cites as the key feature of neoliberalism, it also encouraged a wave of privatization of formerly state enterprises, including hydroelectric projects. Under PT governance, this privatized structure has been joined by a discourse of democratic development in which infrastructure projects are built more responsibility and with citizen involvement. However, in this confluence, the state's logic and concern for citizen well-being comes into conflict with the logic of the private company, which seeks to build Belo Monte in the most efficient and cost-productive way.

Activists and everyday citizens making claims in the face of large development projects such as Belo Monte undoubtedly encounter a host of challenges in the contemporary political context of Brazil. Their efforts are largely fractured while the government, committed to development through infrastructure development, and private companies, with economic interests in these projects, are powerful forces pushing a common agenda. Nevertheless, the increasing opportunities through which people can make demands provide opportunities for activists and citizens to create surprising alliances that can mobilize the state to act for the public good. These alliances, built upon different yet overlapping visions of development, will need to be maintained and strengthened if Brazil's development trajectory is to change.

Acknowledgements

The author would like to thank Rebecca Tarlau and Anthony Pahnke for the invitation to submit a manuscript for this special collection and for providing thoughtful comments on this research. Thanks also go to the anonymous reviewers of this contribution.

Funding

Thanks go to the National Science Foundation and the IIE's Fulbright Program for funding.

References

Abers, R.N. 2000. *Inventing local democracy: Grassroots politics in Brazil.* Boulder, CO: Lynne Rienner.

Abers, R.N., and M.E. Keck. 2009. Mobilizing the state: The erratic partner in Brazil's participatory water policy. *Politics and Society*, 37, no. 2: 289–314.

Baiocchi, G. 2005. *Militants and citizens: The politics of participatory democracy in Porto Alegre*. Stanford: Stanford University Press.

Baiocchi, G., P. Heller, and M. Silva. 2011. *Bootstrapping democracy: Transforming local governance and civil society in Brazil*. Stanford: Stanford University Press.

Baviskar, A. 1999. *In the belly of the river: Tribal conflicts over development in the Narmada Valley*. New Delhi: Oxford University Press.

Carlos, E. 2011. Movimentos sociais: revistando a participação e a instituionalização. *Revista de Cultura e Política Lua Nova* 84: 315–48.

Clark, D., J. Fox, and K. Treakle, eds. 2003. *Demanding accountability: Civil-society claims and the World Bank inspection panel*. Lanham, MD: Rowman & Littlefield Publishers.

Dagnino, E., ed. 2002. *Sociedade civil e espaços públicos no Brasil*. São Paulo: Paz e Terra.

Dagnino, E. 2003. Citizenship in Latin America: An introduction. *Latin American Perspectives* 30, no. 2: 3–17.

Empresa de Pesquisa Energética. 2013. Balanço energético nacional 2013 – ano base 2012: Relatório Síntese. Ministério de Minas e Energia. Rio de Janeiro. https://ben.epe.gov.br/BENRelatorioSintese2013.aspx (accessed November 5, 2013).

Evans, P. 1996. Government action, social capital and development: Reviewing the evidence on synergy. *World Development* 24, no. 6: 1119–32.

Fearnside, P.M. 2001. Environmental impacts of Brazil's Tucuruí dam: Unlearned lessons for hydroelectric development in Amazonia. *Environmental Management* 27, no. 3: 377–96.

Fearnside, P.M. 2006. Dams in the amazon: Belo Monte and Brazil's hydroelectric development of the Xingu River Basin. *Environmental Management* 38, no. 1: 16–27.

Fox, J.A. 2007. *Accountability politics: Power and voice in rural Mexico*. Oxford: Oxford University Press.

Fung, A., and E.O. Wright. 2003. *Deepening democracy: Institutional innovations in empowered participatory governance*. London: Verso.

Goldfrank, B. 2011. The left and participatory democracy. In *The resurgence of the Latin American left*, ed. S. Levitsky and K.M. Roberts, 162–83. Baltimore, MD: Johns Hopkins University Press.

Heller, P. 1999. *The labor of development: Workers and the transformation of capitalism in Kerala, India*. Ithaca: Cornell University Press.

Hochstetler, K. 2011. The politics of environmental licensing: Energy projects of the past and future in Brazil. *Studies in Comparative International Development* 46: 349–71.

Hochstetler, K., and A.P. Montero. 2013. The renewed developmental state: The national development bank and the Brazil model. *The Journal of Development Studies* 49, no. 11: 1484–99.

Johnson, C. 1982. *MITI and the Japanese miracle*. Stanford: Stanford University Press.

Khagram, S. 2004. *Dams and development: Transnational struggles for water and power*. Ithaca: Cornell University Press.

Kingstone, P. 2004. *Critical issues in Brazil's energy sector: The long (and uncertain) march to energy privatization in Brazil*. The James A. Baker III Institute for Public Policy. Houston, TX: Rice University.

Lavalle, A.G., A. Acharya, and P.P. Houtzager. 2005. Beyond comparative anecdotalism: Lessons on civil society and participation from São Paulo, Brazil. *World Development* 33, no. 6: 951–64.

McCormick, S. 2006. The Brazilian anti-dam movement: Knowledge contestation as communicative action. *Organization & Environment* 19, no. 3: 321–46.

McCormick, S. 2007. The governance of hydro-electric dams in Brazil. *Journal of Latin American Studies* 39: 227–61.

McCully, P. 2001. *Silenced rivers: The ecology and politics of large dams*. New York: Zed Books Ltd.

Ministério de Minas e Energia. 2011. Em seminário, ministro diz que Brasil não pode abrir mão de hidrelétricas, 25 October. http://www.mme.gov.br/mme/noticias/destaque1/destaque_284.html (accessed November 2, 2013).

Ministério do Planejamento. 2013. Casa de governo tem nova sede em Altamira (PA), 30 October. http://www.planejamento.gov.br/conteudo.asp?p=noticia&ler=10532 (accessed November 2, 2013).

Moya, C.A., H.C. Franco, and P.F. Rezende. 2007. *AHE Belo Monte – evolução dos estudos. XXVII Seminário Nacional de Grandes Barragens*. Belém, Brazil: Comitê Brasileiro de Barragens.

de Oliveira, A. 2007. Political economy of the Brazilian power industry reform. In *The political economy of power sector reform: The experiences of five major developing countries*, eds. D. Victor and T.C. Heller, 31–75. Cambridge: Cambridge University Press.

PACa. 2013. Geração de energia elétrica. Ministerio do Planejamento. http://www.pac.gov.br/sobre-o-pac (accessed November 2, 2013).

PACb. 2013. 8° balanço completo do PAC 2. Ministerio do Planejamento. http://www.pac.gov.br/sobre-o-pac/publicacoesnacionais (accessed November 2, 2013).

Ribeiro, G.L. 1994. *Transnational capitalism and hydropolitics in Argentina: The Yacyretá high dam*. Gainesville, FL: University Press of Florida.

Ribeiro, A., and F. Redondo. 2011. Os nômades de Belo Monte. *Epoca* 9 July. http://revistaepoca.globo.com/Revista/Epoca/0,,EMI247824-15223,00.html (accessed 19 December, 2014).

Rothman, F.D. 2001. A comparative study of dam-resistance campaigns and environmental policy in Brazil. *The Journal of Environment & Development* 10, no. 4: 317–44.

Schatz, E. 2009. *Political ethnography: What immersion contributes to the study of power*. Chicago: University of Chicago Press.

Schneider, B.R. 1999. The *desarrollista* state in Brazil and Mexico. In *The developmental state*, ed. M. Woo-Cumings, 276–305. Cornell: Cornell University.

Secretaria-Geral da Presidência da República. 2011. *Democracia participativa: nova relação do estado com a sociedade, 2003–2010*. Brasília: Secretaria-Geral da Presidência da República.

da Silva, Luiz Inácio Lula. 2010. *Discurso do presidente da república, Luiz Inácio Lula da Silva, no ato por Belo Monte e pelo desenvolvimento da região do Xingu*. Presidência da República, Secretaria de Imprensa, Discurso do Presidente da República, 22 June. http://www.biblioteca.presidencia.gov.br/ex-presidentes/luiz-inacio-lula-da-silva/discursos/2o-mandato/2010/1o-semestre/22–06–2010-discurso-do-presidente-da-republica-luiz-inacio-lula-da-silva-no-ato-por-belo-monte-e-pelo-desenvolvimento-da-regiao-do-xingu/view (accessed November 2, 2013).

Tarlau, R. 2013. Coproducing rural public schools in Brazil: Contestation, clientelism, and the landless workers' movement. *Politics & Society* 41: 395–424.

Wampler, B. 2007. *Participatory budgeting in Brazil: Contestation, cooperation, and accountability*. Pittsburg: Penn State University Press.

Welch, C. 2006. Movement histories: A preliminary historiography of the Brazil's landless laborers' movement (MST). *Latin American Research Review* 41, no. 1: 198–210.

Wolford, W. 2010. Participatory democracy by default: Land reform, social movements and the state in Brazil. *Journal of Peasant Studies* 37, no. 1: 91–109.

Wright, A., and W. Wolford. 2003. *To inherit the earth: The landless movement and the struggle for a new Brazil*. Oakland, CA: Food First Books.

Education of the countryside at a crossroads: rural social movements and national policy reform in Brazil

Rebecca Tarlau

This contribution explores the strategies used by popular movements seeking to advance social reforms, and the challenges once they succeed. It analyzes how a strategic alliance between the Brazilian Landless Workers Movement (MST) and the National Confederation of Agricultural Workers (CONTAG) transformed the Ministry of Education's official approach to rural schooling. This success illustrates the critical role of international allies, political openings, framing, coalitions and state–society alliances in national policy reforms. The paper also shows that once movements succeed in advancing social reforms, bureaucratic tendencies such as internal hierarchy, rapid expansion and 'best practices' – in addition to the constant threat of cooptation – can prevent their implementation.

Introduction

From 10 to 14 February 2014, the Brazilian Landless Workers' Movement (MST) held its VI National Congress in the capital of Brasília, with 15,000 peasant-activists participating. On Wednesday 12 February 2014, 500 children – the *sem terrinha*, or sons and daughters of the families living in MST settlements and camps across the country – rode in buses to the Ministry of Education (MEC). Several MST activists who had been waiting by the Ministry jumped in front of the doors as the first bus arrived, allowing dozens of children to run into the front lobby before the guards tried to shut the doors. Hundreds more children soon followed. The message of this protest was clearly written on banners the children were holding: '37 thousand schools closed in the countryside'. 'Closing a school is a crime!' '*Sem Terrinha* against the closing and for the opening of schools in the countryside!' Meanwhile, inside the MEC, the recently appointed head of the 'Education of the Countryside' office – a long-time ally of the MST – tried to convince the Minister of Education to meet with the children.[1] Eventually, after three hours of occupation and protest, the Minister came downstairs to address the children, promising them that the federal government was committed to providing quality education in the countryside.[2]

How do popular movements advance social reforms? And, in what ways do state actors, social movements and the goals themselves transform through this process of national

---

[1]Informal conversation with Edson Anhaia, 12 February 2014.
[2]This vignette comes from participant observation in this protest.

advocacy? In this contribution, I analyze how MST activists engaged in a national campaign that transformed the federal government's 'official' approach to rural schooling. I first explore the strategies that MST activists utilized to advocate for these educational reforms, which included developing a network of powerful allies, taking advantage of political opportunities created by previous social mobilizations, strategically framing their struggle to build a broad coalition and developing alliances across the state–society divide. I analyze the tensions inherent in this process of national coalition building, and how social movements transform each other through united struggle. Then, I outline the administrative and bureaucratic barriers activists face implementing social reforms in practice, within the context of a changing political economy. Finally, I reflect on the dangers of 'cooptation', or, in other words, attempts by powerful actors to take on the language of successful policy reforms to promote their own goals.

Background: the educational initiatives of the MST

By the early 1990s, the MST was already deeply embedded in the sphere of public education. Activists were collaborating with municipal governments to improve educational practices, helping to train teachers to work in schools on MST settlements and camps, publishing texts elaborating on the movement's educational ideas and partnering with local universities to run literacy campaigns. The MST's educational approach, known as the 'pedagogy of the MST', garnered increasing recognition among left-wing groups.

While the MST's educational practices were similar to the popular educational programs that many other Latin American social movements incorporated into their political struggles (Arnove 1986; Hammond 1998), the MST was unique in its concern about public schooling. During the first half of the 1990s, MST activists won legal recognition from several municipal governments in Rio Grande do Sul to administer their own teacher-training programs. Then, in 1995, activists founded their first 'movement' school – a private secondary school independent of the public school system, that the MST could oversee with almost complete autonomy.[3] Hundreds of activists received their high school degree through this educational institution. Nonetheless, despite these achievements, the reach of these initiatives only extended to the families living in MST settlements and camps. Activists had not yet discussed the implications these practices held for the entire rural public school system.

Over the next decade, the MST's educational initiatives transformed from a set of isolated practices in agrarian reform communities to a nationally recognized proposal for all rural areas, known as *Educação do Campo* (Education of the Countryside). A national coalition of grassroots movements, union federations and university professors came together to support these ideas, and actively worked with the government to implement them in practice. By 2004, *Educação do Campo* was not only nationally recognized but also become institutionalized within the Ministry of Education. In 2010, President Luiz Inácio Lula da Silva signed a presidential decree giving *Educação do Campo* more legal force, and, in 2012, President Dilma Rousseff announced a multi-ministry federal program that would put this presidential decree into practice. The MST still actively participated in these debates but was now only one of the dozens of groups laying claim to the meaning, content and purpose of these educational ideas. How had the MST succeeded in building a coalition that had such an impact on national educational policy?

---

[3]The Institute of Education Josué de Castro (IEJC), also informally known as ITERRA.

Social movements, social reforms and national advocacy

In a recent collection of successful national-level advocacy campaigns in the Global South, Gaventa and McGee (2010, 10) note a major contradiction: on the one hand, the literature on 'how citizens and civil society organizations (CSOs) interface with national policy' does not acknowledge social movements as a critical base for advocacy; and on the other hand, 'the vast literature on social movements and collective action focuses on explaining the hows and whys of these movements themselves, but not necessarily the policy changes to which they contribute'. This lack of emphasis on national advocacy campaigns may stem from the definition of a social movement itself. McAdam (1999, 37), writing from the 'political process' approach, defines social movements as 'rational attempts by excluded groups to mobilize sufficient political leverage to advance collective interests through non-institutionalized means'. Following this definition, collective efforts to advance social interests can only be described as a social movement if they occur through non-institutional channels. This discounts the fact that many social actors engage in a range of strategies to pursue their goals, simultaneously inside and outside of the state.

In contrast to this literature, the relationship between social mobilization and state reform has long been a focus among agrarian scholars. In analyzing the success of a state-initiated food reform in Mexico, Fox (1992, 6) argues that 'Most explanations of distributive reform tend to emphasize one-way causation, relying on static distribution of power, and they rarely capture the dynamic interaction between state and society'. Fox (1992, 8) proposes an interactive approach to state–society relations, arguing that successful implementation of social reforms depends on a 'sandwich strategy' wherein reformists strategically situated within the state create opportunities for social mobilization. Borras (2001, 548) also advocates for this interactive approach in analyzing land reform initiatives in the Philippines, arguing that state-centered and society-centered approaches cannot explain 'why societal actors attempt to influence and transform state actors, but in the process are themselves transformed – and vice versa'. Similarly, in literature on the Latin American feminist movement, Alvarez (1990, 21) promotes a 'dual strategy', whereby activists engage in contentious politics while also working with policy makers and politicians.

Following these previous scholars, in this paper, I analyze the strategies MST activists utilize to promote social reforms at the federal level. I draw extensively on Gaventa and McGee's analysis of national advocacy campaigns in the Global South, and the strategies these authors highlight as critical to successful reform. Then, in addition to examining the process of policy reform, I outline the 'long chain of actions and reactions that runs from a change in or adoption of a law or policy, to the actual implementation on the ground' (Gaventa and McGee 2010, 31). An analysis of the administrative and bureaucratic barriers that social movement activists face during policy implementation, and how these groups navigate the threat of cooptation, offers important nuance to previous studies of social movements, advocacy and social reform efforts.

Focus and methods

This contribution focuses specifically on the process of policy reform in the Ministry of Education (MEC). Despite the decentralization of public education in Brazil, which devolves authority over schooling to state and municipal governments, the MEC remains one of the most important educational authorities in Brazil. First of all, the MEC is charged with developing general policies and laws for education, which municipal and state governments are legally obligated to follow. Although the implementation of these policies does not always occur in practice, these laws can become important tools for local social actors attempting to implement reforms at the municipal and state level. Secondly, the MEC can influence municipal and state governments through conditional funding and federal–state/municipal partnerships around specific programs. Finally, the MEC also has shared authority over the provision of higher education. These three methods of influence – federal laws and policies, conditional funding and partnerships with municipal and state governments, and higher education provision – in addition to MEC's large budget, make it a powerful agency in the educational landscape in Brazil. Therefore, the recent national campaign to implement *Educação do Campo* in the MEC merits particular attention.

Research for this essay was conducted over 17 months, between 2009 and 2011. I conducted 70 semi-structured interviews with MST activists, including both statewide leaders and local educational activists. I also conducted semi-structured interviews with 60 government officials, five activists from the National Confederation of Agricultural Workers (CONTAG) and 10 members of other rural social movements. Since the focus of this paper is on policy changes in the MEC, I draw extensively on 13 interviews I conducted with government officials in Brasília, and information about national-level politics I collected in interviews with MST and CONTAG activists. In addition, I coded data from the Pastoral Land Commissions (CPT) database on yearly agrarian protests to identity numbers of protests concerning education between 2002 and 2012. I also participated in several educational events organized by the federal government, and examined federal documents concerning *Educação do Campo*.

Strategies for success: from local experiments to national recognition

Powerful allies: UNESCO, UNICEF and university supporters

The MST's investment in educational access in the countryside in the early 1990s stood in stark contrast to the Brazilian government's historical disregard for rural schooling. In the 1990s, the rural public school system in Brazil was still in a dire state. During the two decades of military dictatorship (1964–1985), there had been heavy investment in secondary and tertiary education, levels of schooling that were seen as critical for Brazil's economic development and urban industrialization. Primary education – which was the only level of educational access in rural areas – was largely ignored (Plank 1996, 174–75). The constitution of 1988 brought important structural reforms to the public school system, such as the devolution of schooling to states and municipalities; however, educational improvement was difficult due to the impoverished condition of local governments. Consequently, throughout the first half of the 1990s, rural education did not significantly improve (Plank 1996). Rural schools were seen as an embarrassment to a 'modern' Brazilian state: a system that still contained multi-grade classrooms, teachers without higher education and collapsing infrastructure.

At the same time, in the early 1990s, international organizations such as the United Nations Educational, Scientific and Cultural Organization (UNESCO) and United Nations Children's Fund (UNICEF) were becoming dominant voices in global educational debates (Samoff 1999). These organizations primarily focused on eradicating illiteracy and providing universal access to primary education in high-poverty countries. International program coordinators often criticized the priorities of national governments and tried to circumvent 'inefficient' bureaucracies by working directly with local communities. In this context, UNESCO and UNICEF began to directly fund the MST's educational initiatives, simply because MST activists – in the absence of the state – were organizing the most impressive educational programs in the countryside.

The head of the educational unit of UNESCO-Brazil in 2014, Maria Rebeca Gomes, explains the funding relationship that the agency developed with the MST during this period:

> The MST was the only group working in the settlements ... it is hard to work in these areas if you are not connected to the MST, these are very poor areas. The MST created the infrastructure for these programs, and the families living in the settlements already had a relationship with the MST.[4]

As this official suggests, the imperative for expanding educational access in high poverty areas led the program coordinators of several international organizations to begin establishing programs in areas of agrarian reform. Given the MST's organizational networks in these communities, it made sense to ask the movement for help in the agency's literacy campaigns. Similarly, in 1996, the University of Brasília convinced the Ministry of Education to sign a contract that would allow the MST to train 7000 literacy agents to attend to these agrarian reform areas (Carter and Carvalho 2009, 309). In 1995, the MST even received a prize in 'Education and Participation' from UNICEF for the teacher certification courses MST activists developed for rural teachers (Caldart, Pereira, and Frigotto 2012).

As Gaventa and McGee (2010, 17) argue, 'successful reform campaigns depend on careful navigation to link international pressures with differing and constantly changing local and national contexts'. The support of elite public universities, UNESCO and UNICEF was critical in legitimizing the MST's educational initiatives during the mid-1990s. Nonetheless, although these educational programs were significant, they were still isolated initiatives; it was only with a new political opening in the late 1990s that the MST was able to build a broader national coalition.

Contestation and political opening: education 'free rides' to the capital

The mid-1990s brought both violent conflict and some of the largest social mobilizations of the MST's history. On 9 August 1995, military police killed 11 landless people who occupied a rural property in the poor northwestern state of Rondônia. Less than a year later, on 17 April 1996, 19 MST activists were killed by military police in a march in the northern state of Pará. Perversely, this massacre created a political opportunity, as there was a general public dismay at these government actions that increased national sympathy for the MST and the agrarian reform struggle (Ondetti 2008). In commemoration of this latter massacre, in April of 1997, the MST organized a National March on Brasília for Agrarian Reform, with 100,000 people participating. This march succeeded in pressuring President Cardoso to make many concessions to the movement, including land reform. By the end of Cardoso's first term, 260,000 families had received access to 8 million hectares of land – almost double the amount given in the previous 10 years (Branford and Rocha 2002, 199).

It was within the context of this new political opening – caused by these previous social mobilizations – that MST educational activists began to push the movement's education

---

[4]Interview with Maria Rebeca Otera Gomes, 24 February 2014 (via Skype).

proposal into the national debate. According to MST activist Edgar Kolling, the movement's educational proposal gained national recognition because of its '*carona*' (slang for 'free ride') with the larger movement for agrarian reform.[5] In July of 1997, a few months after the national march, MST educational activists organized a National Meeting of Educators in Areas of Agrarian Reform (ENERA). This meeting was encouraged and financed by UNICEF and UNESCO, in recognition of the MST's educational initiatives. The original plan was for 400 people to attend the first ENERA meeting, but in the end, over 700 people participated (Caldart, Pereira, and Frigotto 2012, 503). Out of these discussions came a proposal for a federal program that would provide educational access specifically for the children, youth and adults living in camps and settlements, known as the National Program for Education in Areas of Agrarian Reform (PRONERA).[6] Gaventa and McGee (2010, 15) argue that 'while political opportunities create possibilities for collective action for policy change, these openings themselves may have been created by prior mobilization'. The fact that the MST's educational proposal was able to 'free ride' on the movement's larger mobilization for agrarian reform supports this assessment.

Framing and coalition building

Up until this point, all of the MST's educational initiatives, including PRONERA, were directed towards populations in 'areas of agrarian reform'. However, during the ENERA meeting in 1997, representatives from UNICEF and UNESCO encouraged MST activists to expand their educational initiatives to include other populations, such as indigenous groups, black communities and rural workers. The MST activists perceived this as a strategic opportunity to receive more financial support for their educational initiatives. They began referring to their educational proposal as *Educação do Campo* (Education of the Countryside).

The MST's new use of the term *Educação do Campo* in the late 1990s – and the quick disappearance of the phrase 'education in areas of agrarian reform' from the movement's public discourse – was a top-down and conscious process of framing (Benford and Snow 2000). The MST's choice of frame resonated with dozens of rural movements, nongovernmental organizations (NGOs) and individuals who were not connected to the agrarian reform struggle. For example, a year after the first ENERA meeting in 1997, the MST hosted the first National Conference for a Basic Education of the Countryside. Participants at this conference included 19 federal and state universities, several government agencies and a dozen other rural social movements and grassroots NGOs.[7]

The 'education of the countryside' frame was also strategic because the federal government was beginning to acknowledge the extreme inequality among public schools, especially between rural and urban areas. In 1998, there was a reform in the financing of primary education, and the federal government began to guarantee a minimum level of spending per student through the National Education Fund (FUNDEF). This provided a surge in financial support for schools that could not reach this minimum (Schwartzman 2004), and increased national attention to the issue of rural education. The MST's strategy

---

[5]Interview, Edgar Kolling, 18 November 2010.
[6]PRONERA was put in the Ministry of Agriculture Development, and has had a very different institutional trajectory than the programs in the Ministry of Education.
[7]All of these organizations and institutions are listed in the final conference document.

to 'frame issues carefully, adjust to changing circumstances and audiences, and draw upon a wide repertoire of strategies' (Gaventa and McGee 2010, 29) succeeded in bringing dozens of new actors into its educational coalition.

Unstable alliances: reconciling labor-peasant tensions

Despite these advances, there was one group conspicuously absent from the National Conference for a Basic Education of the Countryside, which was preventing the coalition's ability to move forward: the Confederation of Agricultural Workers (CONTAG). This union confederation is made up of dozens of rural federations that consist of thousands of unions representing more than 15 million rural workers.[8] The absence of an organization representing millions of rural workers meant that convincing the federal government to support an educational policy for the entire countryside would be unlikely.

In order to understand CONTAG's absence from this national conference in 1998 – and its decision to join the coalition for *Educação do Campo* several years later – it is necessary to trace the history of rural activism back to the years prior to the 1964 coup. During these two decades, Communist Party members and left-leaning Catholic activists were organizing rural workers through the formation of peasant leagues and rural associations. These two groups were often in competition for the allegiance of rural workers, as the Communist Party took a 'quasi-revolutionary approach' and the Catholic organizers 'a moderate, but persistent and firm, demand for the extension of already codified urban workers' rights to their rural counterparts' (Maybury-Lewis 1994, 68). With the passage of the Rural Labor Statute in 1963, CONTAG was founded and, to the dismay of rural elites, communist activists were able to elect themselves to the leadership of the confederation (Houtzager 1998). This situation was short-lived, however, as the military coup took place on 31 March 1964.

When the military government took power, it systematically expelled communist activists from the ranks of the union movement. Simultaneously, the military government *stimulated* the growth of the rural unions in an attempt to foster national integration (Houtzager 1998). Especially important was the Fund for Assistance of the Rural Worker (FUNRURAL), which was established in 1971 to provide medical and dental services for rural populations. CONTAG experienced its biggest growth during this period, from less than 300 unions in 1963 to over 2000 in the 1980s. Almost all of these rural unions had partnerships with FUNRURAL, illustrating CONTAG's 'huge role in dispensing, organizing and managing the regime's rural medical and dental services, in accord with the military government's plan' (Maybury-Lewis 1994, 41).

However, this is not the whole story. While most rural unions functioned as social service providers, some unionists took advantage of the limited space they had to wage a national campaign for workers' rights. These were primarily the activists who had been organizing closely with the Catholic Church prior to the coup. In contrast to the communist party activists, who had largely been purged from their unions, these unionists

> understood that excessive provocation of rural elites and the authorities, given the power relations in the countryside, would hurt them and set back their organization drive ... They learned the value of respecting the law. Indeed, the unionists became champions of the law,

---

[8]Maybury-Lewis (1994, 56) groups rural workers historically connected to CONTAG into three groups: small holders and sharecroppers (people with modest access to land they use to plan subsistence and cash crops), wage workers (with no autonomous control over land) and *posseiros* (homesteaders or squatters).

pushing for enforcement of legislation on the books ostensibly to protect their rights. (Maybury-Lewis 1994, 73)

In 1968, a group of unionists that came out of this organizing tradition took control of CONTAG. Under this new leadership, CONTAG became a progressive force in the countryside, winning concrete legal gains for workers during a highly repressive period. Although most local unions continued to follow a service-oriented path, CONTAG activists helped to develop the leadership of many progressive unionists during this period (Maybury-Lewis 1994). However, the downside of this strategy was that an entire generation of CONTAG labor activists became accustomed to non-confrontational approaches to unionism (Houtzager 1998).

In 1979, in the context of a more general political opening, CONTAG initiated a series of annual strikes in Pernambuco and began calling for large-scale agrarian reform (Maybury-Lewis 1994, 76; Welch 2009). By this time, other rural organizations were also beginning to engage in direct action in the countryside, such as the Pastoral Land Commission (CPT), founded in 1975. The CPT was critical in helping workers occupy land in the early 1980s, the first actions of the soon-to-be MST. These rural activists joined with other urban movements, neighborhood associations and militant unionists to found the Workers Party (PT) in 1979 and the Central Union of Workers (CUT) in 1983. In contrast, the CONTAG leadership 'made a virtual religion out of its autonomy from political parties' (Houtzager 1998, 135).

The relationship that developed between CONTAG leaders and the emerging landless movement is complex. Many unionists developed strong connections to the CPT, the MST and the PT, and with the help of these social movements they took over their local unions (Maybury-Lewis 1994, 173–97). In Pernambuco, in the early 1990s, local union activists actually hosted MST activists in their headquarters and helped the movement organize its first land occupations in the sugar cane region. This eventually led the state union federation in Pernambuco to organize its own land occupations in the mid-1990s – despite a deeply embedded culture of 'following the law' (Rosa 2009, 471–72).[9]

At the national level, however, there were serious ideological divides between CONTAG and the CUT, CPT and MST leaders. First and foremost, CUT believed 'that a rapid separation of the union movement from the money, flows, job sinecures, and state policy orientation' was necessary, while CONTAG 'felt that this would create tremendous organizational difficulties, given the extreme poverty of the workers they were representing' (Maybury-Lewis 1994, 242). In many local unions, a competitive relationship developed between CONTAG and CUT, as CUT activists – often in tandem with the MST and the CPT – ran their own candidates in local union elections. This fed into a general mistrust between the national MST leadership and CONTAG, in addition to other ideological disputes.

A critical moment occurred in 1995, at CONTAG's VI National Congress, when CUT activists won enough local elections that they tipped CONTAG's internal power balance, leading CONTAG to affiliate with CUT. At this congress, delegates also began to discuss a proposal for broader social policies in the countryside, which became known

---

[9]However, CONTAG's choice to occupy land is not necessarily in contradiction to its historical tendency to 'follow the law', as occupations are generally conducted on land that is arguably subject to the land reform law. Therefore, CONTAG can be seen as promoting one dimension of the rule of law.

as the Alternative Project for Sustainable Rural Development and Solidarity (PADRSS). A leader in the CONTAG federation in Pernambuco explained how these developments related to education:

> In 1995 there was a national congress of the rural workers, and we discussed the alternative project we were trying to construct for society ... we wrote the PADRSS proposal ... it was a document that discussed the public policies we wanted for the countryside, and education entered there.[10]

At CONTAG's 7th National Congress, in 1998, the delegates passed the PADRSS proposal. Two years later, CONTAG was at the forefront of the national coalition for *Educação do Campo*.[11]

A legalistic turn

Between the first National Conference for *Educação do Campo* in 1998 and the second National Conference that took place in 2004, CONTAG became one of the most important participants in the national alliance for *Educação do Campo* – at times surpassing the role of the MST itself. CONTAG activists did not take on this struggle simply because MST activists changed how they 'framed' their coalition; there were also internal shifts occurring within the confederation itself. CONTAG's new focus on social and cultural demands in the countryside – emphasized by the PADRSS program – facilitated the decision to make *Educação do Campo* a political goal of the organization. Subsequently, these interactions between social movements – with distinct histories and relationships to the state – transformed the MST's original educational goals. Decades of experiences with a legal approach to workers' rights were still engrained within CONTAG. Consequently, as soon as CONTAG activists joined the coalition for *Educação do Campo*, they sought out ways to work with the government to pass a federal resolution ensuring rural workers' legal rights to this educational proposal.

CONTAG's strategy succeeded, and on 3 April 2002, the National Education Advisory Board[12] approved a federal resolution supporting the 'Operational Guidelines for a Basic Education in the Schools of the Countryside'. Rosali Caldart, a national MST educational activist, admits: 'We participated very little in writing the guidelines, the union movement was closer ... This is not because we decided not to participate, but because this was not our world' (quoted in Marcos de Anhaia 2010, 84). According to Caldart, the MST's 'world' was one of protest, while CONTAG had decades of experience working with the Brazilian state to enforce workers' rights. The coalition between the MST and CONTAG resulted in a slightly different version of Fox's (1992) 'sandwich strategy' and Alvarez's (1990) 'dual strategy', in which MST activists continued to engage in disruptive protest, while CONTAG activists lobbied government officials and drafted public policy. Another factor in this legal victory was the government official who wrote the *Educação do Campo* guidelines, Edla Soarez, a long-time educational activist in Pernambuco with deep connections to many rural social movements. Edla acted as a supportive state reformer

---

[10]Interview with Sonia Santos, 2 March 2011.
[11]The first two coordinators of the *Educação do Campo* within CONTAG also confirm that the PADRSS proposal solidified the importance of public education (Costa Lunas and Novaes Rocha n.d.).
[12]*Conselho Nacional da Educação/Câmara da Educação Básica* (CNE/CEB).

(Fox 1992) within the Cardoso government, contacting grassroots groups to help her write the educational proposal.[13]

After the *Educação do Campo* guidelines were passed, both the MST and CONTAG claimed it as a huge victory; however, President Cardoso took no further actions to put the resolution into practice. The rural social movements also stopped advocating for actions to be taken in respect to this educational proposal; it was an election year and these social movements were focused on bringing the PT to power.

New state–society alliances

When President Lula came to office in 2003, he had a long list of promises to fulfill for the activists who had mobilized his support. One of these demands was the implementation of the *Educação do Campo* Guidelines. In 2004, the President created a Secretary for Continued Education, Literacy and Diversity (SECAD), which included a Department for Diversity and Citizenship. Within this department, an office for *Educação do Campo* was created and an Advisory Board for *Educação do Campo* was established, which included dozens of civil society groups (MEC 2004). While the national coalition for *Educação do Campo* had succeeded in transforming federal law during an antagonistic government, it was only in the context of a more supportive administration that these reforms were implemented. Gaventa and McGee (2010, 16) seem correct in their assertion that 'competition for formal political power is also central, creating new impetus for reform and bringing key allies into positions of influence'.

As soon as the federal government created an *Educação do Campo* office in the Ministry of Education, the historical conflicts between the MST and CONTAG began to reemerge. Both of these organizations demanded that the coordinator of the *Educação do Campo* office come from their own ranks. Under these circumstances, Professor Antonio Munarim – an academic not overtly associated with either movement – was approached by the new Secretary of SECAD to become the first coordinator of the *Educação do Campo* office. In August of 2004, Munarim attended the Second National Conference for *Educação do Campo*, to be vetted for the position.[14]

Unlike the First National Conference, which the MST and a few allies organized, over 38 groups officially sponsored the Second National Conference in 2004.[15] Despite the plurality of voices present at the second conference – ranging from university professors to NGOs and members of the Brazilian legislature – the MST and CONTAG continued to drive the reform process in a tenuous and contentious partnership. When Munarim arrived at the conference, he immediately ran into Rosali Caldart, a national activist in the MST Education Sector. Caldart exclaimed, 'Professor Munarim, what are you doing here?' In response, he told Caldart that he had been tapped as the coordinator of *Educação do Campo*. Laughing, Caldart said that she had better not give him a hug yet, because if CONTAG activists saw him interacting with her, his name would be rejected from the nomination.[16] On 6 August 2004, Munarim became the first coordinator of the *Educação do Campo* office, representing an institutional compromise between CONTAG and the MST, mediated by the MEC. While the Second National Conference for *Educação do*

---

[13]Edla Soarez went to every state to collect data for these guidelines, and she also admits it was difficult to convince MST activists to be part of this process. Interview with Edla Soarez, 6 April 2011.
[14]Interview, Antonio Munarim, 28 November 2011.
[15]These groups are all listed on the conference's official final document.
[16]Interview, Antonio Munarim, 28 November 2011.

Campo in 2004 represented the pinnacle of hope for the national *Educação do Campo* coalition, frustration quickly followed. The rest of this contribution analyzes the challenges and barriers to implementing successful policy reforms.

Policies in practice: administrative and bureaucratic barriers to social reform

Hierarchy: 'a tiny door that opened to a waiting room'

From the beginning, the creation of an office for *Educação do Campo* within the MEC was followed with intense debate and contestation. Armênio Bello Schmidt, a PT activist from the state of Rio Grande do Sul, became the new director of the Department of Education for Diversity and Citizenship in 2004. He recalls the significance of this re-structuring within the MEC: never before had an educational department thought about the diversity of the Brazilian population. 'Now all of these debates were on the table, and there was an institutional space for civil society to participate'.[17] In contrast to Schmidt, Edgar Kolling remembers these institutional changes with anger: the MST had supported the creation of a Secretary of *Educação do Campo*, reporting directly to the Minister of Education. Instead, the minister made *Educação do Campo* an office, within a department within a secretariat.[18] The decision-making power of the coordinator of the *Educação do Campo* office would be highly restricted.

Indeed, from the beginning, Antonio Munarim faced huge barriers sparking interest in *Educação do Campo* within the Ministry of Education, given the low status of his office. Even though dozens of MEC officials worked with rural education, they had little interest in adhering to the new rural educational proposal. Munarim's inability to change the rest of the Ministry led to the increasing isolation of the *Educação do Campo* office. He explains:

> We needed an organizational structure that was strong, with professionals that were competent, and this never happened. The contracting of more people never happened ... In that moment the MEC showed what it really was, a heavy infrastructure. SECAD was an opening, a tiny door that opened to a waiting room, but it never let anyone into the kitchen.[19]

Munarim refers to the 'heavy infrastructure' in the Ministry of Education, or in other words, the hierarchical structures and bureaucratic processes that became barriers for carrying out institutional change. Even though *Educação do Campo* was now part of Brazilian law, the location of the *Educação do Campo* office in a lowly position within MEC's bureaucratic structure meant that influencing other departments was almost impossible.

Munarim says he waited two years for someone with power to enter the waiting room and hear the demands of the social movements participating in the advisory board. No one ever showed up. In beginning of August of 2006, he wrote a letter denouncing the MEC:[20]

> The creation of the Coordinating Committee for *Educação do Campo* signaled a strong commitment within the Ministry of Education for these proposals, and opened up the possibility of inviting civil society groups into these discussions ... The opposite has happened. Representatives from rural social movements have been the only effective presence in these meetings.

---

[17]Interview, Armênio Bello Schmidt, 10 November 2010.
[18]Interview, Edgar Kolling, 18 November 2010.
[19]Interview with Antonio Munarim, 28 November 2011.
[20]Antonio Munarim emailed the author a copy of this letter.

In his letter, Munarim expresses frustration with the weak relationship between the Advisory Board for *Educação do Campo* – where civil society was participating – and the MEC decision-making structure. Although Munarim himself participated in the advisory board, the office of *Educação do Campo* did not have the power to implement any of the advisory board's decisions. In August of 2006, only two years after he was appointed coordinator of the *Educação do Campo* office – as an explicit MST–CONTAG compromise – Munarim was fired.

Rapid expansion: 'losing everything it was supposed to be'

It was only after Munarim left the MEC that a few of the *Educação do Campo* programs were implemented, most likely as a response to the protests that followed his firing. One of these new MEC programs was a bachelor-level certification course for teaching high school in rural areas, known as LEDOC (Bachelor Degree in *Educação do Campo*). This program was adapted from a bachelor-level pedagogy program that the MST had created in 1998, through the federal program PRONERA. The first LEDOC program was administered in one of the MST's own secondary schools, the Institute of Education Josué de Castro (IEJC), in Rio Grande do Sul. The fact that the pilot program took place in one of the movement's own spaces – where activists had been implementing their pedagogical approach for over a decade – meant that the program adhered closely to the MST's previous educational practices. Although the LEDOC pilot program was implemented through an official partnership between the MEC and the University of Brasília, MST activists were the daily participants, administrators and directors of the program (Caldart et al. 2010).

The first year of the LEDOC pilot program had not even ended when MEC bureaucrats launched a second program located at the University of Brasília's own campus. The MEC officials also proposed that the university make LEDOC part of its internal structure, so the MEC no longer had to administer the program. By 2007, LEDOC had been implemented in three more universities.[21] Four years later, in 2011, there were 32 universities that had LEDOC degree programs[22] – with assigned staff, tenured professors, a standardized curriculum and an annual application process. The MEC officials were thinking about quantity: they wanted to reach the largest number of rural students possible. Unfortunately, with this rapid expansion, MST activists lost much of their ability to monitor these educational programs. The MST's lack of capacity to participate – given the fact that the MEC bureaucrats and university professors running the program often lacked a real commitment and knowledge of the original spirit of the proposal – resulted in a significant dilution of the movement's original intentions.

The MST's perspective on these university degree programs is mixed. On the positive side, activists acknowledge that the mere existence of the LEDOC program is a huge advance over traditional urban-centric bachelor-degree programs in pedagogy. Furthermore, the LEDOC courses increase the general level of access to tertiary education for populations in the countryside, since spots are reserved for students from rural areas. On the negative side, activists feel they have lost their ability to be protagonists in the implementation of these courses.

Vanderlúcia Simplício, an MST activist attempting to oversee the LEDOC programs at the University of Brasília, laments that 'the proposal is expanding, but it is losing

---

[21]Interview, Monica Molina, 10 November 2010.

[22]Interview, Antonio Munarim, 28 November 2011; Luiz Antonio Pasquetti, 17 November 2011.

everything it was supposed to be'.[23] When I spoke to Simplício she was observing the fifth LEDOC degree program offered by the University of Brasília. She said that each year it is harder for the program to adhere to the original proposal. Part of the problem is that students may come from the countryside, but many have no previous connection to any social movement. Therefore, these students are more resistant to the collective orientation of the program, such as the shared housing, student collectives and group chores. Simplício attempts to intervene, and remind students about the principles of *Educação do Campo*. Nevertheless, it is difficult, and she fears the situation is worse where LEDOC is being implemented and there are no MST activists present. The case of LEDOC illustrates that even when social movement activists help develop a pilot program, the scale of implementation the MEC hopes to promote as a large government bureaucracy hinders the movement's ability to participate.

Best practices: closing down other experiences

The most far-reaching program that MEC officials implemented through the *Educação do Campo* office was *Escola Ativa*, a program designed to support teachers in multi-grade classrooms. In 2011, *Escola Ativa* was functioning in almost every state in Brazil, with over 1.3 million students enrolled.[24] However, *Escola Ativa* has a very different history than the LEDOC program. The Brazilian government adapted *Escola Ativa* from an internationally renowned educational program first implemented in Colombia in the 1970s, *Escuela Nueva*. In May of 1996 – a few years before the MST's educational initiatives began to receive national recognition as *Educação do Campo* – the World Bank invited a group of MEC program directors to Colombia to participate in a seminar about this program. Impressed, Brazilian officials decided to implement it in Brazil. Renamed *Escola Ativa*, the program was placed under the administration of the National Fund for Educational Development (FNDE), the financial arm of the MEC. In 2007, the program was relocated to the *Educação do Campo* office, due to its similar focus on rural education (MEC 2010).

Activists from both the MST and CONTAG were furious about the imposition of *Escola Ativa* in the office they had mobilized to create. For them, this office was an institutional space for grassroots movements to implement *their* educational ideas. Although there was still disagreement between the movements about the content of *Educação do Campo*, neither the MST nor CONTAG wanted their own educational practices to be replaced by a Colombian program, particularly one sponsored by the World Bank. The MST and CONTAG joined together to oppose the program. Despite their protests, MEC officials insisted that *Escola Ativa* fit into the goals of the *Educação do Campo* office. These officials claimed that the movements were ideologically opposed to the program simply because the World Bank was involved.[25]

Nonetheless, the MEC officials who supported *Escola Ativa* could not completely ignore the MST's and CONTAG's united critiques. In an attempt to appease the activists, they allowed them to help re-write the program's curriculum. The result was a new hybrid curriculum that included aspects of both the Colombian program and the major philosophical underpinnings of *Educação do Campo* (MEC 2010). Thus, *Escola Ativa* represented

---

[23]Interview, Vanderlúcia Simplicio, 9 November 2010.
[24]Interview, Armênio Bello Schmidt, 10 November 2010.
[25]Interview, Armênio Bello Schmidt, 10 November 2010.

both an imposition – of an external program on an office that had been created by rural social movements – and a process of accommodation – state actors allowing activists to reform the program. Nonetheless, the MST and CONTAG continue to denounce *Escola Ativa*.[26]

Increasing levels of education protests

Despite these challenges in the administrative and bureaucratic realm, the social movements in the national coalition for *Educação do Campo* still engaged in Alvarez's (1990) 'dual strategy' and Fox's (1992) 'sandwich strategy' throughout this period: working with MEC officials while also mobilizing contentious actions to support their educational demands. Figure 1 illustrates the rising number of protests concerning education during this period.

As Figure 1 indicates, between 2002 and 2012, the number of MST protests concerning education rose significantly, from less than 5 percent to around 20 percent. The number of total educational protests in the countryside shows a similar trend. This suggests that successful policy reforms actually increased levels of social mobilization around education.[27]

Eventually, this combination of internal and external pressure led to a new development: On 4 November 2010, two months before he left office, President Lula signed a Presidential Decree in support of *Educação do Campo*. This type of legal recognition from a President was unparalleled, and gave activists significant leverage to continue pushing forward their educational demands. Immediately following the signing of the decree, President Lula shook the hands of Fernando Haddad, the head of the Ministry of Education, José Wilson, the head of the Social Policies Department in CONTAG, and MST activist Vanderlúcia Simplício. These three people represented the main protagonists in the process of implementing *Educação do Campo*, up until that point: the MEC, CONTAG and the MST.

Educação do Campo at a crossroads

When PT candidate Dilma Rousseff became the president in January of 2011, the supporters of *Educação do Campo* were at a crossroads. On the one hand, there was no turning back; *Educação do Campo* was now the Ministry of Education's official approach to rural education. There were dozens of universities with *Educação do Campo* departments, hundreds of master's and doctoral students conducting research within this new disciplinary concentration and several massive federal programs attempting to implement *Educação do Campo* in practice. On the other hand, the institutionalization of *Educação do Campo* in the MEC was a far cry from what the MST, CONTAG and other civil society actors had wanted. Activists were frustrated with the dominance of *Escola Ativa* and the expansion of LEDOC without a concern for quality and movement participation. For many, this educational proposal was no longer linked to a socialist development model for the Brazilian countryside. To the contrary, many of the new social actors supporting *Educação do Campo* were interested in *reinforcing* capitalist modes of production.

---

[26]These protests eventually resulted in the termination of Escola Ativa in 2012. It was replaced with a program called '*Escolas da Terra*' (Interview with Edson Marcos Anhaia, 7 February 2014).
[27]These mobilizations were not usually focused entirely on education. For example, only an average of 30 percent of the MST educational protests between 2002 and 2012 were single-issue protests.

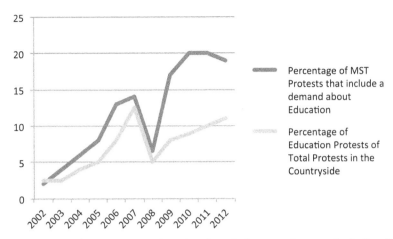

Figure 1.   Percentage of protests concerning education relative to total protests in the Brazilian countryside (2002–2012).
MST: Landless Workers Movement.[28]

Agribusiness, agrarian reform and the PT

The current challenges that the campaign for *Educação do Campo* faces cannot be understood without analyzing the overall agrarian context in Brazil, and particularly the rising importance of agribusiness in the 2000s. The 1980s were a period of transition for the Brazilian economy, when the previous golden-age levels of growth began to slow and social mobilizations increased. Agribusiness groups began to realize a need for more coordination, in the face of economic crisis and land conflicts. The Democratic Rural Union (UDR) was created in 1985 to represent a diversity of elite rural interests. Despite the shifting political and economic context, 'the UDR showed that it had strength to make its interests prevail in the face of new development conditions' (Bruno 1997, 63). In 1993, agribusiness sectors founded the Brazilian Association of Agribusiness (ABAG), in order to 'raise the consciousness of the nation about the importance of agribusiness' and to create 'an institution representative of the common interests of all the agents of the agricultural production chain' (Bruno 1997, 36).

It was during Cardoso's second term in office (1999–2002) that the government began to invest heavily in agribusiness sectors, especially in feed grains such as soy (Delgado 2009, 107). In terms of agrarian reform, while Cardoso had expropriated an unprecedented amount of land during his first term – primarily due to the fallout after the massacres of landless workers in 1995 and 1996 (Ondetti 2008) – the administration shifted to supporting market-based agrarian reform approaches. The justification was that 'market mechanisms will provide access to land without confrontations or disputes and therefore reduce social problems and federal expenses at the same time' (Sauer 2006, 182). This form of market-based agrarian reform was supported by the expanding agribusiness sector.

When President Lula took office in 2003, there was a general assumption that he would reverse these policies and implement a program of agrarian reform based on expropriation.

---

[28]Figure 1 was created by the author, using the Comissão Pastoral da Terra (CPT) database on rural mobilizations (http://www.cptnacional.org.br). I went through the databases from 2002 to 2012 and marked all of the protests (MST and other movements) that included a demand about education. I started in 2002 because the CPT protest database pre-2002 does not indicate the type of demand.

Consequently, right before Lula took office, thousands of families moved into camps to take advantage of the new agrarian reform program. However, once in office the Lula administration did not take any actions concerning agrarian reform, and social movements had to mobilize to pressure him on this issue. In response, the government recruited Plinio Sampaio, one of the most prominent agrarian specialists in Brazil and founding member of the PT, to develop a program for agrarian reform. Sampaio created a plan that would settle approximately one million people in one year. However, according to Miguel Rosseto, the head of the Ministry of Agrarian Development, Sampaio's proposal was not realistic, 'given the actual correlation of social, economic, and political forces' (Branford 2009, 423). In other words, rural social movements were fairly weak and agribusinesses were becoming increasingly stronger. Sampaio was fired shortly after presenting his proposal.

Instead of breaking with Cardoso's previous policies, Lula continued many of Cardoso's market-based agrarian reform initiatives while also publically supporting the PT's previous position on agrarian reform:

> In other words, without criminalizing the struggle for land and still counting on the support of the agrarian social movements and unions, the Lula government was able to operate in a type of 'accommodation' between constitutional agrarian reform and the loan programs for buying land that were supported by the World Bank. (Pereira and Sauer 2006, 198)

Lula also began to incorporate agribusiness allies into his governing coalition during his first term. Consequently, there was a huge expansion of soybean, corn and sugarcane production, also partially driven by an increased investment by international capital in Brazilian agriculture (Sauer and Leite 2012). By 2005, agriculture represented 42 percent of Brazilian exports, and became the principal source of income for the federal government to pay off external debt (Carter 2009, 68). A comprehensive program of agrarian reform through expropriation never moved forward.

Rural sectors, for their part, continued to follow a two-decade long strategy of building up influential networks in congress. As Bruno (1997, 85) argues, 'Although the UDR [Democratic Rural Union] despises the rules of party politics, it recognized the importance of these political-constitutional spaces and bet on the electoral road as a means of increasing its representation'. Between 1995 and 2006, the average representation of landowners in congress was 2587 times more than the representation of landless workers and small peasants (Carter 2009, 62–63). Delgado (2009, 108) argues that this 'powerful political representation – the Rural Block – is structured in various political parties and has between one fourth and one third of all congressman and senators voting in Congress'. This congressional power has also resulted in a series of judicial attacks against the MST and other rural social movements over the past decade, in the form of Parliamentary Inquiry Commissions (CPIs).[29] These developments are in addition to a general criminalization of rural social movements in the media, and attacks through other judicial bodies such as the Public Ministry and the Federal Court of Audits (TCU).[30]

The administration of President Dilma Rousseff has seen a continuation of this support of agribusiness sectors, and currently powerful congressional representatives – such as

---

[29]These include the 'CPMI da Terra' in 2005, 'CPI das Ongs' in 2009 and the 'CPMI do MST' in 2010. (Note: A CPMI, as opposed to a CPI, is a mixed inquiry between both the congress and senate).
[30]The state Public Ministry in Rio Grande do Sul initiated a series of cases against the MST (from 2009 to 2011), and the TCU was the judicial body responsible for preventing the federal educational program PRONERA from functioning for two years (2009–2010).

Kátia Abreu, a senator and president of the National Confederation of Agriculture (CNA) – are part of the PT governing coalition. The current context for *Educação do Campo* cannot be understand independently of these PT-agribusiness relations and the current economic context.

Navigating cooptation

On 20 March 2012, President Dilma Rousseff formally launched a new inter-ministry program, the National Program for Education in the Countryside (*Programa Nacional da Educação do Campo*, or PRONACAMPO), which would dedicate unprecedented funds to rural schooling. At the ceremony announcing PRONACAMPO, a new combination of actors were invited to participate, illustrative of the rising influence of agribusiness in the PT governance coalition. These included Aloizio Mercadante, the new Minister of Education; José Wilson, a CONTAG activist; and Senator Kátia Abreu, one of the most vocal public advocates of agribusiness and infamous among MST activists for her hatred of the movement. While many MST activists were also present in the audience that day, they were not given a chance to speak.

The speeches given at the PRONACAMPO ceremony were illustrative, first, of the tremendous success the campaign for *Educação do Campo* has had transforming national policy; second, of the principal role social movements have played in this campaign; and third, of the contemporary conflicts that exist over the future content of these educational practices. The speech of the new Minister of Education, Aloizio Mercadente, was representative of the success of the national coalition:

> We are sure that this program will contribute to the value placed on the populations of the countryside. Rapid urbanization is not the way forward. We need to value these populations, their stories and culture, and the huge contribution of rural workers to this country.[31]

Less than a decade before, statements such as these from prominent public officials were few and far between. Quality education was considered universal education, which did not differentiate between urban and rural populations. Now, in 2012, the Minister of Education was referring to a 'social debt' the government of Brazil owed populations living in the countryside, and their right to an education that addresses their particular needs.

The choice of a CONTAG activist to speak at the ceremony demonstrates the critical role social movements have played in this national campaign, but also the tensions that still exist. For the Brazilian government, CONTAG is a more moderate organization with a long history of collaboration with the state. Allowing the MST to speak at such a prestigious ceremony might have been controversial, especially given the PT's support of agribusinesses. Nonetheless, CONTAG activist José Wilson gave the MST a space to participate during his speech, pausing to allow MST activist Vanderlúcia Simplício to recite a poem about agrarian reform and deliver a recently published book, the *Dicionário da Educação do Campo* (Caldart, Pereira, and Frigotto 2012), to President Rousseff. The choice of CONTAG to formally represent the social movements, and the informal inclusion of the MST, exemplifies the compromises these two organizations continue to make.

Finally, Kátia Abreu was invited to the podium amid loud hisses from the MST activists in the audience. Abreu began her speech by emphasizing the lack of investment in the countryside: 'There have been decades of abandonment of the countryside ... there are schools without

---

[31]These speeches can be watched online: http://www.youtube.com/watch?v=hPtcdDSqcgk.

internet, without infrastructure, principals of schools absent, teachers earning much less than in the city'. Up until this point, Abreu's speech could have been given by any one of the MST activists in the audience. However, ideological differences quickly appeared. Abreu exclaimed, 'Education is extremely important, so agribusiness can be stronger ... The youth in the countryside need to be qualified workers, whether they are farm laborers or bosses'. In this statement, Abreu claimed *Educação do Campo* as an educational proposal that could support agribusiness, in direct contrast to the origins of these educational ideas.

This situation is indicative of agribusiness' long-standing practice of attempting to influence and manipulate the policy direction of the Brazilian state. As Bruno (2008, 92–93) explains, the preoccupation of the rural elite and agribusiness sectors with poverty emerged at the end of the 1990s, when the quality of life of poor populations began to be considered a 'principal tool' of Brazilian society, due to their potential as consumers. The emphasis agribusiness sectors are currently putting on education is a similar attempt to ensure that any investment in schooling in rural areas adheres to their vision of a qualified workforce that can support the agribusiness sector. This is in contrast to the MST, whose conception of rural education is explicitly tied to a socialist development model centered on workers' ownership of their own means of production, and collective agricultural practices. Six months after the PRONACAMPO ceremony, the National Forum for *Educação do Campo* (FONEC) – an alliance of civil society groups – publically critiqued PRONA-CAMPO, arguing that agribusiness had an 'interest in appropriating a discourse that defends the education of rural workers, in order to affirm (and confuse) society into believing that agribusiness is also interested in overcoming inequality' (FONEC 2012, 8).

MST activists are well aware of the challenges they face, as *Educação do Campo* becomes more entrenched in the federal state apparatus. National MST educational activist Rosali Caldart even admits she is unclear about *Educação do Campo*'s future:

> The fact is that *Educação do Campo* was originally developed by the social movements, and was only practiced within these movements. Now, *Educação do Campo* exists in relationship to the governments, to the universities, or in other words, it exists in a relationship with these other actors ... Those that defend agribusiness are going to have one vision of *Educação do Campo*, and those that defend peasant agriculture are going to have another educational project.[32]

As Caldart articulates, the concept of *Educação do Campo* no longer belongs to the social movements that first developed the proposal; now dozens of groups are laying claim to these ideas. MST activist Edgar Kolling expresses a similar sentiment:

> The creation of SECAD in the MEC was a huge victory; now there is clearly more focus on education of the countryside .... But the MST has been swallowed up in a lot of these SECAD programs, subsumed. We never thought we could create something so big.[33]

According to Kolling, MST activists never imagined the degree of expansion that *Educação do Campo* would achieve. However, this success has also created a new set of challenges, as MST activists must now expend energy navigating both the administrative and bureaucratic barriers of the Ministry of Education, and the potential threat of elite sectors coopting and transforming their proposal.

---

[32]Interview, Rosali Caldart, 17 January 2011.
[33]Interview, Edgar Kolling, 18 November 2010.

Conclusions

The national campaign for *Educação do Campo*, which began in the late 1990s and continues into the current moment, is not a typical story of policy reform. It involves a coalition of social movements that are pushing forward educational policy reforms through both political negotiation and social mobilization. The success of this national advocacy campaign, and the challenges activists still face, offer several lessons for scholars analyzing contemporary state–society dynamics in the Brazilian countryside. First, this study follows Gaventa and McGee and other scholars in illustrating the critical role that social movements play in promoting national policy reform. The MST – through the movement's alliance with CONTAG and other social actors – has been able to publically condemn the trend towards closing down rural schools, legitimize the idea that rural schools should have a unique educational approach and convince the federal government to create dozens of new educational programs designated for rural populations.

Second, this paper outlines the strategies that social movement activists utilize in promoting national-level policy reform. The strategies that MST activists employed to implement *Educação do Campo* in the MEC align with Gaventa and McGee's six propositions concerning successful national advocacy in the Global South. Initially, powerful international and domestic allies were central to the legitimatization the MST's educational initiatives (proposition 3). Then, MST activists capitalized on a political opening caused by previous social mobilization, to bring their proposal to the national level (proposition 1). Through a strategic process of framing and coalition building, the MST garnered significant recognition for their educational reforms (propositions 4 and 6). However, this national coalition was only successful in transforming federal policy once CONTAG embraced these initiatives, which was a complex and historically contingent process. Finally, although the national coalition changed federal law during an antagonistic government, putting this law into practice was only possible once the PT took power at the national level, allowing for influential allies to enter the federal government and new state–society relations to form (proposition 2 and 5).

In addition to illustrating the strategies social movements utilize in national advocacy campaigns, a third contribution of this paper is outlining the many administrative and bureaucratic barriers movements face in actually implementing these reforms. Once inside the Ministry of Education, the MST and CONTAG faced a hierarchical structure, the mass implementation of educational programs, and the imposition of global 'best practices' that competed with the movements' own educational proposals. In response, the national coalition for *Educação do Campo* increased their levels of social mobilization throughout the 2000s.

Fourth and finally, this essay illustrates how the political and economic context affects policy reforms. As soon as *Educação do Campo* became the MEC's official approach to rural schooling, and began receiving significant funding, agribusiness sectors took on the language of the reform to promote their own goals. This is directly related to the surge in the agricultural export sector in the 2000s, and the incorporation of agribusiness interests within the PT governing coalition. Currently, if the federal government takes a stance on the future of the countryside – even if it is only about education – agribusiness interests are invited to participate. Nonetheless, and perhaps this is the most important take-home point, the current terrain of educational struggle is not the same as it was in the early 1990s. Agribusinesses lobbies can certainly deny the relationship between *Educação do Campo* and socialist models of rural development; however, they can no longer publically contest the importance of expanding the access to and the quality of education in the countryside. MST activists have clearly remade the Brazilian state; however, they did not remake it just as they pleased.

Acknowledgements

I would like to thank my committee, Harley Shaiken, Peter Evans, Michael Watts, Zeus Leonardo and Erin Murphy-Graham, for commenting on previous versions of this paper. Multiple colleagues have also given me feedback, including Ellen Moore, Chela Delgado, Nirali Jani, Kimberly Vinall, Alex V. Barnard, Gabriel Hetland, Rasjesh Veeraraghavan, Edwin Ackerman, Krystal Strong, Jon K. Shelton, Khalil Johnson, Robert Gross and Laura Enriquez's Latin American Writing Group.

References

Alvarez, S. 1990. *Engendering democracy in Brazil: Women's movements in transition politics.* Princeton, NJ: Princeton University Press.

Arnove, R. F. 1986. *Education and revolution in Nicaragua.* New York: Praeger.

Benford, R. D., and D. A. Snow. 2000. Framing processes and social movements: An overview and assessment. *Annual Review of Sociology* 26: 611–39.

Borras, S. M. Jr 2001. State-society relations in land reform implementation in the Philippines. *Development and Change* 32: 545–75.

Branford, S. 2009. Lidando com governos: o MST e as adminstrações de Cardoso e Lula. In *Combatendo a desigualidade social: o MST e a reforma agrária no Brasil*, ed. M. Carter, 409–32. São Paulo: Editora UNESP.

Branford, S., and J. Rocha. 2002. *Cutting the wire: The story of the Landless Movement in Brazil.* London: Latin America Bureau.

Bruno, R. 1997. *Senhores da terra, senhores da guerra: a nova face política das elites agroindustriais no Brasil.* Rio de Janeiro: Editora Forense Universitária.

Bruno, R. 2008. Agronegócio e novos modos de conflictuosidade. In *Campesinato e agronegócio na América Latina: a questão agária atual*, ed. B. M. Fernandes, 83–105. São Paulo: Expressão Popular.

Caldart, R.S., A. R. Fetzner, R. Rodrigues, and L. C. de Fretias, eds. 2010. *Caminhos para transformação da escola: reflexoes desde praticas da licenciatura em educação Do Campo.* São Paulo: Expressão Popular.

Caldart, R.S., P. A. Pereira, and G. Frigotto, eds. 2012. *Dicionário da educação do campo.* São Paulo: Expressão Popular.

Carter, M. 2009. Desiqualidade social, democracia e reforma agrária no Brasil. In *Combatendo a desigualidade social: o MST e a reforma agrária no Brasil*, ed. M. Carter, 27–78. São Paulo: Editora UNESP.

Carter, M., and H. M. de Carvalho. 2009. A luta na terra: fonte de crescimento, inovaçao e desafio constante ao MST. In *Combatendo a desigualidade social: o MST e a reforma agrária no Brasil*, ed. M. Carter, 27–78. São Paulo: Editora UNESP.

Costa Lunas, A. and E. Novaes Rocha. A luta dos trabalhadores e trabalhadoras rurais por educação. http://www.contag.org.br/imagens/f307A_luta_dos_%20trabalhadores_Por_Educacao.pdf (accessed August 16, 2013).

Delgado, G. C. 2009. A questão agrária e o agronegócio no Brasil. In *Combatendo a desigualidade social: o MST e a reforma agrária no Brasil*, ed. M. Carter, 81–112. São Paulo: Editora UNESP.

FONEC. 2012. *Notas Para Análise Do Momento Atual Da Educação Do Campo. Seminário Nacional.* Brasília: Fórum Nacional de Educação do Campo.

Fox, J. 1992. *The politics of food in Mexico: State power and social mobilization.* Ithaca, NY: Cornell University Press.

Gaventa, J., and R. McGee. 2010. Introduction: Making change happen - Citizen action and national policy reform. In *Citizen action and national policy reform: Making change happen*, eds. J. Gaventa and R. McGee, 1–43. New York: Zed Books.

Hammond, J. L. 1998. *Fighting to learn: Popular education and guerrilla war in El Salvador.* New Brunswick, NJ: Rutgers University Press.

Houtzager, P. 1998. State and unions in the transformation of the Brazilian countryside, 1964–1979. *Latin American Research Review* 33, no. 2: 103–42.

Marcos de Anhaia, E. 2010. Constituição do movimento de educação do campo na luta por políticias de educação. Thesis (PhD). Universidade Federal de Santa Catarina.

Maybury-Lewis, B. 1994. *The politics of the possible: The Brazilian rural worker' trade union movement, 1964-1985*. Philadelphia, PA: Temple University Press.

McAdam, D. 1999. *Political process and the development of the black insurgency, 1930–1970*. Chicago, IL: University of Chicago Press.

MEC. 2004. *Referências para uma política nacional de educação do campo. Caderno de Susbsídos*. Brasília: Ministério da Educação.

MEC. 2010. *Escola Ativa: projeto base*. Brasília: Ministério da Educação.

Ondetti, G. 2008. *Land, protest, and politics: The landless movement and the struggle for agrarian reform in Brazil*. University Park, PA: Duke University Press.

Pereira, J. M. M., and S. Sauer. 2006. História e legado da reforma agrária de mercado no Brasil. In *Capturando a terra: Banco Mundial, políticas fundiárias neoliberais e reforma agrária de mercado*, eds. S. Sauer and J. M. M. Pereira, 173–206. São Paulo: Expressão Popular.

Plank, D. 1996. *The means of our salvation: Public education in Brazil, 1930-1995*. Boulder, CO: Westview Press.

Rosa, M. C. 2009. Para além do MST: o impacto nos movimentos socias Brasileiros. In *Combatendo a desigualdade social: o MST e a reforma agrária no Brasil*, ed. M. Carter, 461–77. São Paulo: Editora UNESP.

Samoff, J. 1999. Institutionalizing international influence. In *Comparative education: The dialectic of the global and the local*, eds. R. F. Arnove and C. A. Torres, 55–86. Lanham, MD: Rowman & Littlefield Publishers.

Sauer, S. 2006. The World Bank's market-based land reform in Brazil. In *Promised lands: competing visions of agrarian reform*, eds. P. Rosset, R. Patel, and M. Courville, 171–91. Oakland, CA: Food First Books.

Sauer, S., and S. P. Leite. 2012. Agrarian Structure, Foreign Investment in Land, and Land Prices in Brazil. *Journal of Peasant Studies* 39, nos. 3–4: 873–98.

Schwartzman, S. 2004. The challenges of education in Brazil. In *The challenges of education in Brazil*, eds. C. Brock and S. Schwartzman, 9–39. Oxford, UK: Symposium Books.

Welch, C. 2009. Camponeses: Brazil's peasant movement in historical perspective (1946–2004). *Latin American Perspectives* 36, no. 4: 126–55.

Learning as territoriality: the political ecology of education in the Brazilian landless workers' movement

David Meek

In this contribution, I explore the importance of agroecological education in the Brazilian Landless Workers' Movement (MST). I analyze how certain MST educational programs are based in a critical place-based pedagogy. This type of pedagogy can serve as a form of *territoriality*, influencing individuals' interactions with the land. Drawing upon a political ecology of education perspective, I conclude that MST educators can serve as Gramscian 'organic intellectuals', by using a critical pedagogy of place as a form of territoriality to: (1) create a conception of place that is not discrete, but instead relational, and (2) advocate counter-hegemonic land usage.

Introduction

Why do Brazilian agrarian social movements institutionalize education? The institutionalization of education seems paradoxical, because institutionalization often results in the state's co-optation and subsequent neutralization of a progressive agenda. Yet 'Movement institutionalization does not always entail the risk of deradicalization, depoliticization, or demobilization of collective action' (Suh 2011, 444). Social movements institutionalize their vision of education because it is part of a long-term strategy of state transformation (Poulantzas 1978; Jessop 1990; Boden 2011).

Antonio Gramsci refers to this long-term struggle as the 'war of position', and sees activist educators as playing a fundamental organizing role as 'organic intellectuals' (Coben 1995; Mayo 2008; Yogev and Michaeli 2011). Although Gramsci originally applied the organic intellectual concept to the industrial proletariat (Gramsci 1978, 8), I build upon Feierman (1990) and Del Roio (2011) by exploring its relevance to peasant movements. Following the introduction, I justify this application of Gramsci. I draw upon Gramsci's organic intellectual concept to explore the role of educator-students in Brazil's Landless Workers' Movement (*O Movimento dos Trabalhadores Rurais Sem Terra*, or MST).[1] I analyze how these educator-students learn about agroecology, which is the integration of ecological principles into agricultural systems (Gliessman 2006), and the opportunities and constraints they face in disseminating this agroecological knowledge. I shed light on

---

[1] I use the phrase 'educator-students' because the individuals I follow in this research are *both* educators, working as teachers in a primary school within an MST settlement, *and* students in a graduate certificate program.

how the MST is using institutionalized education to strategically reconfigure state-society relations in the Brazilian countryside by exploring the training of these educator-students as Gramscian organic intellectuals.

The MST is Brazil's largest agrarian social movement, and its activists pursue agrarian reform by occupying land deemed 'unproductive' (Wright and Wolford 2003). This tactic evolved in response to the Brazilian constitution, which states that the government can expropriate land if it is not 'socially productive' (Wolford 2006).[2] Both agroecology and transformative education are central to the MST's ideology and struggle (Branford and Rocha 2002; Rosset and Martinez-Torres 2013).

The MST's engagement with agroecology is explicitly political, consisting of a critique of capital-intensive agribusiness and its support of a new model of cooperative production (de Molina 2013). Owing to Green Revolution-era agricultural modernization in Brazil, the country's dominant agricultural model is large-scale, and based in high inputs of petro-chemical fertilizers and pesticides, as well as expensive machinery (Callou 2007; Troian and Eichler 2012). The MST critiques this model as socially and environmentally unjust, and advocates small-scale family farming that can help achieve food sovereignty (Patel 2009; Altieri and Toledo 2011; van der Ploeg 2012). The MST has institutionalized agroe-cological education by collaborating with other social movements and state entities in a broad-based educational movement known as *Educação do Campo*.

The *Educação do Campo* movement seeks to develop pedagogies and opportunities that are attentive to rural realities (Munarim 2008). This umbrella movement has helped create a new emphasis within Brazilian educational policy on locally relevant rural edu-cation as opposed to homogenous national courses that do not attend to local diversity in geography, culture and history (Breitenbach 2011). One way the *Educação do Campo* movement has attended to rural realities is through what Gruenewald (2003) terms critical place-based education. Critical place-based education is a synthesis of critical pedagogy's attention to transforming systems of oppression, and place-based education's focus on learning's historical and geographic context. Critical pedagogy helps students learn 'to per-ceive social, political, and economic contradictions, and to take action against the oppres-sive elements of reality' (Freire 1973, 17). Place-based education draws on 'the power of place as a context for diverse experiences … using diverse communities as "texts" for cur-riculum development … ' (Gruenewald 2008, 143). In certain contexts, the MST has insti-tutionalized critical place-based education by creating agroecological programs funded by the National Program of Agrarian Reform Education (PRONERA) (Araujo 2004). PRONERA, which was launched in 1998, offers funding for institutional partnerships between agrarian social movements and educational organizations (Molina 2003).

I posit two interrelated arguments in this paper. First, I argue that the MST institutiona-lizes critical place-based agroecological education through PRONERA-sponsored certificate programs, which helps transform educator-students into Gramscian organic intellectuals who are capable of influencing agricultural production by advancing counter-hegemonic agricultural practices. Second, I hold that these educator-students face difficulties in institu-tionalizing the agroecological learning they have gained through the certificate program because of the daily politics of settlement life. As a result of these difficulties, these educators are engaged within the school in a Gramscian struggle of territorial dispute.

---

[2]A voluminous scholarship exists on the MST, its origination and its complicated relation to the Brazilian state. See Branford and Rocha (2002), Wright and Wolford (2003) and Wolford (2010) for key perspectives.

In the next section, I explicate Gramsci's concept of the organic intellectual, and what modifications are needed to utilize it in an agrarian reform context. I then draw upon a political ecology of education framework to disentangle the complicated relationships that exist between territory, territoriality, hegemony and counter-hegemony in the MST. I introduce the research site through an ethnographic vignette. In the first of two ethnographic sections, I analyze critical place-based agroecological education in a graduate certificate course offered through a MST-federal partnership. I then explore, in the second ethnographic section, the opportunities and challenges educator-students in this program face in disseminating their critical place-based learning.

I gathered these data during 17 months of ethnographic fieldwork the MST's 17 de Abril settlement located in southeastern Pará state, Brazil. One focal point of the research was a two-year graduate certificate program entitled 'The Agrarian Question, Agroecology, and *Educação do Campo*', which was established for social movement activists of the 17 de Abril and other agrarian reform settlements. I followed the experiences of two educator-students, whom I call Diana and Lucinede, who participated in the program.[3] I gathered data by participating in the program and accompanying these educator-students on their field-based research. I conducted participant observation at the 17 de Abril settlement's school during events that were intended to disseminate the results of this course to the community. I contextualized these data by conducting a survey of 47 percent of all household heads in the community ($n = 330$), which addressed political participation, landscape change and agroecology. Five trained research assistants administered this survey by first dividing the settlement into five areas of comparable population, and then conducting a convenience sample by going house-to-house. I present descriptive statistics derived from these data to buttress the perspectives of Diana and Lucinede.

Gramsci and the question of organic peasant intellectuals

Central to Gramsci's thought is the belief that subaltern subjects have the capacity to both understand and change the world. To do so, there needs to be an 'intellectual and moral reform' involving the critiquing of hegemonic ideas, and the advancement of popular ideologies. Popular 'common sense' can only gain the ability to become hegemonic through a long-term process of movement building that Gramsci termed the 'war of position' (Carrol and Ratner 1994; Meek 2011). Integral agents in this war of position are activist educators, whom Gramsci termed 'organic intellectuals' (Mayo 2008).

Gramsci articulates the concept of the organic intellectual in opposition to the traditional intellectual. Traditional intellectuals were once organic to a rising class, but became complacent, and detached from life's social concerns. In the *Prison notebooks*, Gramsci typifies traditional intellectuals as the 'man of letters, philosopher, artist' who see themselves as classless, and that their knowledge is apolitical (Gramsci 1978, 9). Traditional intellectuals won't advance the struggle of the proletariat, because they are not ideologically committed to this class. Organic intellectuals differ because they arise from within, and are passionately connected to, the subaltern class. These educators play a pivotal role in the 'war of position', by engaging in counter-hegemonic activity.

Gramsci believes that 'every social group, coming into existence … creates together with itself, organically, one or more strata of intellectuals' (Gramsci 1978, 5). For

---

[3]All names used in this paper are pseudonyms in order to protect individuals' identities.

Gramsci, the origination of intellectuals results from class needs. Intellectuals organic to subaltern movements arise from class-based needs, because 'in the modern world, technical education, closely bound to industrial labor, even at the most primitive and unqualified level, must form the basis of the new type of intellectual' (Gramsci 1978, 9). By understanding how industry functions technically and administratively, subaltern groups can overthrow the bourgeoisie (Jones 2006, 85). Yet these organic intellectuals must do more than simply possess technical knowledge; they must also turn this specialist knowledge into political knowledge. Organic intellectuals use this political knowledge to move beyond the 'eloquence' of traditional intellectuals, and become defined by their 'active participation in practical life, as constructor, organizer, permanent persuader and not just a simple orator' (Gramsci 1978, 10). However, Gramsci's particular vision of the organic intellectual requires modification before being applied to the MST.

Gramsci held what are seemingly oppositional accounts of the formation of intellectuals. On one hand, he writes, 'every social group has its own stratum of intellectuals' (Gramsci 1978, 60). On the other, he holds that 'the mass of the peasantry, although it performs an essential function in the world of production, does not elaborate its own "organic" intellectuals' (Gramsci 1978, 6). This disjunction can be explained, because, following Marx, Gramsci believed the peasantry cannot become a class 'for itself' (Shafir 1980, 224).

Although Gramsci himself arose from peasant society, he saw peasant culture as archaic and fragmentary.[4] Gramsci is very critical of the political potential of subaltern movements. He saw Italy's southern peasants as accepting hegemonic values and trying to emulate the characteristics of ruling classes (Gramsci 1994, 327), yet he also believed that these peasants can inject non-hegemonic values into the dominant worldview (Gramsci 1978, 420; Cirese 1982, 226). In sum, Gramsci saw Italy's peasants as a class that maintained itself in subordination through internal weaknesses and its acceptance of the social, political and moral leadership of the ruling classes. Yet it could also become a revolutionary class through alliances with workers, and the development of a class consciousness (Arnold 1984, 158–59).

In arguing for a non-orthodox application of Gramsci's organic intellectual concept, it is important to acknowledge his changing views on the peasantry. Following the success of fascism in 1922, Gramsci realized that (1) any revolution must take into account the peasantry, (2) without understanding their world view there would be no chance of their mobilization, and (3) only intellectuals who were committed to the class could help solidify its worldview (Davidson 1984, 145). The justifications of peasant movements themselves, who have long applied Gramsci's concepts to function in disparate historical and geopolitical contexts, are particularly instructive.[5]

---

[4]Perhaps as a result of his own lack of sustained focus on the peasantry, or the difficulty of extracting a coherent narrative from his *Peasant notebooks*, Gramscian scholars tend to focus on his writings about the industrial working class in comparison with the peasantry (Arnold 1984); but see Davidson's (1977) *Antonio Gramsci: towards an intellectual biography* for a critical analysis of this neglect, and Davis's (1979) *Gramsci and Italy's passive revolution* for a consideration of how the peasant problem is interconnected with his larger corpus of ideas.

[5]Gramsci's concepts have been applied to peasant groups in diverse international contexts. Feierman (1990) translates Gramsci's concept of the organic intellectual to rural Tanzania, exploring the formation of peasant intellectuals. Karriem (2009) draws on a Gramscian approach to understand the interplay between space, ecology and politics in the Brazilian MST.

In the late 1970s, Brazilian movements began drawing upon Gramsci to inform their political-pedagogical activities (Semeraro et al. 2011). Paulo Freire, for example, an influential Brazilian critical pedagogue, cites Gramsci as having a direct influence on his thinking about the role of peasant intellectuals (Freire et al. 1986, 68). In situating the historical and political translation necessary to make Gramsci relevant to Brazilian movements, Marcos Del Roio (2011) first describes how Gramsci himself needed to translate Lenin into Italy's national context and Machiavelli to that time period, to understand the development of capitalism and the State, and strategies for socialist revolution. Del Roio considers one of the principal problems of translating Gramsci into a Brazilian context the form of popular social movements. Yet he concludes that if one questions the nature of these movements from the position of the actors themselves, then we see how these movements understand themselves as

> a moment in the construction of the people, of the unification of subaltern classes, of the realization of a moral and intellectual reform, of a new hegemony that will result in the construction of a new State, and the translation of Gramsci will coalesce in praxis. (Del Roio 2011, 81; my translation)

This revolutionary process, Del Roio goes on, is slow, and requires the type of moral and intellectual reform that can only come about through the accumulation of organic intellectuals that are organizing on behalf of syndicates, parties and movements. The MST, Del Roio argues, is a paradigmatic example of peasant organic intellectuals. He qualifies this by describing how the MST's Escola Nacional Florestan Fernandes has built alliances between the rural areas and the city, restructuring Gramsci's vision of the nature of the alliance between workers and peasants (2011, 82). In the context of the MST, I believe organic intellectuals are characterized by technical training they've received through the movement, and their active involvement in the everyday politics of the community as knowledge producers, disseminators and mediators.

Education as a tactic for influencing relations to the land

I employ a political ecology of education framework to explore how MST educator-students function as Gramscian organic intellectuals, and struggle to advance counter-hegemonic agroecological production. A *political ecology of education* framework analyzes how reciprocal relations between political-economic forces and pedagogical processes mediate resource access, control and landscape change (Meek 2010). The linkages between territory, territoriality, hegemony and counter-hegemony are key to understanding the potential for MST educator-students to affect agricultural production.

Territory consists of a combination of material and immaterial aspects (Fernandes 2009). Material territories are natural landforms and infrastructure (Rosset and Martinez-Torres 2013). Immaterial territories, by contrast, are the ideologies connected with landscapes, including ideas about what constitutes appropriate land use. Immaterial and material territories are intrinsically interconnected through territoriality, which is a process of exerting control over territory (Sack 1986; Storey 2012).

Educators can employ education as a form of territoriality. One way educators exert control over territory is by promoting the dominant ideologies and associated forms of land use. An alternate way educators can influence territory is by critiquing the dominant ideas about land use, and advocating counter-hegemonic forms. MST educators can function as organic intellectuals due to their ability to raise awareness about contradictions inherent in dominant economic, cultural and material systems, and advance counter-hegemonic forms of agricultural knowledge and practices.[6]

'The struggle is over and now there's just victory'

The MST's 17 de Abril settlement was created in 1996, following the Brazilian paramilitaries' massacre of 19 MST members outside the city of Eldorado dos Carajas on 17 April 1996. The settlement consists of 690 families, who each have a small plot of land in a peri-urban village, and a larger plot of land (~25–50 hectares) in the surrounding rural area. The original settlers were largely from the Northeastern state of Maranhão. These individuals came from disparate backgrounds, having worked as miners, laborers on others' farms and urban merchants. All of these respondents described joining the movement out of desire for 'a piece of land to work, and a place to live'. Since that point, due to failed credit projects, poor health and the general difficulty of living off the land, many settlers have left the community for the neighboring urban centers of Paraupebas, Curionópolis and Eldorado dos Carajás. The majority of respondents (64 percent) have lived in the settlement since its origination. Newcomers (36 percent of respondents) have flocked to the settlement because it is 'good, and really cheap land'. Many of these new inhabitants were landowners in the south, and have replicated their ranching activities in this settlement. Given the emigration of many original MST members, and arrival of non-MST affiliated individuals the community has an incredibly heterogeneous feel, as described in the following ethnographic vignette:

> I groan as the settlement's radio begins crackling at 5:30 a.m. Samio, in charge of the radio, can't be starting this early I think. However, it's the unmistakable first notes of the Brazilian Landless Workers' Movement's anthem that causes to me to sit straight up in bed. Although I've lived here for 17 months, it's the first time I've ever heard the MST anthem played on the community radio of the 17 de Abril settlement.
>
> I'm surprised as the song ends and begins again. I reach the central square, watching what would appear to be another normal morning unfold, if not for the fact that the anthem finishes and then begins yet again. After the fourth repetition, I hear a voice break the radical monotony, but it's not Samio's. Rather, it's Arnoldo. Arnoldo is the settlement's president, and extols the importance of the settlement's anniversary in the historical context of regional agrarian reform. Three other MST activists follow Arnoldo and each give speeches on the importance of the movement. Following these speeches Samio turns on a popular Forró song, whose refrain is 'the struggle's over and now there's just victory'.
>
> As if on cue, a vehicle lurches into the central square. It is a flatbed truck carrying sets of giant speakers bolted together. A man with a cowboy hat stands on a stage that rests atop the speakers. He does a sound check, and proceeds in what is clearly a professional rodeo announcer's voice: 'Wellllllllllllllcome Amazonian horsemen from all neighboring cities. Wellllllllllcome to Amazonia's biggest rural horse parade!' As his call bellows out, from all six intersecting streets men and women on horses begin to trickle out into the central square.
>
> Douglas is orchestrating the horse parade. He is a large-scale dairy producer and in this role has become one of the settlement's most powerful individuals. While directing the riders he tells me, 'My dream is for this event to be on the scale of the big cattle expositions. Complete with lasso events, and a real rodeo'. Douglas hands out flags to several horse riders: a state of Pará flag, a Brazilian flag, and one from the MST. Aside from this morning's radio announcements, it would be the only sign of the MST on this settlement's 17th birthday. Sometimes absences are more telling than presences.

---

[6] As I discuss below, educators do not always fulfill this role as intended.

Conspicuously *absent*, to me, are the signs of the movement on this day, such as the MST flags, banners, t-shirts and hats. Also absent are the members of the MST's state secretariat, or activists from other encampments or settlements. The colorful horse parade, a manifestation of regionally dominant cattle culture, occupies this negative space, and draws attention to the forces transforming the settlement.

This anniversary used to be a harvest festival, marking the importance of various subsistence crops. However, a number of factors ranging from credit incentives, to urban migration, to the failure of agricultural projects and cooperatives have resulted in the large-scale transition away from agriculture and towards milk production. As a result, what used to be a harvest festival is now a rodeo. Yet despite the dominance of cattle culture at this event there are still some indications of the settlement's agrarian roots. José Batista is one vestige.

Although I had never met him, I knew of José Batista from attending a recent presentation about agroecology at the settlement's school. MST educator-students, who had recently finished a graduate certificate program in agroecology, were presenting results to their students from their field-based research in the community. As part of this research, they had analyzed environmentally destructive forms of production, such as intensive cattle production within the settlement. But they had also researched emerging alternative forms of production, such as agroforestry, that were environmentally sustainable. With an image of José Batista projected on the school's wall, one teacher described to an auditorium of students how through this research she found that the settlement was not solely comprised of cattle ranching but, rather, was a diverse universe of production. José Batista exemplified this rich universe.

On the settlement's anniversary, José Batista stands in front of a cart decorated with hanging fresh mangoes, papayas, squash and various other agricultural products. On the cart's edge sits a homemade water wheel, which is fed by a tank in the back of the cart. 'People hardly remember these', José Batista admonishes, 'but they should. I'm here to remind them, to represent the settlement, to represent the small family farmers and the various products they produce'. José's reminder is of the traditional agroecological experimentation that characterizes the community's agrarian history.

I use this ethnographic vignette to introduce the contest between hegemony and counter-hegemony in the 17 de Abril settlement. In what was once an agrarian reform settlement of politically committed activists, it is now an anomalous event to hear the MST anthem on the community radio. Political participation on the settlement's anniversary consists of the perfunctory carrying of an MST flag; as the song's refrain reminds us, 'the struggle's over and now there's just victory'. Agriculture is also a constraint: subsistence production has been replaced by export-oriented dairy production. The harvest festival, which was timed to coincide with the settlement's founding, has been replaced by a rodeo, paying homage to the importance of regionally dominant cattle culture. This political and agricultural context is antithetical to the MST's agroecological vision, and calls into question whether 'the struggle's over and now there's just victory'.

While the original struggle for land might be over, the struggle over the meaning of land continues. The settlement's inhabitants are largely divided concerning how they see the transition from subsistence agriculture to cattle ranching. Two responses typify this division. The first is that 'without cattle ranching I wouldn't be able to live here'. Those holding this perspective have abandoned subsistence agriculture for one reason or another (most frequently due to lack of forest to convert to swidden agriculture, difficulties with non-mechanized production or lack of transportation for agricultural goods) and see ranching as the best available option. The opposing perspective is typified by sentiment that 'the transition from subsistence agriculture signals the death of the MST'. Those holding this view lament the degree to which the settlement has moved away from the movement's ideals of self-sufficient smallholder agriculture. This perspective is encapsulated by the oft-heard critique among respondents that 'we won this land in order to farm it, not to rely upon store-bought foods'. These two perspectives signal the competition between common sense and popular common sense in the 17 de Abril settlement. It is a contest between those ideologically aligned with the MST who originally expected, and continue to believe, that the land should be used for subsistence agriculture, and those who have adopted the hegemonic system of cattle ranching, and are trying to emulate the characteristics of ruling classes (Gramsci 1994, 327). As one MST activist described this ideological struggle, 'these days everyone wants to be a *fazendeiro* (rancher)'.

Despite, or perhaps because of, these constraints, critical education has incredible potential to transform the settlement's political and agricultural milieu. The settlement's teachers use José Batista as an example of an unseen universe of agroecological production to help their students challenge the hegemonic system of ranching. These critical education practices have the potential to help maintain students' ideological commitment to the movement, engagement with its agroecological practices and role in a long-term movement towards social change.

Institutionalizing critical pedagogy of place: experiencing place as relational

Problematizing as creating relational place

The certificate program in 'The Agrarian Question, Agroecology, and *Educação do Campo*' exemplifies the institutionalization of critical place-based agroecological education. The course format consisted of alternating 3-week segments in the home settlement and distant school community. During school time, students participated in field research trips, consisting of visits to sites representing opposing forms of production. My informants described these opposing forms of production as hegemonic and counter-hegemonic. They told me that these trips provided clear examples of the abstract concepts of hegemony and counter-hegemony. While in their home community, they conducted critical field research and gained an appreciation for how these hegemonic and counter-hegemonic practices were manifest at a smaller scale.

Professors in the certificate program described the objective of the program's first section as problematization. Problematization took three forms: conflict mapping, 'dislocating' exercises (field trips) and field research. Professors used conflict mapping to graphically depict contradictory forces, such as the infrastructure of exploitative industries and the spaces of social movement resistance. A professor indicated that conflict mapping was chosen as a form of problematizing in the course for several reasons. First, southern and southeastern Pará is a region historically defined by territorial land conflict; mapping helped students visually understand the political geographical nature of these conflicts (Simmons 2004). Second, conflict mapping advanced the course's 'political objective' to map 'strategies'. As a professor described it,

> We were trying to map the capitalist strategies that confront the students' own territoriality; we decided to focus not on industrial capital in an abstract sense ... or the personification of the enemy, i.e. the large land owner, but rather on these strategic processes.

Student activists from dispersed settlements mapped these strategic processes by placing their communities on the map and discussing what geographic features spatially, politically, economically and ideologically link their communities together. For example, a railroad line operated by Vale, a transnational mining corporation, runs from Paraupebas to Marabá, and then into the neighboring state of Maranhão. The students discovered that each community had been affected by Vale's strategy of extractive capitalism, whether through environmental pollution, land grabs or exploitative labor. Students also mapped their resistance activities, such as where they had occupied the railroad in 2009, and agroecological practices such as agroforestry. Students visualized how community-level and industrial sites of extractive capital are interconnected. These mapping exercises helped to solidify the abstract concept that scale is socially constructed, meaning it is not ontologically fixed as local *or* regional, but rather 'a contingent outcome of the tensions that exist between structural forces and the practices of human agents' (Marston 2000, 220). Critical mapping can be understood as training these educator-students as Gramscian organic intellectuals, providing the tools to critically analyze the contradictions of hegemonic land use, and the spaces for popular resistance.

In addition to conflict mapping, the first section's focus on problematization consisted of 'dislocating' exercises. Dislocation pedagogy is 'the removal of students from what has become familiar by disrupting their geography … and their assumptions about … the authority of academic knowledge' (Godlewska 2013, 385). In this program, dislocation consisted of several 2–3-day field research trips to aluminum smelters, cattle ranches and mines, as well as farmers' markets, agricultural cooperatives and land occupation encampments. During one of these dislocation exercises, Diana, who is a long-term educator from the 17 de Abril settlement, critically recognized a contradiction between an industrial mineral extraction operation and its claims of environmental conservation:

> In Serra dos Carajás we went to the areas where they deposit the mine tailings. There (in the forest) we realized the contradiction that they say that this is an area of environmental preservation, but what type of environmental preservation is this, where they destroy the rainforest to discard the tailings that they don't want? What type of preservation is this? Was it set up this way to preserve, or was it to distance people from what they were doing, so they could be free to do what they want? These were the types of questions that we asked, *why* are they preserving?

Scholarship highlights how contradictory moments, like Diana's recognition of the contradiction between mining and claims of environmental conservation, are when learning happens (Mezirow 1990).

Diana subsequently elaborated that her learning during this trip was not simply about *those* mines, but about larger questions of how scale constitutes place:

> The experience involved the questioning of, 'What space are we living in? What is happening here?' Because we live here, and lots of people say, 'Oh this is happening over theeeere in Amazonia.' But hell, Amazonia is here, Amazonia is right here. They say, 'Oh that is happening way over theeeere in the mines of Serra dos Carajás,' but you know what? Serra dos Carajás is right here.

Diana's scalar reflection typifies moving 'away from thinking of scale as an area or a circumscribed space – we should think of scale as a network, or a strategy linking local struggles to regional, national, or global events' (Jones 2006, 26). This conception of place as a product of scalar interactions is fundamentally relational.

A relational perspective of place opposes the conception of places as coherent and distinct locations (Massey 1999, 14). Diana's narrative exemplifies Massey's (1999) description of place as 'the sphere of juxtaposition, or co-existence, of distinct narratives, as the product of power-filled social relations … . This is place as open, porous, hybrid' (21–22). By asking, 'What space are we living in?' Diana called into question accounts of environmental change that see place as discrete and environmental devastation as distant. Rather, by seeing place as 'open, porous, hybrid' and constituted through interconnected scales, Diana visualized how the strategies of extractive capitalism connect places in a 'sphere of juxtaposition' (Massey 1999, 22). Describing her understanding of how contesting scales constituted a relational place, Diana emphasized how *that* 'Amazonia [the one of extractive capital] is right here ... Serra dos Carajás is right here'. Through dislocating pedagogy, students gained further training as organic intellectuals by developing a relational understanding of place as grounded in struggle at multiple scales.

Intervention through problematizing

While the certificate program's first section dealt exclusively with problematization, the second section was structured around intervention. As one student explained,

> One of the grounding principles, for the professors and for us, in the course was that this cohort can't be simply another group of researchers who go to farmers' lands, take away an understanding of the problems, but never return. Rather, the proposal was that we conduct the research and then do projects that can help the farmers.

The program's grounding principle of applied research trained these educator-students as organic intellectuals who could advance counter-hegemonic forms of production. Bernardo, another participant in the certificate course, conducted research on one of the most prominent local strategies of extractive capitalism: sand extraction. He sought to apply his research finding in the community, using critical place-based research as a form of territoriality.

Bernardo stood on the undercut bank of the *Rio Vermelho* with a community-managed agroforestry grove to his back. Motioning with his hand over the muddy river on which floated makeshift barges coughing black smoke into the air, he reflected upon his research, and its transformative potential for the settlement:

> The river is dying here. When you extract the sand you're going to take out all of the land from over there, and so all of the soil will continue sloughing off from the bank, filling in the river, and the river will dry up. While discussing this process in the course, we've been talking about how rural farmers become involved in these processes. Many rural farmers are having difficulties meeting their subsistence needs through agriculture, and see the best way to make a living as the easiest, which in these areas is the extraction of sand. But, they don't recognize that they're negatively affecting the environment. So, this research has enormous potential within the community. The community could control the sand extraction if they had the knowledge, and this is what we tried to do in the research, talking with farmers, but it's difficult, because they see it as the easiest way to make money.

Bernardo exemplifies how critical place-based education serves as a form of territoriality, by advocating for alternative forms of land use. Bernardo's usage of agroecological education as a form of territorial communication underscores Sack's perspective that 'human interaction, movement, and contact are ... matters of transmitting energy and information in order to affect, influence, and control the ideas and actions of other and their access to resources' (1986, 26).

As part of the certificate program, the students conducted a final project that sought to unify the course objectives by both problematizing and creating solutions. Diana and Lucinede, the educator-students in the course who came from the 17 de Abril settlement, focused on related topics – milk production and pasture – that underscored the historical transformation of the region's economy and ecology. Although community members largely engaged in subsistence agriculture following their original settlement, in the last 15 years they have almost entirely transitioned to dairy production. The majority of survey respondents indicated that they transitioned to cattle ranching due to economic necessity, because 'it's more income' (57 percent). Lucinede, an educator-student, adds context to these data with an exasperated sigh:

> The tendency in our settlement is ranching! In a little while, we won't be producing anything; we're simply going to be raising cattle for milk. I've realized that there are many families that have abandoned the practices of working the land for agriculture, and have become dependent upon working with cattle.

I found the settlement's inhabitants were divided as to whether or not this agricultural transition was positive: 38 percent of respondents saw the settlement's transition away from subsistence agriculture as a bad change, 40 percent viewed it as a good thing and 22 percent saw it as both. Those seeing it as a good thing shared the view that it was 'less work and more money', whereas those who saw it as a lamentable change expressed a similar view that 'we shouldn't have to buy foods with pesticides, but instead be able to plant healthy food to eat'. Lucinede's frustration with the transition to cattle ranching motivated her research project:

> I began to observe this [the predominance of ranching] 2 years ago, and I talked about it with my friend, and it annoyed me. When I began the course, I wrote several assignments about this, talking about this irritation I had, and at the end of the course's first section, our professors proposed some assignments for our time in the communities, and the assignment was for us to work to identify the form of production in the settlement and through researching it, work to understand both hegemony and counter-hegemony.

When Lucinede's professors encouraged her cohort to focus on conflicting commodities, she chose milk production, because of its cultural, economic and political value (Hoelle 2011). Once she settled on milk production, her professors urged her to research an alternative product, asking her, 'What is it that you have in your settlement?' In response, Lucinede

> discovered so many things, so many lovely things, so many interesting things. I discovered that there is a farmer who has milk cattle, but he also has a mandala agricultural planting, he works with agroforestry, he has *cacau* trees. There's another farmer who works doing beekeeping, but he also produces *cupuaçu* fruit. There's another farmer who is working with agroforestry and is working to reforest his land with native Amazonian forest species. And so I discovered within the settlement a universe that I had never seen, which was so broad, so vast and so marvelous, filled with diverse experiences of production.

Although Lucinede had lived in the community for 17 years, she had largely internalized the narrative that one finds circulating both within the settlement and the surrounding cities, which is that in the settlement 'it's only extensive cattle ranching'. Lucinede's narrative highlights how researching the relations between hegemonic and counter-hegemonic forms of production led to a critical place-based rediscovery. This realization pushed Lucinede towards action:

> When I discovered this I thought, 'I need to do something to show these people the potential that exists within the settlement, the potential for the diversification of production', and I realized that someone has to do something, and I thought that I can do this.

I argue that Lucinede, by working to communicate these realizations, becomes positioned as a Gramscian 'organic intellectual' in several ways. First, she is dedicated to working in the community and not leaving it after she gets her degree. Second, as an educator in the settlement's school, she can share the counter-hegemonic examples of land use she uncovered, engaging in the long-term work of transforming territory from monoculture to agroecological polyculture.

Lucinede and Diana began to take action by discussing the creation of an agroecological 'brigade', which would study agroecology in the settlement. This brigade would ideally work in the school, but also engage with the farmers in the community. However, as I show in the next section, Lucinede and Diana faced significant obstacles in the endeavor.

Transforming the school?

The school where Lucinede and Diana work is defined by the ebb and flow of political participation among both educators and the larger community. The micropolitics of educators' political participation serves as both an impediment to and opportunity for agroecological change within the settlement. I first explore how educators themselves perceive their political participation as both an opportunity and obstacle for curricular change. I then examine how Lucinede and Diana, the educator-students from the certificate program, communicate their critical place-based research to influence their students' perceptions of land use.

The school's namesake points to its origination in struggle: Oziel Alves Perreira was a 17-year-old MST leader who, as local lore tells it, screamed 'Long live the MST' before being fatally shot during the 1996 Eldorado dos Carajás massacre. Following the creation of the settlement, the community was extensively involved in the building and staffing of a rudimentary wooden school. As infrastructure within the settlement began improving, the community demanded a 'quality' school. Students, parents, teachers and administrators actively participated in the politics of the new school's creation. Between 2007 and 2009, the community held a variety of protests to pressure the government for the resources to construct the new school. Several times in 2006 and 2007, students, parents, teachers and administrators occupied highway PA-275 in the southeastern Amazon, transforming it into a makeshift two-lane classroom. Government officials eventually arrived and began a discussion, asking the MST activists what were their priorities in terms of creating a new school. Although building materials were promised at these protests, the months dragged on and there were no signs that the government would make good on its promises. When the school material failed to arrive, these MST activists and those from other encampments and settlements occupied the train tracks of the Vale Corporation for over a month. This action led to more substantial and formal dialogue with the state and municipal governments, which ultimately resulted in the construction of the school.

In the five years since the occupation of the railroad, both the school and the politics of its educators have undergone a transformation. When I first conducted pilot research in 2009, the school consisted of classrooms separated by flimsy wood walls, and a cafeteria under a thatched-roof. At that point, construction on the new school complex was just beginning. When I returned in 2010, the old school was leveled, and the new school had been completed. At that time, I was struck by the visual absence of the movement in the new school in comparison with the old one. In 2012, when I returned for a year of fieldwork, the school had once again metamorphosed, taking on the appearance of a more radical space. MST posters now adorned the administrative wing. Two large spray-paint stencils of Che Guevara adorned the exterior walls. Slogans from revolutionary intellectuals like Rosa Luxemborg, and critical pedagogues such as Paulo Freire, graced the wall, formed from cut-out letters created in a children's art class project. When asked about this trans-formation in the school between 2010 and 2012, Edison, an MST activist who works at the school, pointed to the change in the coordinating council: 'When the coordinating council of the school changes, it wants to put on a face that represents itself'. The coordinating council of this municipal school is made up of six teachers who are elected by all of the teachers, and who function as the administration. Edison continued,

> This year is the first year that the coordination of the school has been completely composed of activists. As such, we've decided to give the school a makeover, to give it the face of what we represent. It happens all the time, if you have a coordination that is Evangelical, you'd expect to have an Evangelical 'face' at the school.

Edison's perspective points to the complicated daily politics within the larger settlement. The political participation of educators mirrors that of the larger settlement, in the sense that neither is a homogenous space of movement activism. The fact that all inhabitants of an MST settlement are not MST activists might be surprising, given the community's political victories. However, Wolford (2003) reports similar findings from multi-sited research on the MST: 'Settlers in both places struggle over their understanding of what community means to them – they struggle inwardly and they struggle with each other. Sometimes the everyday experience of "community" is not very communal at all' (501). Edison underscores this discord in his community:

> It's really quite complicated, this relation between the school and the movement. The school tries to work by following the organizational principles of the movement. But the problem is that not everyone who works in the school belongs to the movement. It's one thing to live in an MST settlement; it's another to belong to the organization. These two things are quite different, and it's difficult to reconcile these two aspects.

Part of this inconsistent participation is related to the influx of non-MST inhabitants, as many of the settlement's original inhabitants have left, due to failed agricultural efforts, lack of resources or other problems. Another part of 'the problem', as Edison called it, stems from the inhabitants' complacency, as they no longer see the need for political engagement. Whereas 19 percent of inhabitants described their participation in the MST as 'high' at the time of the settlement's creation, at present only eight percent see them-selves as being actively involved in the MST. As one MST activist opined, 'Everyone's got their television, their house, their motorcycle. They have enough food, and can go and hang out with their cell-phone in the central square. What need do they have for the movement?' Other activists within the community continue to see the MST as a fundamen-tal organizing force of the community, arguing, 'The struggle is continual. When one struggle ends, another begins'. These differing perspectives support Wolford's (2010) point about how people flow into and out of the movement, participating in a protest one year, and disassociating from the movement soon after. This ebb and flow of participation was manifest in the production of knowledge within the school. As Edison explained:

> There are a number of teachers that disagree with the MST's principles, and because they dis-agree with these principles it becomes difficult to direct this process [of integrating MST prin-ciples into the school]. For example, a teacher arrives, and he's from São Paulo, but he grew up in Goiana, and he grew up with a completely different reality than ours. And he arrives and wants to work with the principles that he brought from there. I'm not saying that the principles that he brought are wrong, and those that we have are right. We have an ideology, and so we want to preserve our ideology, and work with the grassroots, in the manner in which we think is correct. But it's really quite complicated.

Edison's description of these internal politics draws attention to the importance of geography, because, in his analysis, where a teacher is from structures whether or not they will defend the MST and its pedagogical principles of advancing alternative forms of knowledge.[7]

The geographic question inherent in educators' political participation is not simply about who is or is not from the settlement. Rather, *place*, and its attendant opportunities and constraints, is relational. Whereas all of the school's teachers give lessons in the same space, their understanding of that place, its transformation and their role in it is structured by their life trajectory of experiences. As Edison remarked, 'They may have lived in the settlement for several years, but so what?' The following example highlights Edison's point.

Luana is a biology teacher who works in several schools, spending three months in each of five communities on a rotational basis. At one class I observed, she showed pictures of the larval stages of bee development, and a student remarked, 'How in the world can we see it that close?' Luana informed the student that they would use magnifying glasses, which they should all have. 'Magnifying glass?' the student remarked. 'Where are we going to get money for a magnifying glass?' Luana jokingly responded with a tasteless play-on-words: '*Sem Terra*, I swear, it should be *Sem Nada* [those that have nothing]'. Luana's comment was an insult to the students and the movement as a whole and illustrates how far removed Luana is from the realities of her students in the settlement. Although she occupies the same educational space, her relational sense of place has an entirely different referent than those like Diana and Lucinede, who have literally grown up, and developed as activist educators, within the movement's ideological spaces.

The relation between school micropolitics and those of the larger political sphere became pronounced in the weeks after the October 2012 municipal election. Municipal elections strongly shape settlement politics, because the winning candidate typically appoints his supporters to leadership positions in various institutions, such as the municipal school in the settlement. While elections strongly shape the school's political composition, MST activists disagreed about the ultimate impact of electoral politics on the school. The next vignettes explore these MST activists' differing perspectives, and yield insights about how the politics of place mediate curricular change.

The week after the election, Joata and Francisco, two of the settlement's most active MST leaders, met with Genilda, a state-level MST leader, to discuss the potential impact of the local elections on the school. Upon greeting them, Genilda, the state MST leader, asked what would happen to the school's director. Joata said resignedly, 'We haven't decided yet who it will be'. 'She's definitely going to be out, that much we know', interjected Francisco, the other representative from the settlement. Genilda sighed in frustration, and added,

> This business with the elections has to end. Every time we have an election we have a change in administration in leadership within the school. Look at Palmares [a nearby MST settlement];

---

[7]Although Edison describes these internal politics as geographical, not all agree. Salete Campigotto, who is known as the first educator of the MST, indicated in a personal communication to MST education scholar Rebecca Tarlau that whether a teacher is from the MST community, or not, is not important. According to Campigotto, some teachers born in MST settlements will refuse to use the movement's pedagogy, and some teachers from outside the community will become the biggest activists. Therefore, the MST should try to engage with all teachers equally. Following Campigotto, one's 'place' (or where they come from) does not necessarily translate to how they defend or do not defend counter-hegemony in a particular territory.

they have direct elections in the settlement for the schools coordinating collective. Perhaps that could work in the 17 de Abril …

As she trailed off, Francisco and Joata both sighed and seemed uncertain. Genilda continued, looking Francisco, the older and more influential of the two activists, directly in the eye: 'We *need* to retain our presence in the school. Who is being considered for the position?' Francisco responded, spitting as he said, 'Daniel, most likely. He has no interest in the MST. No interest whatsoever'.

As these comments make clear, these MST leaders saw having an MST-oriented school principal as crucial. As the MST-supportive party lost the election, the school's director will be replaced. From their perspective, this was a very important and unfortunate consequence, because, as Edison previously remarked, the director's personal politics affect the 'face' of the school and its curricular programming. Losing the current director would thus result in a transformation in the coupled ideological and material terrains of the school and, consequently, the larger settlement.

Yet not all MST activists placed such importance on municipal politics. Lucinede, one of the certificate program participants and a long-term teacher in the school, reflected,

> Sure, you can put a representative of the movement in the school, but if you don't have a consensus among teachers that education is a fundamental part of our transformation, than that opportunity is lost, because we won't be able to get it done [the integration of MST politics in the curricula].

Lucinede's understanding of the importance of electoral politics contrasts markedly with that shared by Genilda, Joata and Francisco. Lucinede emphasized that having a school director 'who walks with the movement', as she often described it, is important. Yet she exemplified a Gramscian perspective on the importance of advancing popular consent, arguing that if there is not a larger consensus among educators about the political role of education, then it can be difficult to create mobilization in spite of the director's MST sympathies.

Instead, Lucinede explained, the politics of the teachers themselves would ultimately determine whether or not the MST's pedagogy could be implemented:

> We have a series of problems in the school. Even within a settlement of the movement we have significant difficulty implementing the pedagogy of the movement, but it's not because the municipal government doesn't allow it. Having worked as the director of the school, I never had a directive from the municipal government or the secretary of education saying, 'You can't work with this. You can't work with that'. So, if the school doesn't function as it should, then the problem is us – the settlement – in implementing this work [the pedagogy of the movement]. The problem is not the municipal government.
>
> I've suggested at various times in various meetings that what we have to change is the mentality of the educators who are here within the settlement. Teachers used to be trained by the movement, and understand the importance of debate within the school. These days, frequently, you invite the teachers to come to a discussion and they don't come.

As Lucinede sees it, the crux of the problem is the educators' lack of consensus regarding the place of the MST in the school.

Lucinede's perspective draws attention to how educators' participation is *political* in two senses. First, the act of educating is a form of political participation because – depending upon one's intent – it either transmits or omits particular ideals supported by the movement. Those who function as 'organic intellectuals', such as Diana and Lucinede, participate daily in the movement by communicating its ideology to other teachers and students. By contrast, those who either actively denigrate or simply do not acknowledge the importance of the movement practice their own ideological resistance, and simply drag their feet. Their daily resistance to the MST's efforts to advance a counter-hegemony within the school took the form of not showing up at teachers' meetings where political projects were being discussed, and not encouraging their students to participate in movement events. Second, MST educators' teaching is political because it sees the school as a mechanism for achieving social and environmental transformation. MST teachers are thus organic intellectuals who have the power to advance a counter-hegemonic project. Each of these manifestations of educating *as* political participation is intimately linked to the act of territoriality, because the terrain of ideas and land is interwoven.

Lucinede went on to describe how being an educator committed to the movement is at variance with the interests of the majority of educators in the school.

> Working with the movement is something that demands a significant focus: more availability on behalf of the people, in terms of being available to come to the school to discuss things, to become involved in activities. What we [MST educators] have found is that people don't want to be more available; they don't want to be present at the school more than is required in their little contract. For example, if my contract says you need to be there for 6 hours, I'm only going to be there for 6 hours. And so people are really caught up in the question of salary. And it's a salary that says I need to work for a total of 100 hours, and so at the end of that hundred hours I'm finished. The rest that needs to be done, such as extraneous projects, oh, just leave it to the side. But in the pedagogy of the movement it's more than this. The person needs to really be able to make time available to plan and organize, to propose activities, to involve the community in these discussions, and these discussions go forward veryyyyyy slowly within the community. You have to find methodologies that bring the community to the school, and we've not been able to achieve this because this requires time, it requires resources.

Particularly important, from Lucinede's perspective, is educators' personal political commitment to the MST's mission. This intrinsic motivation is necessary for transcending everyday concerns about the number of hours worked to achieve the larger objective of social transformation. If there is going to be a lasting agroecological 'face' at the school, consisting of anything from posters and slogans, to substantive curricular content and applied student research projects, it needs to arise from the educators' own political commitments.

I have presented two different perspectives regarding the links between the settlement's formal educational curriculum and the larger political environment: one represented by Edison, Genilda, Joata and Francisco, which views the dominant political party, and subsequently the school administration, as the key driver of school politics; and one represented by Lucinede, who asserts that regardless of the larger political leanings of the school leadership, educators' individual commitments to the movement are crucial. This exploration highlights how the politics of place – and in particular, a lack of consensus in the community and in the school regarding the importance of MST electoral and internal politics – can be a constraint for integrating the pedagogy of the movement within the school; the next section explores the flipside of how these politics provide opportunities for institutionalizing agroecological education.

Territory as opportunity

Various MST activities take place in the school throughout the year, and the driving forces behind them are usually the organic educators dedicated to the MST and its ideals of transforming the social and ecological relations of production. For example, in April, there is an annual 10-day 'pedagogical encampment' where MST activists study agroecology and agrarian theory; during June an agroecological student garden is annually planted; and every August the MST Youth Journey (*Jornada de Juventude*) takes place.

The Youth Journey is a week when normal classes in the school are canceled, and in their place MST activist youth and teachers lead lectures and workshops. One afternoon during the 2012 Youth Journey, which was devoted to the topic of agroecology, Diana and Lucinede led a session about the individual research projects they conducted during their certificate program. Lucinede told the students, 'Sometimes you'll hear people talk about certain things as if they were something that only took place really far away; but sometimes, those things are actually occurring quite close to you, it's just you're not able to realize it'. Lucinede's description is strikingly similar to Diana's previous comment about discovering a 'marvelous' and 'unseen universe' of agroecology in her own community. The students began to learn about these invisible landscapes as Lucinede continued, 'Through this research we were able to learn many things, to discover many things here in the settlement. We observed a variety of agroecological initiatives going on, which are barely known by the population of the settlement'. Lucinede had lived in the settlement for 17 years without knowing of these agroecological areas. Her narrative ensured that the students learned about this unseen geography through a critical lens. Lucinede continued,

> I'm going to tell you all two stories from our research. One is an inhabitant in our settlement, and another is from another MST settlement, which we visited as part of our course. When we were visiting that other settlement, Mede took us on a tour of his lot. And he told us that the first thing he did was try to use the lot to raise cattle, but the lot was very small and didn't work to raise cattle, which is often the case here.

None of her students had been to this other settlement, but Lucinede painted a relational vision of place by using a description of *that* place to educate students, critically, about *this* place. She indicated that MST settlers in both settlements face similar constraints with the land and its small lot sizes, which are inadequate for cattle ranching. Lucinede also used her experience to instruct the students about how both settlements' inhabitants had similar agroecological opportunities available to them. Lucinede went on to describe, in exquisite detail, the ecological richness of Mede's land and how he was able to sustain his family through the agroecological products he and his wife sold from it, ranging from fruit pulps to natural cosmetics to orchids. Challenging the students, she asked rhetorically, 'Now, where is all this taken from?' 'From their lot', a student interjected. 'From their lot is correct', Lucinede responded, 'from Nature, exactly correct, so it's a different form of producing. They are able to survive without destroying the rest of Nature that still exists there'.

Lucinede's presentation then shifted from illustrating a relational geography, to explaining her critical place-based learning about hegemonic and counter-hegemonic production in the 17 de Abril settlement. 'Another place where we did research was here in our settlement. And sometimes you see someone's land plot that has a lot of forest, really high forest, and you say, "Man, he's lazy", right?' She stopped and emphasized her next statement, meting her words out slowly to describe a hapless imagined individual: 'We ... see ... them ... as ... *lazy*, now don't we?' Without missing a beat, the auditorium responded in unison, 'We do'. 'Right,' Lucinede continued, 'we see them as lazy because they've been on the land for 15 years, and you can see from the beginning to the end, it's just forest, *just* forest'. Lucinede drew upon the students' own experiences, setting them up for a problem-posing moment by asking what seeing '*just*' forest on someone's lot indicated about that person. 'But I don't think this is laziness', Lucinede offered. 'You know what it is? It's a choice to engage in a new form of production. Do you think economically he just survives on cattle and pasture?' 'No', the students all replied again. Lucinede probed further, 'Do you think that he just knocks down the forest to burn so that he can then plant crops?' 'No!' shouted the students. 'There are other forms of production', Lucinede said in a voice that was reserved, yet forceful, and filled in the other half of the contradiction she had created: having one's land comprised completely of forest is not laziness, as the community tends to believe; rather, it can be tactical. Lucinede continued,

> What is lacking amongst us is knowledge of how to do this, and so for that reason we did this research, and through this research we learned that this individual has *açai*, mahogany, cedar, *cupaçu, cacau, goiaba*, he has a huge list of tree species, including *castanha do pará*. He has planted more than 5000 trees on his land.

Lucinede's research was, as she described it, a process of discovering the counter-hegemonic forms of production that constituted place.

Lucinede's talk took on an increasingly normative nature as she described the threats that this forest farmer faced:

> He needs to have a lot of courage, because many people say it's deplorable, they'll say that he hasn't *even* been able to generate any income from this. And what happens to all of the effort and resources he invests in this, traveling hours away to acquire native transplants, and then someone comes with a nice sharp machete and makes a trail to hunt the animals that are living in the forest, or to cut down the trees, right?

The students unanimously agreed, 'They will be cut down.' With statements like these, Lucinede's narrative directly confronts the competing ideologies and forms of land usage in this MST settlement. Drawing on this sole example from her research, she described to the students the conflict, within their own settlement, between the proponents of hegemonic forms of cattle production who see agroecology not merely as lazy, but as something that needs to be stopped, and counter-hegemonic forms of production based in agrecological diversity. Lucinede concluded her presentation by telling the students,

> These examples can be a way of encouraging our parents to work in a type of production that is not simply ranching. This new form of agroecological production is in equilibrium with nature, because one thing depends on the other. This type of work is gratifying because it creates in the person a perspective of a future that is more healthy, and that is much better than having someone take a land plot and mechanize the entire thing and plant pasture across it. And so these two experiences bring for us a new hope for life.

Lucinede's conclusion epitomized how education can serve as a form of territoriality, encouraging particular uses of and relations to land through production.

Conclusion

I have explored in this contribution the opportunities and constraints facing the MST's institutionalization of critical place-based agroecological education. I found that both these opportunities and constraints were geographic in nature.

The certificate program in The Agrarian Question, Agroecology, and *Educação do Campo*, exemplifies the successful institutionalization of critical place-based agroecological education. This program was created by a partnership between the MST and the Federal University of Pará, funded by PRONERA. This institutionalization of critical place-based agroecological education was an *opportunity* for the MST, because it enabled the training of its educator-students as Gramscian organic intellectuals. This Gramscian training was structured into the certificate program, which was designed around the problematizing of hegemonic forms of production, and applying research findings about those system's contradictions and emerging counter-hegemonic forms of production. In this case, institutionalizing a critical place-based agroecological education helped the MST advance its struggle within the framework of the state. It thus became a way to begin transforming the state from within (Poulantzas 1978; Jessop 1990; Boden 2011).

The critical place-based education helped educator-students develop a relational conception of place. Diana's description of the ties between Serra das Carajás and her settlement, Bernardo's analysis of sand extraction and Lucinede's description of extensive cattle ranching all illustrate an understanding of place as relational, comprised 'of distinct narratives, as the product of power-filled social relations' (Massey 1999: 21), yet also interconnected. This cohort learned through dislocating field trips and community research how scales are interconnected, and socially constructed through hegemonic forms of production, such as sand extraction and cattle ranching.

The certificate program also emboldened the educator-students to apply their research findings about alternative methods and systems of production. Bernardo visualized his research as having an 'enormous potential' to communicate the negative effects of sand extraction to his community, filling a knowledge gap and facilitating the control of small-scale mining. Through her research, Lucinede also realized that 'someone has to something' and that 'I can do this'. Together with Diana, they decided to form an agroecological brigade, and bring this debate about hegemonic and counter-hegemonic forms of land use into both the school and larger community. Bernardo, Lucinede and Diana's efforts to use knowledge as a tool to transform material production within their communities highlight how agroecological education can be employed territorially, affecting the relation to and control over land. Although institutionalization provided opportunities for these educator-students to develop as Gramscian organic intellectuals, there were also constraints.

Whereas the educator-students had the training to advance counter-hegemonic knowledge and practices, the geographic politics of the school were a constraint. As Edison cautioned, 'It's really quite complicated, this relation between the school and the movement'. Edison's warning signaled that educators in social movement schools are not necessarily counter-hegemonic agents simply because of their location within an agrarian reform settlement. The example of Luana, the temporary teacher who joked that her students were better characterized as '*Sem Nada*', illustrated how all educators are the products of their particular spatial histories, which are relational and hybrid. The contrast between Luana, the temporary teacher, and Diana and Lucinede, the organic intellectuals, illuminates how educators' spatial histories of political participation mediate whether or not they will support MST counter-hegemonic objectives through their teaching.

Various theoretical perspectives could shed light on these issues. Critical educational scholarship takes as a foundational principle the idea that all education is political (Giroux and McLaren 1994; Apple 1995). Similarly, the political economy of education draws attention to how public policies shape curricula (Carnoy 1985). Political ecologists draw attention to how struggles over power shape ecology (Neumann 2005). Yet none of these perspectives illuminates the interconnections between politics, economy, education and ecology. The political ecology of education lens provided insight into how mapping strategies of hegemonic and counter-hegemonic forms of production, which one professor described as the 'political objective' of the certificate program, influenced the students' independent and collective understandings and attempts at transforming place. This lens also shed light on how the MST is institutionalizing its projects within the state as part of a long-term war of position. This analysis of *why* movements institutionalize education clarifies the omnipresence of spaces of resistance within the larger project of neoliberalism.

Funding

This work was supported by the National Science Foundation doctoral dissertation improvement grant [grant number BCS# 106088], a Social Science Research Council international dissertation research fellowship, and a Fulbright scholarship.

References

Altieri, M.A., and V.M. Toledo. 2011. The agroecological revolution in Latin America: rescuing nature, ensuring food sovereignty and empowering peasants. *Journal of Peasant Studies* 38, no. 3: 587–612.

Apple, M.W. 1995. *Education and power*. New York: Routledge.

Araujo, S.G. 2004. *O PRONERA e os movimentos socias: Protagonismo do MST. A educação na reforma agrária em perspectiva*. Brasília: PRONERA.

Arnold, D. 1984. Gramsci and peasant subalternity in India. *The Journal of Peasant Studies* 11, no. 4: 155–77.

Boden, M. 2011. Neoliberalism and counter-hegemony in the global south: Reimagining the state. In *Social movements in the global south*, ed. S.C. Motta and A.G. Nilsen. London: Palgrave Macmillan, 83–103.

Branford, S., and J. Rocha. 2002. *Cutting the wire: The story of the landless movement in Brazil*. London: Latin America Bureau.

Breitenbach, F.V. 2011. A educação do campo no Brasil: uma história que se escreve entre avanços e retrocessos. *Espaço Academico* no. 121: 116–123.

Callou, A.B.F. 2007. Extensão rural no Brasil: da modernização ao desenvolvimento local. *Unircoop* 5, no. 1: 164–76.

Carnoy, M. 1985. The political economy of education. *International Social Science Journal* 37, no. 2: 157–73.

Carrol, W.K., and R.S. Ratner. 1994. Between Leninism and radical pluralism: Gramscian reflections on counter-hegemony and the new social movements. *Critical Sociology* 20, no. 2: 3–26.

Cirese, A.M. 1982. Gramsci's observations on folklore. In *Approaches to Gramsci*, ed. A.S. Sassoon. London: Writers and Readers, 212–47.

Coben, D. 1995. Revisiting Gramsci. *Studies in the Education of Adults* 27, no. 1: 34–51.

Davidson, A. 1977. *Antonio Gramsci: Towards an intellectual biography*. London: Merlin Press.

Davidson, A. 1984. Gramsci, the peasantry and popular culture. *The Journal of Peasant Studies* 11, no. 4: 139–54.

Davis, J.A. 1979. *Gramsci and Italy's passive revolution*. New York: Barnes & Noble.

de Molina, M.G. 2013. Agroecology and politics. How to get sustainability? About the necessity for a political agroecology. *Agroecology and Sustainable Food Systems* 37: 45–59.

Feierman, S.M. 1990. *Peasant intellectuals: anthropology and history in Tanzania*. Madison: University of Wisconsin Press.

Fernandes, B.M. 2009. Sobre a tipologia de territórios. In *Territórios e territorialidades: teoria, processos e conflitos*, ed. M.A. Saquet and E.S. Sposito. São Paulo, Brazil: Expressão Popular, 197–215.

Freire, P. 1973. *Pedagogy of the oppressed*. New York: Continuum.

Freire, P., et al. 1986. *Pedagogia: diálogo e conflito*. São Paulo: Cortez.

Giroux, H.A., and P. McLaren. 1994. *Between borders: pedagogy and the politics of cultural studies*. New York: Routledge.

Gliessman, S.R. 2006. *Agroecology: The ecology of sustainable food systems*. 2nd ed. Boca Raton: CRC Press.

Godlewska, A. 2013. Dislocation pedagogy. *The Professional Geographer* 65, no. 3: 384–9.

Gramsci, A. 1978. *Selections from the prison notebooks*. New York: International Publishers.

Gramsci, A. 1994. *Gramsci: Pre-prison writings*. Cambridge: Cambridge University Press.

Gruenewald, D.A. 2003. The best of both worlds: A critical pedagogy of place. *Educational Researcher* 32, no. 4: 3–12.

Gruenewald, D. 2008. Place-based education: Grounding culturally-responsive teaching in geographical diversity. In *Place-based education in the global age: Local diversity*, ed. D. Gruenewald and G. Smith. Malwah, NJ: Lawrence Erlbaum Associates, 139–53.

Hoelle, J. 2011. Convergence on cattle: Political ecology, social group perceptions, and socioeconomic relationships in Acre, Brazil. *Culture, Agriculture, Food and Environment* 33, no. 2: 95–106.

Jessop, B. 1990. *State theory: Putting the capitalist state in its place*. Cambridge: Polity Press.

Jones, S. 2006. *Antonio gramsci*. New York: Routledge.

Karriem, A. 2009. The rise and transformation of the Brazilian landless movement into a counter-hegemonic political actor: A Gramscian analysis. *Geoforum* 40, no. 3: 316–25.

Marston, S.A. 2000. The social construction of scale. *Progress in Human Geography* 24, no. 2: 219–42.

Massey, D. 1999. *Power-geometries and the politics of space-time*. Heidelberg: University of Heidelberg.

Mayo, P. 2008. Antonio Gramsci and his relevance for the education of adults. *Educational Philosophy and Theory* 40, no. 3: 418–35.

Meek, D. 2010. Pedagogies of Brazil's Landless: Towards a Political Ecology of Education. Association of American Geographers annual conference. Washington, D. C. April 14–18.

Meek, D. 2011. The Brazilian Landless Worker's Movement and the war of position: A content analysis of propaganda and collective participation within the Jornal Sem Terra. *Studies in the Education of Adults* 43, no. 2: 164–80.

Mezirow, J. 1990. *Fostering critical reflection in adulthood: A guide to transformative and emancipatory learning*. San Francisco: Jossey-Bass Publishers.

Molina, M. 2003. A Contribuiçã do Pronera na construção de políticas públicas de educação do campo e desenvolvimento sustentável. Centro de Desenvolvimento Sustentável. Brasília, Universidade de Brasília. Dissertation.

Munarim, A. 2008. Trajetória do movimento nacional de educação do campo no Brasil. *Educação* Jan/Abr: 57–72.

Neumann, R.P. 2005. *Making political ecology*. London: Hodder Arnold.

Patel, R. 2009. Food sovereignty. *Journal of Peasant Studies* 36, no. 3: 663–706.

Poulantzas, N. 1978. *State, Power, Socialism*. London: New Left Books.

Roio, D. 2011. A tradução histórica e política de Gramsci para o Brasil. In *Gramsci e os movimentos populares*, ed. G. Semeraro, M. Marques de Oliveira, P. Tavares da Silva, and S.N. Leitão. Niterói, RJ: Editora da Universdidade Federal Fluminense, 69–82.

Rosset, P.M., and M.E. Martinez-Torres. 2013. *La via campesina and agroecology. La via campesina's open book: Celebrating 20 Years of struggle and hope*. La Via Campesina: Jakarta. http://www.viacampesina.org/downloads/pdf/openbooks/EN-12.pdf (accessed November 10, 2014).

Sack, R.S. 1986. *Human territoriality: Its theory and history*. New York: Cambridge University Press.

Semeraro, G., et al., eds. 2011. *Gramsci e os movimentos populares*. Niterói, RJ: Editora da Universdidade Federal Fluminense.

Shafir, G. 1980. *Intellectuals and the popular masses: An historical and sociological study of Antonio Gramsci's 'prison notebooks'*. Sociology. Berkeley, University of California, Berkeley. Doctor of Philosophy.

Simmons, C.S. 2004. The political economy of land conflict in the eastern Brazilian Amazon. *Annals of the Association of American Geographers* 94, no. 1: 183–206.

Storey, D. 2012. *Territories: The claiming of space*. New York: Routledge.

Suh, D. 2011. Institutionalizing Social Movements: The dual strategy of the Korean Women's Movement. *The Sociological Quarterly* 52, no. 3: 442–71.

Troian, A., and M.L. Eichler. 2012. Extension or communication? The perceptions of southern Brazilian tobacco farmers and rural agents about rural extension and framework convention on tobacco control. *Journal for Critical Education Policy studies* 10, no. 1: 315–26.

van der Ploeg, J. 2012. The drivers of change: The role of peasants in the creation of an agro-ecological agriculture. *Agroecologia* 6: 47–54.

Wolford, W. 2003. Producing community: The MST and land reform settlements in Brazil. *Journal of Agrarian Change* 3, no. 4: 500–20.

Wolford, W. 2006. The difference ethnography can make: understanding social mobilization and development in the Brazilian Northeast. *Qualitative Sociology* 29: 335–52.

Wolford, W. 2010. *This land is ours now: Social mobilization and the meanings of land in brazil.* Durham: Duke University Press.

Wright, A., and W. Wolford. 2003. *To inherit the earth: The landless movement and the struggle for a new Brazil.* Oakland: Food First!.

Yogev, E., and N. Michaeli. 2011. Teachers as society-involved 'organic intellectuals': Training teachers in a political context. *Journal of Teacher Education* 62, no. 3: 312–24.

The Landless invading the landless: participation, coercion, and agrarian social movements in the cacao lands of southern Bahia, Brazil

Jonathan DeVore

This contribution draws on Nancy Fraser's concept of 'participatory parity' to analyze the reproduction and contestation of inequalities internal to land reform settlements affiliated with the Landless Rural Workers' Movement (MST) located in the cacao lands of southern Bahia, Brazil. These inequalities are variously manifest in unequal control over land and legal documents, disparities in status and what Fraser calls 'voice'. These circumstances help account for quantitative evidence that shows a strong preference among local landless populations for land reform organizations that are more decentralized and less hierarchically organized. These circumstances also motivate direct actions undertaken by grassroots MST settlers seeking to destabilize the conditions that ground these inequalities. This research highlights the importance of attending to local histories and interactions through which participatory disparities are christened and reproduced; indicates potential methodological consequences; and examines the interplay of transgressive action, dialogue and recognition as settlers struggle to bring about 'participatory parity' – or what they might call genuine 'friendships' – in their communities.

'I am a slave.'
Sharecropper, Settlement 3, November 2009

'Slavery ended, but it's just like slavery.'
Settler, Settlement 5, June 2010

' ... we are convinced that liberty without socialism is privilege, injustice; and that socialism without liberty is slavery and brutality ... '
M.A. Bakunin, September 1867

1. A small rebellion; or expelling the MST

Sometime around 1985, a small group of squatters began to occupy and cultivate the forests on the outskirts of an old cacao plantation in southern Bahia, Brazil (see Figure 1, Settlement 6). Twelve years later, in 1997, these squatters resolved to invite leaders from the Landless Rural Workers' Movement (MST)[1] into the settlement they

---

[1]*Movimento dos Trabalhadores Rurais Sem Terra.*

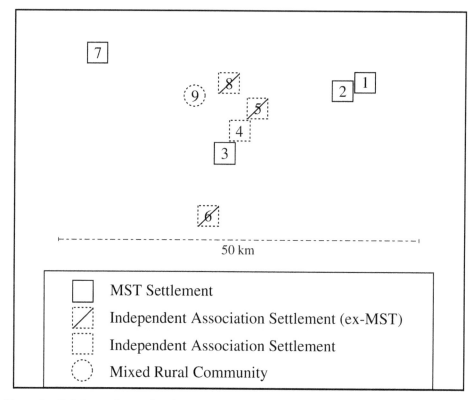

Figure 1. Relative settlement locations.

had established there. Their hope was that affiliating with the MST would enable them to access the elusive bureaucratic channels necessary to formalize their land claims against those of the plantation owner. MST militants (*militantes*)[2] began to establish their presence in the community. With time, however, the families living there became increasingly dissatisfied with the MST's leadership. One farmer suggested the MST's leaders had 'put us [to work] as slaves – more than when we worked on the plantation'. Emboldened by an outspoken farmer named Honório,[3] these experiences culminated in the settlers' decision to expel the MST's leaders. They sought to recover what Honório described as a time when '[we lived] independent with our liberty' in the years after they first occupied the plantation.

Honório recalled the early conversations that occurred as community members became conscious of their situation and reached their decision:

> And then we started to talk, we started to talk, we started to become conscious (*conscientizar*).
> And so it went, becoming conscious:

---

[2]To capture relevant sociological distinctions, I translate *militante* as either 'militant' or 'activist' according to each person's situation in the MST hierarchy, which differentially distributes the exercise of command and control over other members. 'Militants' can be understood as more central to the exercise of command and control, whereas 'activists' are more marginal.
[3]All names in this contribution are pseudonyms.

[Someone asked:] 'Man, could this really work?'

[Honório:] 'Yes, it'll work. We just have to be courageous, but it'll work. We'll send these guys running from here, and we'll take back our freedom'.

Then the people started to believe and said, 'How do we do it?'

I said, 'Do the following … '

At the time, the MST militant living in the region was a man named Osório. Honório's plan was to commandeer the car that Osório had for personal use, precipitate a confrontation and tell Osório to leave:

As soon as the car arrives, we'll take the car [from the driver] and then Osório will come. And when he comes, we'll tell him how it's gonna be. End [the conversation] there and expel him from here – say that we no longer want him here. Period. And if he stays, we'll break him.

On the evening that the settlers went through with the plan, conflict ensued. Osório left but then returned to the settlement later that night, armed with a gun, and backed by people from other MST settlements whom he brought to help suppress the small rebellion. He came looking for Honório, but instead of finding him, he encountered a whole community united against him. The episode ended without any violence, Osório was expelled and the MST's leaders never returned.

As a scholar politically sympathetic to the MST, and broadly interested in the prospects of redistributive land reform, this moment of collective action leads me to critical questions: Why were people so dissatisfied with these particular MST leaders? What was it about this local instance of the MST that compromised the settlers' sense of freedom, requiring them to forcibly take it back? This contribution builds an ethnographic framework for understanding this episode, and draws attention to historically durable structures of unfreedom that have shaped the experiences of families living on several land reform settlements in Bahia's cacao lands. Although redistributive land reform may be understood as part of the emancipatory project following the abolition of slavery, this account from the cacao lands examines ways in which inequalities and domination endure and have reemerged in ostensibly liberatory contexts.

2. Land, power, and participatory parity

Leal's (1977[1949]) landmark study of *coronelismo*, the Brazilian patron-client relationship, argued that understanding social power in rural Brazil requires attention to unequal access to, and control over, land. Clients – whether tenants, sharecroppers or day laborers – found themselves in a position of coercive dependency, subjected to the whims of their patrons, whose social power was partly rooted in disproportionate control over land and access to bureaucratic power. Diverse rural movements that emerged in Brazil over the past century have posed fundamental challenges to this unequal distribution of land and the persistent forms of social domination that it accompanies.

Among these movements and organizations, the MST stands out as one of the largest and most notable (see Wright and Wolford 2003). The MST is a social movement with deep agrarian roots, emerging in southern Brazil in the 1980s and becoming increasingly important on Brazil's national political scene. While the MST has attracted a great deal of attention from activists and scholars from around the world, other researchers have documented the broader diversity of rural movements in Brazil (e.g. Fernandes, Servolo de Medeiros, and Ignez Paulilo 2009, Feliciano 2006, Simmons et al. 2010, Welch 2009), and the diversity of experiences within the MST (e.g. Brenneisen 2002, Caldeira 2009, and Wolford 2010a). This paper contributes to both literatures by examining processes of con- testation internal to the MST, and by exploring the MST's complicated relationships to other land reform organizations in Bahia's cacao lands. Because the cacao lands are rela- tively prosperous and have a long and acute history of struggle over resources (see various accounts by DeVore 2014, Leeds 1957, Mahony 1996, and Wright 1976), the con- flicts described here may not occur with the same intensity – or at all – on MST settlements in other regions.[4]

As has been seen, some settlers draw analogies to plantation life, and even slavery, to make sense of problematic experiences on the region's MST settlements. The persistence of this interpretive frame raises a question: To what extent have the legacies of worker-owner relations been reproduced in some MST locations? The evidence presented here, corrobor- ating some of Eliane Brenneisen's (2002) observations for southern Brazil, suggests that some aspects of the MST's institutional structure have allowed for the reproduction of 'patron-client' analogues internal to these settlements, especially in the diverse relation- ships between settlers and militants at local and regional levels of the organization. This is owed in part to (1) ongoing problems of land tenure insecurity on MST settlements, and (2) settlers' personal dependence on movement leaders who mediate access to govern- ment bureaucracies and resources.[5] My research calls attention to anti-authoritarian chal- lenges to the MST's organizational hierarchy from within, as settlers embrace ideals of free association and show a deep skepticism toward certain expressions of authority. These challenges are not properly understood as oppositional to the MST. Indeed, given the historical movement of which the MST is one expression – aiming to create a more just and equal Brazil – these settlers' challenges are best understood as an unflinching embrace of precisely this same historical spirit.

The problem of reproduced inequality and privilege raises questions about the prospects for participatory democracy in the context of the MST, a question extensively reviewed by Wolford (2010b). In the present analysis and context, I draw on Nancy Fraser's discussion of 'participatory parity' and 'voice' (2001, 36) as a framework for evaluating forms of par- ticipation in the region's MST and other land reform settlements.

Fraser suggests that people cannot participate in social life as 'peers' – with equal voice – so long as they are constrained by asymmetries in social status and access to material resources. Establishing participatory parity requires fulfilling what Fraser calls 'intersubjec- tive' and 'objective' conditions. The intersubjective condition is fulfilled when 'institutio- nalized patterns of cultural value express equal respect for all participants and ensure equal opportunity for achieving social esteem'. For the objective condition to be fulfilled, the 'dis- tribution of material resources must be such as to ensure participants' independence and "voice"' (2001, 36). This latter condition, Fraser explains, is undermined by 'forms and levels of economic dependence and inequality that impede parity of participation. … denying some people the means and opportunities to interact with others as peers' (2001, 36). As Fraser notes (2001, 108), this framework draws on critical feminist research on household gender inequalities that, in turn, develops insights from A. O. Hirschman's *Exit, voice, and loyalty* (1970). Barbara Hobson (1990), for example, explores the ways

---

[4]I owe this observation to conversations with Gregory Duff Morton (personal communication, March 4, 2014).
[5]Wolford (2010b) similarly suggests that the way relationships between MST leaders and settlers are institutionalized may contribute to the reproduction of certain inequalities.

in which women with access to an independent livelihood can exercise greater 'voice' in domestic relations, precisely because exit is a material possibility. When exit is foreclosed by fear of material destitution or social stigma, then 'voice' diminishes in inverse proportion.

Mobilizing Fraser's framework, I draw attention to participatory disparities that have been structured by specific moments of interaction in the community 'assembly' (*assembléia*; section 8), and related disparities in control over material and legal resources (sections 9 and 11). These experiences motivate settlers' analogical references to plantation life and slavery, and give meaning to their notion of 'friendship' as a social form that approximates participatory parity (section 13). While the research presented here raises a series of critical questions for further investigation, general conclusions are untenable in the absence of comparative research that is attentive to potential methodological issues indicated at the end of section 8. In section 14, I briefly develop some of these political and methodological implications, and, finally, contribute a few observations about Fraser's concept of participatory parity.

3. Methodology

This paper draws on 38 months of fieldwork in a number of land reform communities in Bahia's cacao lands. I first met families living in these communities in 2002, and I have maintained these relationships for more than a decade. The research is grounded in extensive participant-observation in farm, family and community life, and draws on over 200 hours of oral histories and semi-structured interviews, research into local historical documents, and survey data.

My earliest relationships were established with families at Settlement 4, when I was studying various dimensions of resource use and agricultural production (see Paulson and DeVore 2006). Over time, I built relationships with people living at neighboring settlements. These relationships were initially mediated by ties of friendship and kinship that crossed settlements, and subsequently as a researcher working independently from local institutions.

The subject population consists primarily of grassroots squatters and settlers. While several of these settlers have held positions in the MST's local leadership, and some self-identify as 'activist' or 'militant', none belong to the MST's extra-local (regional or state) leadership. With some exceptions noted above, a good deal of research on the MST focuses on the most visible movement leaders and official narratives. When settlers present contentious and contrary politics of their own, these are often framed and evaluated in terms of the MST's broader political vision, rather than in terms that draw from settlers' own politics and lifeworlds. This contribution, therefore, gives sustained and systematic attention to settlers' experiences and politics, and suggests that earlier inattention to settlers' lifeworlds may be owed precisely to the asymmetries of representation that I examine here. While I gesture toward the MST's broader political logic at various turns, extra-local MST militants have not been the focus of this study, both because a substantial body of research is already available, and because they were relatively absent from the settlements where I conducted my research. Instead, their relationships to settlements are increasingly bureaucratic, as seen in section 12.

Over time, it became evident that many settlers, and an even stronger majority of landless people, held distinct preferences for participation in non-MST land reform settlements and institutions. Even those settlers who were the most outspoken about their general support for the MST expressed ambivalence, and often opposition, to particular institutional arrangements in local MST settlements. In some cases, those settlers who willingly joined the MST in the early years were among the first to challenge local MST institutions, as seen in section 10. In order to better understand these issues more broadly, I designed a survey instrument to help clarify some key relationships. After exploring the appearance of the MST and other land reform organizations in the region, and briefly examining some key institutional differences, I discuss the preliminary findings from this survey in section 6.

4. Collapse and upheaval in the cacao lands

In the early 1990s, the fungal disease 'witches' broom' (*Crinipellis perniciosa*) was somehow introduced into the Bahian cacao lands.[6] Once home to the most successful large-scale cacao cultivation in South America, producers in Bahia faced an agricultural epidemic that devastated the region's cacao economy. In this same period, Brazil's federal government ended price supports for natural rubber, which had been introduced into the region in the 1950s. This confluence of events led many of the region's plantations into bankruptcy. The plantations laid off numerous workers and let large stretches of land return to fallow.

Following this collapse, diverse land reform organizations proliferated throughout the region. These organizations ranged from:

(1) *Squatter groups*: These decentralized groups, more anti-authoritarian in character, initially lack legal status and are unaffiliated with institutionalized social movements. They tend to be comprised of local rural workers who occupy abandoned plantation lands where competing claims are not forthcoming (cf. Simmons et al. 2010). Squatters' property claims are established through continuous occupation and cultivation of the land, and appeals to squatter rights such as 'usucaption' (*usucapião*). These groups may eventually seek the status of a legal association, as they secure various legal 'documents' (*documentos*) for their landholdings (see section 6). Squatter landholdings are apportioned to individual families, and no one family exerts institutionalized control over any other's landholding.

(2) *Independent legal associations*: These groups are comprised of local workers that have organized into legal associations in order to purchase land directly from large landowners. While these groups are independent, they may acquire occasional support from syndicates or social movements. Their transactions result in the transfer of legal land title, and may be facilitated by state entities. As with squatter groups, landholdings are apportioned to individual families and settlers only exert control over their own landholding. Legal associations have low operational costs, and come in and out of existence as settlers' needs change.

(3) *MST settlements*: These settlements are embedded in two sets of relationships: with the MST's broader organization, and with the federal government. The MST is a centrally coordinated social movement with levels of organization comprising a hierarchically nested set. The direction of command tends to proceed from national, to state, regional and settlement levels of organization. Each level of organization exercises relative autonomy in its local affairs. The organization's grassroots

---

[6]For an account of the contentious circumstances under which the disease may have appeared in Bahia, see Caldas and Perz (2013).

consist of families living in MST encampments and settlements, with locally elected association leaders to represent them. Families on these settlements, ideally, form a disciplined rank-and-file that is responsive to increasingly higher orders of the MST's leadership. Militants at the organization's lower levels are, in theory, accountable to militants at higher levels.

MST settlements are also embedded in a legal relationship with the federal government through the National Institute of Colonization and Agrarian Reform (INCRA), which administers various services and agricultural credits to settlements. Although the MST is not a legally recognized entity, it serves as an intermediary between settlements and INCRA (see Wolford 2010b), and MST leaders and INCRA officials engage in informal relationships at various levels of both organizations.

Property claims on MST settlements are established through legal 'declarations of possession' (*declarações de posse*), through which the federal government grants usufruct to local associations of MST settlers. Landholdings are initially held collectively, but over time, INCRA is supposed to promote land 'emancipation' (*emancipação*) wherein each family would receive legal title to a parcel of land. This would involve a series of transactions through which families buy the land from the state and repay any agricultural credits received. The national MST organization is opposed to such emancipation.

These different settlement types can be broadly differentiated by the types of institutional support they receive and their respective modes of property acquisition and distribution, all of which may shift over time. These distinctions are neither exhaustive, nor always mutually exclusive. Although space prevents me from detailing the institutional trajectory of each settlement depicted in Figure 1, the conventions there indicate some relevant changes.

5.   Land reform institutions and the MST

The MST is among the most historically significant social movements of its kind. Over the past few decades, it has contributed to public debates about the legitimacy of land reform both within Brazil and beyond. But the MST is a large, complex and controversial social movement, not just from the perspective of the landowning classes, but also from the perspective of people sharing political sympathies – most notably, settlers themselves. Both the MST's opposition to emancipation (i.e. each family receiving a legally titled plot of land) and its hierarchical form of organization help to make sense of struggles that sometimes emerge internal to MST settlements, and between the MST and other organizations.

Officially, the MST's opposition to emancipation is explained by various concerns: emancipation could promote negative forms of 'individualism', settlers might sell their land and their access to state resources and representation could become curtailed. Settlers have diverse reactions to the MST's concerns, which some take to be valid but also paternalistic. Many settlers in my study suggest a more immediate reason for why the MST opposes emancipation: its financial dependence on the settlements.

In place of emancipation, MST lands are held, at least initially, in what is called the 'collective' (*coletivo*). Land collectivization is part of the MST's broader strategy to decommodify land. More broadly, the collective also refers to collective work and collectivization of financial resources. Internal conflicts become especially acute in relation to this institution, as will be seen.

The MST's organizational hierarchy is instituted through various methods of implementing 'discipline' (*disciplina*). Discipline refers both to general processes through which settlers learn to inhabit the MST hierarchy (e.g. respecting community rules and militants' orders) and specific practices involved in creating discipline. These practices may be pedagogical in nature, or involve physical instruction including forms of punishment. These latter forms, described in section 8, are a source of tension between some MST settlers and militants.

In what follows, I will explore settlers' critical engagement with questions of discipline, collectivization, and emancipation, as settlers attempt to create more just and non-hierarchical settlements. Before moving on, I would like to specify the general argument: (1) Although the MST is an explicitly socialist and, therefore, ideologically egalitarian movement, its hierarchical organization creates structural affordances through which some members of its leadership may engage in coercive and exploitative activity vis-à-vis grassroots settlers. (2) Despite occasional and unfortunate facts of reproduced exploitation, it does not follow that the MST is essentially exploitative outside of the specific institutions and relationships that its membership builds. (3) This process of institution-building can be usefully examined by looking at how settlers challenge, transform and contribute to building community institutions. This last point has implications for problems of participatory parity and voice, and suggests an empirical thread that I will follow at this point.

6.   Settler participation in land reform

Over the course of my research, I began to discern a distinct preference among rural people in this region for life on non-MST settlements, and I encountered several instances in which people compared life on MST settlements to both plantation life and slavery. By late 2010, realizing that I needed to deepen these observations, I developed a survey instrument consisting of 338 and 575 items, respectively, for landless or settled respondents. Over 2011 and 2012, my research assistant and I administered the survey to 100 respondents across Settlements 3 and 4 and Community 9. Among these respondents, 55 were men and 45 were women; 38 identified as landless; and 62 claimed some form of land ownership, of which 35 were MST settlers at Settlement 3. My research assistant had ties of kinship and friendship to people at all three communities, and, crucially, had never held any position of authority in these communities.

The survey questions presented here asked settlers to rank their preferences among the settlement types identified above, and to provide some account of their choices in a recorded interview. Instead of referencing abstract settlement types, local settlements served as cases for comparison. While this makes comparative research more complex, providing respondents with a tangible reference point helps to account for local experiences and knowledge. For this purpose, I selected Settlements 3, 4 and 8 as these corresponded to MST, squatter, and association settlement types, respectively. Table 1 presents the results in aggregate.

Items 1–4 show very strong interest in land reform generally, with interest in squatter and association-type settlements slightly higher than that in the MST. When participants hierarchize these preferences, a distinct pattern emerges: item 5 indicates a significant preference for association-type settlements over MST settlements, and items 6 and 7 indicate a very strong preference for squatter-type settlements over both MST and association settlements.

Items 8–11 suggest that squatter-type settlements hold top preference among current MST settlers, landless people, women and men. Item 12 suggests a more even spread of second-place preferences among current MST settlers. Items 13–15 suggest a strong second-place preference for association-type settlements among landless people, women

Table 1. Participation preferences.

General questions about participation in land reform (n = 100)

	Yes	No	Maybe
(1) Are you interested in participating in land reform?	94	6	–
(2) Would you be interested in entering an MST settlement?	70	28	2
(3) Would you be interested in entering an association	90	10	–
(4) Would you be interested in entering a squatter settlement?	96	4	–

Ranked participation preferences (n = 100)

	MST	Association	Squatter	No reply
(5) MST vs. association	34	**62**	–	4
(6) MST vs. squatter	13	–	**82**	5
(7) Association vs. squatter	–	19	**71**	10

First place preferences (by subcategory)

	MST	Association	Squatter
(8) Current MST members (n = 34)	7	7	**20**
(9) Current landless (n = 30)	2	2	**26**
(10) Women (n = 36)	4	3	**29**
(11) Men (n = 50)	6	9	**35**

Second place preferences (by subcategory)

	MST	Association	Squatter
(12) Current MST members	**14**	10	10
(13) Current landless	6	**21**	3
(14) Women	12	**20**	4
(15) Men	10	**27**	13

First/second place paired (by subcategory)

	MST/Assn	MST/Squat	Assn/MST	Assn/Squat	Squat/Assn	Squat/MST
(16) Current MST members	3	4	1	6	7	**13**
(17) Current landless	1	1	0	2	**20**	6
(18) Women	2	2	1	2	**18**	11
(19) Men	2	4	0	9	**25**	10

Documents vs. productivity

	Forest with document	Farm without document	No reply
(20) Would you prefer a plot of forest with a 'document' (documento) or a productive farm with no 'document'?	**83**	14	3

Note: The information presented in items 8–19 includes ranked responses for people that responded to all items 5–7. Partial responses (n = 8), no replies (n = 2) or contradictory responses (n = 4) to items 5–7 did not allow responses to be ranked. Consequently, responses for the following number of respondents could not be counted for items 8–19: current MST members (n = 1), current Landless (n = 9), women (n = 8) and men (n = 5). Values in bold indicate the highest values for each question.

and men. When first- and second-place preferences are paired, items 16–19 show the same overall preference for squatter-type settlements.

While the follow-up interviews await further analysis, my initial examination of these and other data suggests that one important factor constituting this pattern relates to land tenure and control over various legal documents that establish a family's claim to the land. In this respect, and in local estimations, families living at locations such as squatter Settlement 4 are understood to have greater tenure security than families at other settlements. Although none of the families on any of the settlement types have secured what is legally referred to as 'definitive title' (*título definitivo*) to their land,[7] the families living at Settlement 4 have been able to legally document their land claims through other means that are available to squatters. These include documents such as the 'Declaration of Ownership of Rural Property' (*Declaração de Posse de Imóvel Rural*), which can be secured through local land registries (*cartórios*) sanctioned by the state government, or the 'Certificate of Registration of Rural Property' (*Certificado de Cadastro de Imóvel Rural*), which is issued through INCRA and enables squatters to pay an annual (and nominal) 'Rural Property Tax' (*Imposto sobre a Propriedade Territorial Rural*) that generates its own paper trail. In the hierarchy of legal forms, these different legal 'documents' (*documentos*) have significant weight for squatters seeking to formalize their land claims. These forms of documentation, which are specific to individual families and individual land claims, are not institutionally available to families on the MST settlements. In the context of the region's history, where earlier generations had been dispossessed of their land because their claims went unrecognized by the state, securing various means to legally document one's land claims has become a pressing social issue (see DeVore 2014, especially chapters 3, 4, and 5). Whereas I have recorded no case of any family being expelled from Settlement 4 after nearly 17 years of residence, the potential for interpersonal conflict with MST leaders (see section 8) appears to ground a sense of insecurity among some settlers on the region's MST settlements. From the perspective of many MST settlers, it is the personal authority of militants, and not the authority of the 'document', that mediates their access to the land (see section 11).

Item 20, as such, brings the social value of 'documents' into greater relief. When asked if they would prefer owning a legally documented plot of uncultivated forest (*mata bruta*) or a fully productive farm (*roça safreira*) without legal documentation showing ownership, the vast majority of respondents opted for a documented plot of forest. Bearing in mind the steep investment of time and labor required to build a cacao farm from uncultivated forest, the document emerges as a highly salient figure.

To develop a more robust understanding of the social meanings of documents, we can examine brief excerpts from follow-up interviews with two women. One woman from Settlement 5 suggested that if someone has a plot of land 'without any document, she still doesn't have anything'. She explained the significance of having a document:

> Nobody takes [the land] because it goes directly into the person's name, and who'll take [that away]?… God is above all – and these days, under God, everything else is the document. Without a document, a person is nothing at all, and doesn't have anything either.

---

[7]Indeed, given the incredible complexity of Brazilian property law, 'definitive title' has an almost mythical status for large and small landowners alike. See Holston (2008) for an important discussion of the complexities and pitfalls of property ownership.

Another woman from one of the MST settlement in the region, who had served as a 'work coordinator' (*coordenadora*) for the local association, suggested:

> A person with a document, he – he's got freedom. And us here in the Movement [MST], we are – we have our plots (*roças*) – we have our plots – but we don't have the same freedom like a person who's really got an individual plot for him [and not others] command and un-command (*mandar e desmandar*).

Here we see some of the ways documents connect issues about land tenure, social agency and autonomy from others' potentially problematic demands. Predictably, control over land titles is a point of contention for settlers on the region's MST settlements. To understand why requires some understanding of the MST's initial appearance in the region.

7.   First contact, first conflict: plantation workers, associations, and the arrival of the MST

Around 1996, a group of rural workers formed a legal association with the intent of purchasing the plantation that later became Settlement 1. The association entered into direct negotiations with the landowner and, shortly thereafter, began negotiations with the former landowners at Settlements 2 and 3. Most association members were workers on these plantations. The association's plan was to divide each plantation into separate lots for each family, each of which would purchase their lot from the former landowners through bank financing and INCRA's institutional support. By January 1997, the plantation at Settlement 1 had been legally disappropriated and placed under INCRA's administrative guidance, while negotiations with the two other plantations were in progress.

Prior to late 1997, the MST had no representation in this part of the cacao lands. Sometime in 1997, MST leaders established ties with a local syndicate to coordinate a series of land occupations. In November 1997, the MST arrived in the region, bringing a number of hopeful settlers from neighboring municipalities. The MST's leaders planned to establish encampments at what would become Settlements 1, 2 and 3 – where the workers' association was already in negotiation. In short, the MST sought to displace the association, whose members felt they had a legitimate prior claim to purchase the plantations.

A couple of weeks after the MST arrived in the region, minutes from an association meeting suggested that the plantation workers were under pressure from MST leaders to either join the MST or abandon the plantation. Many of the workers were afraid of the MST's leaders. The meeting minutes concluded:

> [G]iven what is happening, we are going to lose everything, because we are sheep and we do not want violence ... [One worker reported that] complete control remains under the MST's yoke, and he said that [the workers] do not even have the right to speak or give an opinion. They are living as slaves of this MST. ... [Another person reported that the MST's leaders] took [two] of the old residents [from Settlement 2] and beat them up and left them tied up [to a tree].

The standoff between the two groups continued for several months until April 1998, when the plantation workers ultimately agreed to join the MST for fear of losing everything they had worked for.

Eventually, the association lost all three plantations to the MST. In an interview, a former association leader recalled what he had said to local MST leaders during a verbal confrontation with them:

'We're already here – this is ours. It's our work, three years of work [negotiating] … You're invading the landless themselves. The Landless invading the landless (*O Sem Terra* [i.e. MST] *invadino o próprio sem terra*) – [invading] the workers themselves.'

8. Instituting discipline, fear, and silence

As suggested previously, the MST's national strategy involves organizing rural (and increasingly urban – Simmons et al. 2010) workers into the rank-and-file of its command hierarchy. Such a well-coordinated organization, presumably, would be able to confront potential violence from landowners and more easily secure resources and support from the state.

Forming a disciplined rank-and-file involves different methods of instituting respect and deference to the MST's hierarchy of authority. This begins with consolidating encampments (*acampamentos*). Encampments are built near plantations targeted for occupation and consist of huts covered with iconic black plastic tarps (*lona preta*). Building an encampment transforms a legal claim into a materially tangible presence, curtailing competing claims from other groups.

In the cases of Settlements 1, 2 and 3, MST leaders faced the prior claim of the workers' association that sought to directly purchase the plantations. This created significant tension between the plantation workers and prospective MST settlers. At Settlement 1, where negotiations were more advanced, the tensions between the old plantation workers (*os velhos*) and prospective MST settlers (*os chegantes*) who took control of the plantation are still salient and a lingering source of pain.

One woman from Settlement 3, who had resided on the former plantation there, recalled the pressure she felt to join the MST:

> They made threats … They said if we didn't go to their side … we'd be expelled. There'd be evictions – there would be evictions, and so many things, so much talk … We didn't feel very happy, no – because the threats were too much.

Those making the threats, she recalled, were 'the big' (*os grandes*) movement leaders, 'these [guys] who come from outside [the region] … the big [leaders] who come from the outside'. She and her husband had been working at the plantation before the MST arrived. Her husband was 44 years old, and, at his age, it was hard to find work on plantations that preferred hiring young men. Their family depended on the income from the plantation, especially as they were raising two children, so they could not easily exit in search of work elsewhere. Since the local association had been supplanted, she figured that they had no option but to join the MST.

Her husband similarly recalled pressure that the MST militants put on the workers to either join the MST or abandon the plantation. His family had planted a vegetable garden behind the house where they lived, and this was destroyed by some more aggressive MST militants:

> We had a tomato garden … They got their hands into it and collected the tomatoes, broke the corn, pulled up the beans, and took it. I didn't say anything [but I said to myself]: 'I won't run'.

Because of pressure like this, many former workers left the plantation frustrated, and some resentfully so, as they had hoped to purchase the old plantation through the association. While many settlers were recruited in neighboring municipalities and joined the MST willingly, and some plantation workers merely followed the shifting tides of fortune, others felt they had no choice but to join, as they had no materially viable exit.

Processes of disciplining also involved subduing those outspoken personalities that came into conflict with the MST's leadership. For example, some settlers remember people who, as a punishment, were told to stand in the community square for an entire day holding up an MST flag. One settler from Settlement 6 explained that the authoritarian form of command exercised by the militants was no better than life under plantation management, while another man from Settlement 7 explained that if you crossed the leadership, you risked being told 'to leave without a right to anything'. A person from Settlement 5 elaborated on this widespread fear:

> You have to do whatever the militant wants. Either you do what they want, or they take [your stuff] and drop it out on the road … There's none of that business of complaining, no, because otherwise, they put – they take your things, bed, whatever little things you have, and put it all out on the road.

This fear is not a mere impression. On several of the settlements that were, or still are, associated with the MST, people remember individuals who had disputed the authority of some militant only to be met with coercive discipline or even expulsion. At Settlement 6, for example, Honório remembered one man who had publicly disagreed with a militant who later ordered the other settlers to remove the man's personal belongings from his house: 'they took his things, put them on top of a truck, and took them away. And when they arrived out on the road (*pista*), they tossed it all [out on the road]'.

Another case involved a woman named Izaura who lived at Settlement 5, which had also been occupied by the MST against the plantation workers' wishes after the absentee owner had ceded the land to the workers. One man recalled the situation. In a community assembly, the MST militants who had taken control instructed the workers to leave their homes for the encampment where they would 'stay under the *lona* [*preta*] just like everybody else'. The militants warned that they would remove the ceramic roofing tiles from the old plantation houses, which would force the workers to either leave the plantation or go to the encampment. Izaura, who had been a long-time resident in the house where her family lived relatively independently, was upset with these demands and she spoke out angrily. The man narrating this story recalled how Izaura's confrontation with the militants ended:

> Well, the woman confronted them. She wouldn't let them [remove the tiles from her house] and wouldn't go [under the *lona preta*] … . And well, they got into a fight, and [the militants] really went [to her house]. They took 30 people up there, took the roofing tiles off the house, and took them away.

He explained the outcome:

> [The militants] made an example [of her] for the others to see. And, well, [the workers] started to obey. [The militants] did something to give the others an incentive [to obey], so that whatever they said, the others would obey.

While actual cases of abuse like these are a quantitatively rare matter, their social psychological effects generalize in a quantitatively different manner. As stories about such events quickly circulate, singular events take on a generalized semiotic life. In Izaura's case, she was offered as a public exemplar that modeled a particular relationship between militant and settler, as a relation of command and submission, which demonstratively represented one range of possible actions and their social consequences. This transgressive action christened new participatory norms on the settlement, solidified the militants' voices, and transformed an instance of mere physical coercion into a complex social fact. This grounded the authority, if not the legitimacy, of these MST militants in experiences of anxiety and fear.

Relationships inflected by fear raise question about settlers' voices in the community meetings. One young activist – a frequent participant in MST congresses, marches and protests – reflected on this problem in community assemblies:

> Those who speak the loudest are the people from the leadership ... The [settlers] don't have complete freedom to say, 'That's wrong'. For example, [if a militant] makes a point, [a settler can't say,] 'Oh no, that's wrong, we won't accept that, no'. If [a militant] says it, you have to accept it. Even having a majority [against some proposal], you have to accept it.

This activist understood the militants as having a stronger voice, while settlers had a weaker voice that they often self-censor out of fear (e.g. the fear of expulsion) that some people living in and around these MST settlements express.

A woman at Settlement 3 elaborated on the problem of voice in forums like the assembly, suggesting that many people are 'afraid of what they say' because they are 'afraid of the threats'. In an interview marked by an occasionally hushed voice, she explained what happened to settlers who were too outspoken:

> [They] go through one of their disciplines ... [They] are punished ... [The militants] keeps perturbing them ... They don't let that person participate in things ... They don't like when the person speaks the truth, and they keep hammering [those people] – [The leaders say,] 'No!' They don't want to help that person, they don't want to let that person grow ... Often times they make threats: 'Oh, you did this? So there's none of that [for you]. We won't free up this [resource] for you because you are like this, this and that.'

She concluded that 'in the MST, there are good people and bad people', but if someone with power does something wrong, 'oftentimes [people] don't say anything because they're afraid. They're afraid to speak'. She characterized a prevalent attitude on the settlements: '"I don't know anything, I didn't see anything"'.

These observations raise critical questions about participatory parity on these MST settlements where power disparities have been institutionalized. These observations also raise important methodological questions for researchers who need to attend to their own position in the settlements, as perceived alignment with, or dependence upon, local leaders may contribute to problems of self-censoring among settlers in the research process, thereby further reproducing potential disparities of voice and representation. I address this problem further in section 14.

9. Inequality, privilege, and the extraction of wealth

While the MST's land occupations largely depend on volunteer labor from prospective settlers, and various kinds of support from the federal government, such as food baskets (*cestas básicas*) that are sent to encampments, one of the ways in which the MST's larger organization finances itself is through monetary contributions that settlers are obligated to pay to the organization.[8] Settlers refer to these payments as 'percentages'

---

[8]Gregory Duff Morton (personal communication, March 4, 2014) suggests that in Bahia's arid interior, such contributions proceed more voluntarily, partly because settlers have fewer resources from which to draw.

(*percentagens*) or 'discounts' (*descontos*), invoking analogies to forms of extraction that they are familiar with through plantation work and sharecropping. The MST's public position on these contributions is that they are crucial to advancing the broader campaign for land reform, and that those families who have already been settled have a duty to contribute financially to the cause. Such contributions may come from agricultural production. For example, each family is obligated to pay between 10 and 20 percent of their rubber harvest, which is automatically deducted at the point of sale, as the rubber harvest is sold collectively. Other financial contributions come from percentages that settlers pay out of agricultural loans they receive through federal agricultural credit programs. Practices such as these are well known in Brazil (Ondetti 2008, 187–188; Wolford 2010a, 173).

While many settlers acknowledge the importance of advancing the cause of land reform, few trust that the MST's local and regional leaders use the money in good faith. One important factor contributing to this sense of discontent, widely shared by settlers in MST Settlements 1, 2 and 3, are memories of significant financial hardship during the first years of collectivized work, before settlers forced the division of the collective (see section 10). Many recalled receiving a pittance after each week of collective work. Two settlers from Settlements 1 and 2 reported that they had kept record of the harvests – presumably a habit from their days as plantation workers and sharecroppers. According to their own calculations, the settlers' share of the harvest was not adding up. Even if no one had thought to keep records, it is plausible that settlers had developed a practical sense for returns on labor after years of plantation work, and they knew when they were being cheated. Indeed, many of the settlers seemed to suspect that they should have been receiving more than they did. One man explained: 'When our day wages should have been more than a [minimum] salary, it only came out to less than a salary'. In the end, 'nobody knew' where the money went, and there was a widespread feeling that the leadership 'lied, they lied'.

These were not mere intuitions, however. Many settlers perceive militants as occupying a privileged position, not only in terms of authority, but also in terms of material well-being. One young man, who worked as a sharecropper for a militant who eventually 'retired' at Settlement 3, explained that the militant sold his rubber harvest separately from the collective. Unlike the settlers, the militant did not pay any percentage of his production to the MST's organization. At Settlement 3, furthermore, those settlers who were more closely integrated into the MST leadership during the initial occupation ended up living in the nicer buildings located near the old plantation headquarters. None of them had settled down in the 'avenue' (*avenida*) that used to house plantation workers.

One afternoon while I was working with some families living at Settlement 4, one of the people present pointed to an MST militant on the other side of the river. He was there with a group of young men who were helping him wash his new car. The person observing this commented: 'Yesterday there were five people washing his motorcycle, today there are four people washing his car'. Another person at Settlement 7, referring to the division between the settlers and the militants, suggested that 'those who work the hardest [i.e. the settlers] aren't the owners, those who work least [i.e. the militants] are the owners'. Observations like these hold deep resonances with the inequalities that constituted plantation life.

10. Counting the trees, dividing the collective

As a result of their experiences of inequality and financial hardship in the collective, some of the more courageous people living in the MST Settlements 1, 2 and 3 began to secretly organize other settlers who would be willing to confront the MST's leadership and force the division of the collective. These were tense moments between MST settlers and their local leaders, on the one hand, and the MST's regional leadership, on the other.

One elderly man living at Settlement 1, who had spent decades as a plantation laborer, had eagerly joined the MST in the 1990s, and traveled to the region on the promise of land. He had been active in the local settlement association, having served various leadership roles including association president. He explained that, with time, he felt deeply betrayed by the MST's promise of land, and he was particularly frustrated with his experiences in the collective. He came up with a plan to count and divide all of the rubber and cacao trees from the old plantation, and distribute them equally among the families:

> And then one day I made a decision … I invited the people … [I invited] the [other] settlers [to talk]. Then they supported [the plan, and said,] 'Yes'. And I already knew how we would go about it, so, I say – I say – I took a position like this:
> 'If I can find 10 companions here, really strong ones, who know what we are going to do [and can support it in the assembly], then I'll step forward.'

He expected that most settlers would support his proposal to divide the collective, but that many would be too afraid to speak up in the assembly alone. Therefore, he went about secretly organizing the most courageous settlers, who agreed to support him against the MST's regional leaders. When their group was established, they called the regional leaders to an assembly:

> They came and we sat down and argued (*discutiu*). Oh! It was an ugly discussion, see? Then the 10 … unfortunately [for the MST's leaders], the 10 really guaranteed [their word] and supported me on this point, and we put our foot down.

After a contentious debate, the MST's regional leadership eventually ceded to the settlers' demands. The settlers counted all of the rubber and cacao trees on the old plantation, divided them into equal lots, and held a lottery to distribute them among the families.

Identical processes eventually occurred on Settlements 2 and 3 around that same period. In each case, the collective fell apart because of settlers' experiences with tangible and widespread injustice on the settlements, and not because of abstractions like 'individualism' or 'capitalism'. While it was not always clear who exactly was taking their money or where it was going, it was clear enough to the settlers that they were being cheated. Their initial fear of confronting this situation was compounded given that they knew the MST leaders had a vested interest in the settlements' financial contributions. They had seen, in earlier years, the consequences faced by lone individuals like Izaura who questioned the militants' authority. Dividing the collective through clandestine organizing – making a collective claim by creating a new 'we' and a socially distributed voice – was one means of mitigating the hardships that settlers experienced, but dividing the trees did not solve all of their problems.

11. Titles, documents, and land tenure insecurity

In early 2011, a young MST activist related an encounter he had witnessed at the MST's regional headquarters. A woman, who lived on an MST settlement in the region, had come to inquire if the settlers would eventually receive titles for their land. Dismayed with what ensued, the young activist recalled that the militant 'got angry' and yelled at the woman: 'You guys can forget about that!' This brief scenario succinctly illustrates an ongoing source of tension between many settlers and some members of the MST's regional leadership, and even among some militants themselves. One self-identified militant, for example, who had been a base organizer with the MST for over a decade, evoked a theme of insecurity related to land titles on the settlement. He explained that the settlement technically belonged to the government, and that, at any time, some official could 'show up and wave their hand [in the air] and send me [away]'.

One man at Settlement 3 gave a series of analogies to describe settlers' land tenure. He variously compared their landholdings to those of renters,[9] and also described their holdings in terms of an incomplete transaction between the settlers and the government. Throughout the interview, he reiterated that the settlers wished to complete the transaction, pay for the land and finally receive 'documents' (*documentos*), or title, for the land.

I asked him why, if this was such a crucial matter for the settlers, they did not pursue the land titles. He explained, emphatically, that it was risky business to make such claims:

Settler: No! *Who's that crazy*, man!?
Jonathan: No? But who – is there someone who's against that?
Settler: *Who're against it, against it*!
Jonathan: But who?
Settler: Everybody – a bunch [of people] in the Movement [MST].

He explained that the MST was financially dependent on the settlements, and identified a conflict of interest between MST leaders and settlers. Providing settlers with land titles would afford them more say in terms of what, and how much, they contributed to the MST or other social movements. Conversely, and in inverse proportion, the MST leaders would exert less financial, social and political control over the settlements.

For the settlers, however, control over land titles would mean a more definitive exit from the vestiges of plantation life. This observation was brought home as the person I was interviewing drew these reflections on documents together with a description of the militants:

Those who don't have courage to work, [who] want to live all mixed together (*no bolo*), working in the collective … Eating and drinking and sleeping and walking around like a militant – walking around the world [running] after the government. No, we don't want that, nobody wants it.

By drawing on tropes typically applied to plantation managers – who appear to him as idle, well fed, living from others' sweat – the speaker invites further analogies between plantation life and life on the MST settlements. Indeed, rural workers can gain access to land on settlements and plantations alike, even if, as sharecroppers, the share of the harvest they keep is significantly less. The problem that settlers identify with respect to control over land titles, however, is less an argument about percentages and proportions.[10] What seems to matter to settlers is that control over the harvest is mediated by uncertain and sometimes precarious relationships with movement leaders, who also appear to have unwarranted say over who gets access to land, who can stay, and who must go. Insecurity in personal relationships – whether patrons, plantation managers or movement leaders – is connected to insecure access to land. This problem is crystallized in access to legal documents like land titles.

---

[9]Wolford (2010a, 199) presents similar evidence.
[10]While there is a clear quantitative shift in the percentage that settlers keep, settlers' observations highlight an insufficient *qualitative* shift in their control over the means of production.

12. Loan documents, agricultural credit, and coercion

Struggles to control other kinds of documents emerged when federal planting credit became available to families living at Settlements 1, 2 and 3. In early 2009, an independent agricultural cooperative approached settlers with a proposal to cultivate and market non-traditional agricultural products through the cooperative. Each family that opted to participate would receive a federal project loan totaling R$ 21,500. A total of 60 families between Settlements 1 and 3 eventually agreed to participate, eager to improve their lot.

Although the MST's regional leadership was involved neither in elaborating nor in implementing the project, the MST's leaders had to give final approval of the project. For settlers to receive approval, participants would have to pay 7.3 percent of their loan, or R$ 1500 per family, directly to the MST, or else they would be unable to participate. The MST's regional leaders explained, one settler recalled, that the money would be used to pay MST agronomists or 'technicians' (*técnicos*) to help implement the project, although, the settler noted, none ever materialized in the end.

Nobody was happy about this. The families at Settlement 2 opted out altogether, while families at Settlement 3 reluctantly agreed to the terms. The families at Settlement 1 insisted on participating, but refused to pay such a large sum of money to the MST. In response, the MST's regional leadership refused to approve the project by not releasing a crucial document that was necessary for loan approval, referred to as the 'DAP'.[11] After a stalemate that dragged on over a period of several months, the families at Settlement 1 realized that they had no choice in the matter.[12] Either they had to give up their hopes of participating in the project, or else pay the 7.3 percent to the MST. The settlers eventually agreed to the terms. The combined amounted that the MST took from both settlements amounted to R$ 90,000. To get a sense of the magnitude, this was equivalent to nearly 16 years of plantation work, or three months of plantation labor per family.[13]

Similar occurrences involving other planting projects occurred on Settlements 6 and 7. In the former case, settlers recalled figures that would have amounted to between R$ 492,000 and R$ 738,000 of federal loan money that was transferred from their settlement to the MST organization in the late 1990s. Assuming 1999 wage levels,[14] the magnitude of this payment would have been equivalent to between 300 and 450 years of plantation work. Even a fraction of these amounts would have been enough to spark the small rebellion that occurred at Settlement 6. The precise relationships of control and coercion in these situations are not entirely clear, as settlers struggle to decipher their relationships to the larger MST organization that are often opaque. What was clear enough to settlers, though, was that they felt coerced between their desire to participate in these credit programs, and the inflammatory conditions that were imposed upon their participation.

One man living at Settlement 2 concluded that the interest of the MST's leadership was 'just to take, just to take', and many settlers infer that some of the militants (and perhaps some other officials) misappropriated the money that they extracted from the settlers' harvests and loans. What permits these problematic relations to emerge, however, was not something they had to infer: settlers do not have control over certain legal documents.

---

[11]*Declaração de Aptidão ao Pronaf.*
[12]It is unclear how the MST's regional leadership was able to intervene here. Ondetti (2008, 187–88) seems to suggest an institutional understanding between movement leaders and INCRA officials more than anything else.
[13]In 2009, Brazil's minimum monthly salary was R$ 465.
[14]In 1999, Brazil's minimum monthly salary was R$ 136.

Such control over documents would give settlers greater control over their livelihoods, and the ability to exit relationships that they found unacceptable *without* sacrificing their access to various productive resources, such as land or agricultural credit. It is this logic, grounded in important historical experiences, that helps account for the strong desire to participate in non-MST land reform settlements.

13.   Friendship and friends of the MST

> We need very strong ears to hear ourselves judged frankly; and because there are so few who can endure frank criticism without being stung by it, those who venture to criticize us perform a remarkable act of friendship; for to undertake to wound and offend a man for his own good is to have a healthy love for him. (Michel de Montaigne 1958[1587], 825)

People in the region often use the idiom of 'friendship' (*amizade*) to evaluate various kinds and qualities of relationship. Unlike patriarchal, managerial or other relationships grounded in various kinds of coercive authority, the ideal and practice of friendship lacks this important element of coercion. While patron-client relationships are often cast in terms of the idiom of 'friendship' (see Leal 1977[1949], 1–25), people in the region make epistemic distinctions between 'true' (*verdadeiro*) and 'false' (*falso*) friends. The mode of friendship operative in patron-client relationships is grounded in an asymmetrical distribution of social resources, such as land and bureaucratic power, which gives some parties the power to potentially disenfranchise others. This coercive threat stabilizes such relationships, at least in some measure, but it does not make for much of a friendship.

For many people living on and around the area's MST settlements, the structure of life there reproduces too many of the conditions that they had become familiar with on the plantations, whether various 'discounts', different forms of insecurity (e.g. fear of being sent away) or dependence on people in positions of authority whose favor one had to carefully maintain. As a consequence, historically salient idioms of plantation life and slavery continue to resonate with local understandings of the MST. For those who still participate in the MST, but have found means of mitigating some of the asymmetries that underlie persistent inequalities (e.g. by dividing the collective), genuine forms of 'friendship' may emerge in settlers' relationships with one another and with the MST's leadership. Even for people at Settlements 5 and 6, which expelled the MST altogether, their stance is not one of unbridled enmity, but of potential friendship. One man living at Settlement 5 suggested that while he would never work with the MST again, he could still remain friends with the organization. The ability to be friends with the MST, however, meant creating conditions under which no one party had the power to illegitimately disenfranchise the other, and each could support the other in shared work and political struggle. When collective action comes about under circumstances where electivity and mutual recognition can emerge – circumstances approaching participatory parity – the meaning and meaningfulness of 'the collective' and collective action are fundamentally transformed.

For Honório and other families at Settlement 6, their exit from the MST was a moment of conflict that was necessary for them to regain the freedom that gave their quest for social equality substance. Their exit from the MST was not, however, a disavowal of the broader struggle for human dignity that, to be certain, animates the vast majority of people who support the MST. Indeed, Honório himself continues to feel kinship with the MST: 'I left the MST, but I didn't stop helping … in some things, in some visions … If I have to go under a tarp (*lona*) to hold onto a spot for somebody that doesn't have land, I'll help. I'll go, too'.

Honório's sense of freedom does not prevent him from recognizing the importance of collective action, and he recognizes that in the struggle for equality, one cannot ignore the needs of others or the importance of reciprocity. In this case, however, attending to the needs of others is an activity that is grounded in something more akin to a virtue rather than a coercive duty. In the end, Honório draws an important distinction with respect to the MST: '[T]he MST has leaders, right? And those guilty of some wrongdoing is not the MST [in general], it's some leaders in the MST, you see?'

Honório is acknowledging here that movement leaders are never mere extensions of some official movement ideology, however well or poorly conceived. The particular instantiations of the MST that Honório and his companions faced were simply not tolerable to them. More importantly, though, the collective institutions that empowered the MST's militants reproduced experiences of humiliation and hardship that were not so different from those they faced on the plantations. In short, the structure of community under the MST was simply not adequate to their sense of dignity; the way of life that the MST imposed there was not one the settlers could recognize as their own, and this was why the MST had to go.

14. Directions: socialist politics, voice, and the institution of participatory parity

This contribution is meant to foster critical reflection on the role of authority and authoritarianism in socialist politics, and should be of special interest to MST activists as they craft organizing strategies for the coming years. In sections 5, 8 and 9, I gestured toward the MST's strategy as a national political actor – which insists on non-emancipation of settlements, financial contributions and internal discipline – and I explored difficulties that authoritarian instances of this strategy posed for some settlers. While it seems doubtful that the MST could achieve its national political goals in the absence of organizational discipline, if a central goal is to support families seeking secure access to land – and in a way that attends to their sense of justice and dignity – then the evidence presented here, especially in section 6, suggests that in its political exchange with the rural poor, the MST may be expending more political capital than it is creating. This can be seen most poignantly among those settlers who willingly joined the MST with great hope, but later found that the MST's leaders had failed to make good on their promise for land, as was seen in section 10. Finally, this paper highlights other viable pathways that people have taken in the search for land, and draws attention to moments when the MST may have actually undermined those efforts, as seen in section 7.

When examining the egalitarian prospects of social movements and social life more broadly, we can benefit by paying close attention to the ways in which historically particular norms of participation are established in time and through specific interactions. Rather than supposing that institutions like the 'assembly' unproblematically foster equal participation, researchers can benefit by exploring moments of social interaction that christen actual forms of participation, such as those described in section 8. These christening events shaped the prospects for actual participation and voice, and did so in a way that indicated some measure of historical continuity with plantation life and, as Brenneisen (2002) suggests, patron-client relations.

Exploring local histories through which participatory norms are structured can highlight methodological hazards that may emerge from institutionalized asymmetries of representation and voice along the militant-settler continuum. Researchers who rely too heavily upon movement leaders for collecting data, or as mediators to interactions with settlers, may be unintentionally aligned into roles and commitments that are not readily apparent, and this may contribute to self-censoring among settlers. In this way, researchers run the risk of colluding (unwittingly) in the reproduction of representational inequalities. Deliberate empirical attention to these issues can help to mitigate this potential problem. This observation might be considered in other research contexts, as well, whether multinational corporations, non-governmental organizations (NGOs), indigenous organizations, and so forth.

If the christening of participatory norms appeared, in Izaura's and other cases, as a transgression toward the institution of some norm of interaction and participation, this transgressive-institutive character can help us understand the extent to which participation and voice are institutionally conditioned.[15] For Fraser, as suggested earlier, participatory parity and voice require the fulfillment of 'intersubjective' and 'objective' conditions. The present context is one in which neither of these conditions appears fully secure,[16] as can be seen in the reproduction of status distinctions, reproduced at various orders of the militant-settler relationship, and differential control over resources that is part of that relationship. Yet, although these conditions were not fully secured at Settlements 1, 2 and 3, the families living there succeeded, at least in some measure, in *creating* a participatory voice, as seen in section 10. Their voice was not a result of their frictionless participation in the assembly, as an equally available medium for expressing their voices, since the assembly appeared to privilege militants' voices. Rather, it was through creating a distributed voice, and a new 'we' through which they could speak, that settlers were able to stake a legislative claim over their own lives. They were able to create this collective voice through clandestine dialogue in the shadow of the assembly. Through dialogue – by *talking* with one another – the former plantation workers and newly arrived settlers at Settlement 1, who had previously been pitted against one another, were able to acknowledge commonalities in their standing and standpoints. In this way, they *created* an intersubjective condition for joint action, committing to appear *together* in the assembly, where they made a joint claim (subsequently recognized by the MST's regional leaders) to the trees and their own work: the material (or Fraser's 'objective') condition for subsequent life.[17]

In this sense, the emergence of a participatory voice proceeded hand in hand with the settlers coming to recognize one another in a new way, and this was achieved through hazarding the expression of voice itself. While their claims and voices proved to be anything but indefeasible, as seen in the failure of Izaura's and other collective claims in sections 8 and 12, these instances help to clarify the way in which voice emerges and becomes stabilized precisely through hazarding the *expression* of voice, and hoping that it will be acknowledged by another. In this sense, participatory voice, at least at some stages of its historical emergence, does not have an external or prior condition. Rather than being understood as institutionally *conditioned*, such a participatory voice might also be understood as part of the emergent process of *instituting* conditions.

15. Conclusion

Settlers' rejection of collective practices introduced by the MST is not a rejection of collective action, generally. The egalitarian ideals of generosity, sharing, and mutual aid are by no

---

[15] At this point, I have been broadly influenced by J.M. Bernstein's publicly available lectures on G.W. F. Hegel's *Phenomenology of spirit*, which develop a theme of what Bernstein elsewhere calls the 'transgressive-legislative deed' (Bernstein 2003, 421).

[16] This would appear to forestall the operation of what Fraser calls 'cross-redressing' (2001, 83).

[17] Tracking this particular history helps to navigate the complex debate between Nancy Fraser's 'perspectival dualism' and Axel Honneth's 'normative monism' (see Fraser and Honneth 2001), since what Fraser understands as an 'objective' condition for participatory parity appears to be at least initially grounded in (or emerging from) various relationships of intersubjective recognition.

means foreign to people in Bahia's cacao lands (see DeVore 2014, especially chapter 3). Such practices are voluntary and grounded in mutual trust and respect. If conditions grounding collective action fail, then the parties may freely exit the relationship without fear of any one party disenfranchising another. It was precisely the coercive dimension of some MST practices that was offensive to many settlers' sense of what makes for a good life. This is what motivated some settlers to exit the MST altogether, and others to dismantle the collective and create institutions through which they could recognize and affirm one another. Although settlers still do not control vital and sought-after documents – such as the DAP or the almost mythical 'definitive' land title – they continue to make the MST their own.

In *The German ideology*, Karl Marx famously wrote: 'Only within the community has each individual the means of cultivating his gifts in all directions; hence personal freedom becomes possible only within the community' (1998[1846], 86). He goes on to observe that community has often taken on 'illusory' forms in which personal freedom is only possible for the minority owning class, while the majority is excluded, individuated and cut off from real community. In short, the forms that 'individuality' assumes hinge, at least partly, upon the forms of personhood available through particular historical communities. The form that community ought to take – and how such communities could be brought about – informed internal disputes that occurred in the nineteenth century among major socialist figures such as Bakunin and Marx. These questions remain with us today, as the conflict in this ethnographic setting is precisely a conflict over the form of community. Some observers might imagine that the division of the collective and the desire for land titles is derived from an abstract notion of liberal 'individualism' imposed upon people from without, whether by a sweeping appeal to 'neoliberalism', 'capitalism' or some other abstraction. But the notion of 'individuality' in the cacao lands is anything but an abstract or universalizing notion. Whatever else it may be – or may become – it is the result of a long and persistent history of social domination that has recurred in unexpected ways. People's efforts to recover positive notions of 'individuality' are part of a longer process of struggling to wrest dignity from the multiple bonds that subsume it, while creating forms of life together through which individuated 'I's can be realized as affirming 'we's.

Acknowledgements

The author would like to express sincere thanks to José Henrique Bortoluci, Aaron Ferris, Webb Keane, Michael Lempert, Bruce Mannheim, Barbra A. Meek, Gregory Duff Morton, Susan Paulson, Rebecca Tarlau and two anonymous reviewers for critical comments and conversations about earlier versions of this paper. He would also like to express his gratitude to the team at Taylor & Francis for their skill, thoroughness and patience in the process of transforming the original manuscript into its present form.

Funding

Fieldwork for this contribution was supported the following institutions: National Science Foundation, Doctoral Dissertation Improvement Grants (2009); Fulbright-Hays, Doctoral Dissertation Research Abroad (2009); American Philosophical Society, Lewis and Clark Fund (2010); Rackham Graduate School, University of Michigan, International Research Award (2010).

References

Bernstein, J.M. 2003. Love and law: Hegel's critique of morality. *Social Research* 70, no. 2: 393–432.

Brenneisen, E. 2002. *Relações de poder, dominação e resistência: o MST e os assentamentos rurais.* Cascavel: Edunioeste.

Caldas, M.M., and S. Perz. 2013. Agro-terrorism? The causes and consequences of the appearance of witch's broom disease in cocoa plantations of southern Bahia, Brazil. *Geoforum* 47, no. 1: 147–157.

Caldeira, R. 2009. The failed marriage between women and the Landless People's Movement (MST) in Brazil. *Journal of International Women's Studies* 10, no. 4: 237–258.

DeVore, J. 2014. Cultivating hope: Struggles for land, equality, and recognition in the cacao lands of southern Bahia, Brazil. PhD. Thesis. University of Michigan.

Feliciano, C.A. 2006. *Movimento camponês rebelde: a reforma agrária no Brasil.* São Paulo, SP: Editora Contexto.

Fernandes, B.M., L. Servolo de Medeiros, and M. Ignez Paulilo, eds. 2009. *Lutas camponesas contemporâneas: condições, dilemas e conquistas.* Vols. 1–2. São Paulo, SP: Editora UNESP.

Fraser, N. 2001. Social justice in the age of identity politics: Redistribution, recognition, and partici-pation. In *Redistribution or recognition? A political-philosophical exchange*, ed. N. Fraser and A. Honneth, 7–109. London; New York: Verso.

Fraser, N., and A. Honneth. 2001. *Redistribution or recognition? A political-philosophical exchange.* London; New York: Verso.

Hirschman, A.O. 1970. *Exit, voice, and loyalty: Responses to decline in firms, organizations, and states.* Cambridge, MA: Harvard University Press.

Hobson, B. 1990. No exit, no voice: Women's economic dependency and the welfare state. *Acta Sociologica* 33, no. 3: 235–50.

Holston, J. 2008. *Insurgent citizenship: Disjunctions of democracy and modernity in Brazil.* Princeton, NJ: Princeton University Press.

Leal, V.N. 1977[1949]. *Coronelismo: The municipality and representative government in Brazil.* Cambridge, England: Cambridge University Press.

Leeds, A. 1957. Economic cycles in Brazil: The persistence of a total-culture pattern: Cacao and other cases. PhD. Thesis. Columbia University.

Mahony, M.A. 1996. The world cacao made: Society, politics, and history in southern Bahia, Brazil, 1822–1919. PhD. Thesis. Yale University.

Marx, K. 1998[1846]. *The German ideology.* New York: Prometheus Books.

Montaigne, Michel de. 1958[1587]. Of experience. In *The complete essays of Montaigne*, trans. D.M. Frame, 815–857. Stanford, CA: Stanford University Press.

Ondetti, G.A. 2008. *Land, protest, and politics: The landless movement and the struggle for agrarian reform in Brazil.* University Park, PA: The Pennsylvania State University Press.

Paulson, S., and J. DeVore. 2006. 'Feeding the nation' and 'protecting the watershed': Forces and ideas influencing production strategies in a Brazilian agricultural community. *Culture and Agriculture* 28, no. 1: 32–44.

Simmons, C., R. Walker, S. Perz, S. Aldrich, M. Caldas, R. Pereira, F. Leite, L.C. Fernandes, and E. Arima. 2010. Doing it for themselves: Direct action land reform in the Brazilian Amazon. *World Development* 38, no. 3: 429–444.

Welch, C. 2009. Camponeses: Brazil's peasant movement in historical perspective (1946–2004). *Latin American Perspectives* 36, no. 4: 126–155.

Wright, A.L. 1976. Market, land, and class: Southern Bahia, Brazil, 1890–1942. PhD. Thesis. University of Michigan.

Wright, A., and W. Wolford. 2003. *To inherit the earth: The landless movement and the struggle for a new Brazil.* Oakland, CA: Food First Publications.

Wolford, W. 2010a. *This land is ours now: Social mobilization and the meanings of land in Brazil.* Durham, NC: Duke University Press.

Wolford, W. 2010b. Participatory democracy by default: Land reform, social movements and the state in Brazil. *Journal of Peasant Studies* 37, no. 1: 91–109.

The Brazilian quilombo: 'race', community and land in space and time

Ilka Boaventura Leite

More than a century after the abolition of slavery in Brazil, the term 'quilombo' continues to evolve new meanings, not all of them associated with its common definition as a runaway slave community. In this article, I discuss the significance of quilombo in its diverse social, political and historical contexts, demonstrating how changes in the uses and meanings of the term reveal broader trans-historical, juridical, political and metaphorical processes. I argue that quilombola communities should not be conceptualized as a racial category, but rather as a system of social organization and a right. Specifically, I show how the term quilombola is currently a way actors identify with Afro-descendants in order to achieve political recognition. I also describe how contemporary practices involving quilombos reveal historical tensions over land conflicts between historically marginalized rural black communities, private interests and governmental authority. I draw on evidence from field research in southern Brazil to illustrate my understanding of how quilombos work.

Introduction: quilombos in the development of a multicultural Brazil

In the Brazilian national discourse, 'quilombo' is commonly used to refer to a runaway slave community. Currently, the term no longer refers only to this historical process. The increasing prevalence of the word quilombo in Brazilian society marks subtle changes in Brazilians' perceptions of their identities, which have resulted from citizens considering their society less ethnically homogeneous and more multicultural. Brazilians are embracing some kind of African ancestry, often doing so by demanding particular public policies. Statistics from the last two decades of census data published by the Brazilian Institute of Geography and Statistics (IBGE) reveal this trend. In fact, the IBGE has begun to adopt the category of 'self-identification' as the basis for its categorization of 'race/color'.

As Alfredo Wagner B. de Almeida (2008) wrote, claims of quilombo identity are more nuanced. They arise, above all else, from contextually specific, local notions of land access and natural resource use. In addition to context, individuals who adopt the quilombo identity consider its importance connected to struggles over land acquisition experienced by groups in order to find a place to live and reproduce their livelihood. Agriculture censuses from the last 30 years reveal how certain rural identity claims are often connected – almost necessarily – to conceptions of territory. Groups such as foresters, fisherman, peasants – and quilombolas[1] – are connected to local meanings and corresponding territorial designators.

My central argument in this article is that although 'quilombo' and its various meanings have great importance in the struggle for land rights in Brazil today, the term cannot be analyzed solely as connected to territory, an instrumentalist interest in property rights or race. Certain contemporary scholars adopt such interpretations, attempting to show that the quilombo of today is something different from prior understandings (Arruti 2006). In contrast to this theorist, I argue that a necessary point of departure to understand the nature of quilombo identity is to relate it to conceptions of Brazilian nationalism. Over the last 20 years, we have witnessed the meanings linked to the term quilombo changing in response to the creation of a multi-ethnic and multicultural nation. Quilombo, initially regarded as a form of political resistance against slavery and the colonial regime, has become a form of social organization and ethnic category, and a legal category found in public policy.

What I highlight in this contribution is that the various meanings of quilombo, including its appearance in Angola during Portuguese colonization, reinforce the organizational sense and struggle for autonomy that are reconfirmed and highlighted in current struggles for territory. It is also important to recognize the historical, socio-anthropological and political presence of Africans and their descendants in nation-building, and their implications for understanding the development of Brazilian nationalism. It is not my purpose to enter into the history of quilombo formation in Brazil, even though I make extensive use of many studies in anthropology, history, geography, sociology and other fields dedicated to the subject.[2] My intention is to highlight the quilombo's multiple meanings, while demonstrating its importance as a metaphor for Afro-Brazilian identity. I argue, however, that quilombola communities should not be determined as a racial category, but rather as a system of social organization and a right. In addition, I show over the course of this paper how the term quilombo is a way for actors to identify with Afro-descendants in order to achieve political recognition.

I examine the current struggles of quilombola communities in Brazil through an analysis of the legal trajectory of the term 'quilombo', using as a reference my ethnographic research in the state of Rio Grande do Sul since the 1990s. I have coordinated several research projects in this region focused on black territories and quilombos, through the Nucleus of Studies of Identity and Inter-ethnic Relations (NUER) at the Federal University of Santa Catarina. The research that I conducted in the Municipality of Mostardas in Rio Grande do Sul, in the quilombola community of Casca, began in the 1990s as a part of the Plural-Ethnic and Intolerance Project, with the support of the Ford Foundation and National Council for Scientific and Technological Development (CNPq), the latter a Brazilian governmental foundation. In 1998, I coordinated a NUER investigation about the socio-anthropological context of Afro-descendants of Casca. Since then, my students, colleagues and I have completed a variety of archival and field research in this area, under the purview of NUER.

At the end of 2002, NUER published the first edition of the book *O Legado do Testamento: a Comunidade de Casca em Perícia* [*Legacy of testimonies: experts discuss the*

---

[1] The corresponding words in Portuguese are: *castanheiros, ribeirinhos, faxinais, quilombolas*. The translations above fail to capture their meaning in Portuguese. The territorial designations for each group are the following: *'terras soltas'*, *'terras de santos'*, *'terras de preto'*, *'terras de parentes'*. Loosely translated, respectively, these designations are 'land of the free', 'holy land', 'land of the blacks' and 'land of parents'.

[2] In anthropology, for example, these studies include Silva (1996), Almeida (2005, 2008), O'Dwyer (1992, 2002, 2012), Arruti (2006).

community of Casca]. I participated in all of the public hearings that took place in this community in 2005, when the National Institute of Colonization and Agrarian Reform (INCRA) requested a new investigation to clarify some doubts about the territorial limits of Casca. This new investigation took place over a period of one year. From this investigation, in 2006, a supplementary report was published in the Third Informational Bulletin of NUER. This report served as the basis for the territorial regularization of Casca through INCRA, and initiated a series of actions that resulted in the official recognition of this land as the Community Association of Quitéria, which is currently the judicial representation of the quilombola community of Casca. The community of Casca is the first quilombola community to be recognized in the Southern region of Brazil and serves as the basis of the case study in this essay.

The meanings, practices and controversies concerning quilombos

The term quilombo has multiple meanings and uses. In 1994, the Palmares Cultural Foundation, a federal agency that is part of the Ministry of Culture, defined quilombos as 'any Black rural community composed of descendants of slaves, who survive through subsistence agriculture, with cultural manifestations strongly linked to the past' (FCP 2013). Later, the concept was widened to include urban areas, as well as other communities that claim an African cultural legacy. The term has shifted from being a form of opposition to the regime of slavery, to signifying the enjoyment of full citizenship through the inclusion of African descendants in land tenure regularization, housing, health, education and cultural policies. The struggle for Quilombo recognition has become a political project intended to change the present-day situation of African descendants in Brazil.

Elsewhere, I have developed three analytical categories to differentiate the various meanings of the term quilombo: the trans-historical, the juridical–formal and the post-utopian (Leite 1995). The trans-historical quilombo refers to the memories of African-descendants of past struggles against colonial exploitation, and also the relationship between these struggles and those of other African diasporas around the world. The juridical–formal quilombo defines a quilombo as a right, related to affirmative action and compensation policies. The post-utopian quilombo seeks to overcome the nostalgic view of a quilombo related to a colonial past of slavery and oppression, instead promoting a heroic understanding rooted in a radical transformation of society. The post-utopian quilombo represents a deconstruction of color and race as a criterion of exclusion, highlighting the quilombo as a human right (the right to life, education, health and housing for Afro-Brazilian populations).

The lineage of the word quilombo can be traced back to 'kilombo', a word taken from the kimbundo language, the second most widely spoken Bantu language in Angola. The word 'kilombo' has a double meaning – one is toponymic, that is, related to the region, and the other is ideological. War campaigns, more or less permanent, were named 'kilombo', as well as fairs and markets in Kasanji, Mpungo-a-Ndongo, Matamba and Kongo (Parreira 1996).[3] Another appearance of the word quilombo is in the Ultramarine

---

[3]In a recent trip to Angola, I came across a newly created district, close to Universidade Agostinho Neto, named 'Kilamba'. I found that 'kilamba' was the codename of the war leader Agostinho Neto, who understood the term as representative of courage and military leadership. The famous Portuguese historian Cadornega, based in Angola in the seventeenth century, described and made a drawing of a 'quilumbo', a ritual with fire for the initiation of young warriors. From the colonial period to today, this word has been common in Brazil, used to name neighborhoods, city streets.

Colonial Legislation of 1740, which was the colonial law during the period of Portuguese colonialism. Here, quilombo is defined negatively, as 'any meeting of more than five Blacks'. This definition represents the real and everyday potential of slave rebellion during this period, and the fear that the Portuguese had of any gathering of African slaves under the colonial system.

Resistance becomes part of the term's meaning in late nineteenth-century Brazil. In this period, the term 'quilombo' was understood as a practice and discourse pressing for social change and the transformation of Brazilian society. For example, concepts such as 'abolitionist quilombo' and 'disruptive quilombo' are analyzed by Silva (2003, 11–18) to describe the final stages of crises during the period of slavery. Similarly, the 'Brazilian Black Front' (FNB), a social movement that emerged in the 1930s, re-appropriated the idea of the quilombo to denounce both the ideology of whitening and the marginalization of Black people during the Republican project of modernization. In 1948, following the lead of the FNB, Abdias do Nascimento and the Black Experimental Theatre (TEN) group founded the journal *Quilombo*. In the first editorial article of this journal, Nascimento (1948) writes that, 'the black population rejects debasing piety and philanthropy, and instead struggles for the right to rights'. The newspaper edited by the TEN circulated between 1948 and 1950 in Rio de Janeiro, and published in its first editorial that 'the struggle of the quilombo is not specifically against those who deny our rights, but in particular to remember and know that Blacks have the rights to life and culture' (Nascimento 1948, 1). The newspaper covered many subjects, but also urged the need to regularize conditions in the slums.

The Unified Black Movement (MNU) began to use the term 30 years later in order to promote the identification of African-descendants with their historical sense of *pertencimento* (to belong), or membership, in the African diaspora. In this moment, quilombo represented a form of mobilization and struggle for freedom and citizenship, as well as for political, economic and social liberation. With the growth of cities and the expansion of capitalism in the countryside under military rule in the 1960s and 1970s, Afro-Brazilian politics in regard to the quilombo lost salience. This was recovered in the late 1970s and 1980s by the black social movements that challenged discrimination and prejudice based on race. Organizations and major groups such as the MNU sought, and still seek, to foster a broader anti-racist, emancipatory agenda. The marches and actions since 1970 have taken racism as a central issue, while demonstrating their specificities in Brazil. Relevant studies have been produced by and about black intellectuals such as Abdias do Nascimento and the Black Experimental Theater, the activist Beatriz Nascimento and Lelia Gonzalez (Ratts 2007), all of whom were active in the counter-culture movements of the years 1960–1980. Their actions gave visibility to the quilombo as a symbol for racial justice. In the 1970s and 1980s, the claims of these activists in the Black Movement and of several members of the Parliament, such as Abdias do Nascimento, were taken to the National Constituent Assembly. This is the moment when the term quilombo was transformed into a juridical device to promote the effective entrance of Africans descendants from diasporas into the new institutional and national order – '*quilombismo*'.

Another class of meanings blends juridical language with struggle and historical marginalization, emphasizing the intersection of race and territory. The 1988 National Constitutional Assembly passed Article 68, the Temporary Constitutional Provision Act (ADCT n.d.). This text states that 'the definitive property rights of "remanescentes", or "remnants" of quilombos that have been occupying the same lands over time are hereby recognized, and the state shall grant them title to such lands'. Article 215 of the constitution (Constitution of the Federative Republic of Brazil n.d.) goes on to say that the Brazilian state shall ensure the full exercise of cultural rights, and 'shall protect the expressions of popular, indigenous and African-Brazilian cultures, and of other groups participating in the national civilizing process'. Article 216 (Constitution of the Federative Republic of Brazil n.d.) also refers to cultural heritage, proclaiming that:

> The Brazilian cultural heritage is property of an immaterial nature, taken individually or together, and bearing reference to the identity, action and memory of the different groups that form Brazilian society. These cultural heritages include: I- forms of expression; II- ways of creating, doing and living; III- creation of science, art and technology; IV- works, objects, documents, buildings and other spaces intended for artistic and cultural expressions; V-urban complexes, and historical, natural, artistic, archaeological, paleontological, ecological and scientific sites.

These three articles, #68, 215 and 216, represent the most important legal devices in the 1988 Brazilian constitution for protecting the rights of African descendants. It was the 'quilombo movement' in the 1988 Constituent Assembly that urged the granting of land titles to Quilombos communities throughout the Brazilian countryside. The National Coordination of Quilombo Communities (CONAQ) has positioned itself since then as an organization fighting for the defense of Presidential Decree No. 4887 of 2003, which regulates Article 68 of the Brazilian Constitution of 1988. The use of juridical language shows how actors who identify as 'quilombolas' use the term in order to gain political recognition.

In addition to the Constitution, Presidential Decree 4887 of 2003 (Presidência da Republica 2003), signed by President Luiz Inácio Lula da Silva in 2003, states that:

> For the purposes of this Decree, the characterization of the remnants of Quilombo communities is attested by self-definition of the community itself, based on historical background, and they are endowed with specific territorial relationships with the assumption of Black ancestry and their relationship of resistance to the historical oppression suffered.

It is important to note here that the official land title of a quilombo in the decree is *collective*, and a community must go through a process of collective registration to become a quilombola community association. The association is a nongovernmental organization made up of any number of participating, self-identifying members of a given community.

Another definition comes from the 2004 Convention of the International Labor Organization (ILO) on Indigenous and Tribal Peoples, promulgated by Brazilian President Luiz Inácio Lula da Silva through Decree 5051 (Presidência de Republica 2004). This Decree recognizes communities' right to ethnic self-identification, to an education that is appropriate to their local reality, and to the right of return to Africa. Brazil is a signatory of this Convention. Quilombolas' desire for recognition extends the borders of Brazil to the international community.

Researchers have also been involved in crafting definitions and meanings for quilombo. In 1994, the Brazilian Anthropological Association (ABA) stated:

> The term 'Quilombo' has taken on new meanings in the academic literature, as well as in groups, individuals, and organizations. ... They have mainly consisted of groups that developed everyday practices of resistance in order to maintain and reproduce their everyday ways of life, as well as to consolidate their own territory. The identity of such groups is not defined by the size or the number of members, but by the experience and shared visions of their common trajectory and continuity as a group. They constitute ethnic groups that are

respected by the discipline of anthropology as an organizational type that grants membership through norms and means of affiliation or exclusion.[4]

This was ABA's first public expression concerning the term quilombo, and it was endorsed by the president of the association, João Pacheco de Oliveira Filho. This document aimed to synthesize the theoretical and methodological contributions of the anthropological research that had been conducted on black communities in diverse regions of Brazil up until that point. The document emphasizes quilombos as cultural collectives instituted through a complex process of resistance and cultural preservation, which guaranteed the unity and survival of Afrodescendants and their ways of life. It also recognized how quilombos resisted the efforts of society and the state to attack their ways of being. The ABA definition of quilombo de-emphasizes the criteria of genealogical descent or biology (genetic heritage verified through DNA), instead reaffirming that 'the remnant quilombo communities are not defined in biological or racial terms, but rather, by the social organizations they have developed since their first common possession of land and the creation of their own cultural patrimony' (ABA 1996, 82). This notion is important for Brazilian nationhood because it emphasizes its constructed nature, rather than some biological or primordial one.

Emphasizing the rural and agricultural roots of quilombos, the Federal Social Program 'Brazil Quilombola' (PBQ), implemented in 2006, promoted sustainable development policies for quilombola territories, and also required that the communities adhere to historical and cultural needs (SEPRIR 2005). The program also promoted social inclusion in order to combat inequalities. In 2007, Article 3 of the National Sustainable Development of Traditional Peoples and Communities Program was established by Presidential Decree 6040, and reads as follows: 'Traditional territories are necessary spaces for cultural, social and economic reproduction of traditional peoples and communities, whether used permanently or temporarily' (Presidência da Republica 2007).

Recognized in this decree is how quilombola territories have particular forms and rules for transmitting tangible and intangible assets, which constitute a group legacy: a collective memory, a symbolic heritage of the group and a patrimony related to Brazilian culture. In February 2007, through another Presidential Decree, the National Policy for the Viable Development of the Peoples and Traditional Communities (CNPCT) program was created. This program recently announced that it will operate through a forum made up of representatives of the government and 'traditional' communities.[5] The estimates tell us that about one fourth of the national territory is occupied by traditional peoples and communities. This is equivalent to an area of 176 million hectares, in which approximately 4.5 million people live (Mendes 2006). In this sense, the quilombo is a political unit, sometimes falsely racialized, historically rooted in rights, and, at the same time, constitutes a form of agricultural production.

Challenges facing quilombo access to land

For generations, Black populations have been denied official land titles that would have allowed them to stay on the land and cultivate it. The first Land Law in Brazil (a Lei da

---

[4]Document from the ABA Thematic Group 'Rural Black Communities,' published as a letter after a conference realized on 17–18 October 1994, in Rio de Janeiro (ABA 1996, 81–82).
[5]Those considered to be traditional populations are: *quilombolas, ciganos, caiçaras, ribeirinhos, geraizeiras* (inhabitants of the inland), *pantaneiras* and *quebradeiras de côco*, among others.

Terra), in 1850, effectively excluded non-Brazilian, non-white populations in an attempt to distribute public lands to 'Brazilians' and 'Foreigners' (European immigrants). This law inaugurated one of the most effective mechanisms for territorial expropriation: the law denied Africans and their descendants full Brazilian citizenship by placing them in the special liminal category of 'libertos' (Free Africans and their descendants), which under the law gave them only limited rights. This law, which transferred public lands to private ownership, favored large landowners and the accumulation of large properties (Silva 1996).

At this time, large landowners and European immigrants were favored over former slaves. Furthermore, the 1850 Land Law's supposedly universal provisions created numerous legal means for expropriating the land: expulsions, removals, enclosures, the registering of 'vacant' lands, forced and arbitrary divisions of community and the seizure of lands for failure to pay taxes. Their legal status made existent rural black communities invisible, a result of a racist hegemonic legal order that used symbolic violence to subordinate certain groups. This hegemonic order has operated by criminalizing those black populations that have fought to remain on their lands. Violence, as brilliantly demonstrated by Foucault (1999), can be enacted under the guise of protection, through control of access and through legal requirements. The forms of violence practiced in the case of the expropriation of black communities' land rights are directly related to technologies of control and manipulation. What is significant about the current legal moment is that the law incorporates a sector of the Brazilian population that has been de facto disenfranchised for centuries.

Currently, despite all the work on self-representation and organization by quilombos, the development of quilombo recognition has continued to face hurdles. The results of Presidential Decree 4887, and the articles conferring rights to Afro-Brazilians in the 1988 constitution, have been minimal at best. Decree 4887, passed in 2003, established a two-part process by which quilombola communities can receive official land titles. First, the community has to be recognized as a quilombola community by the Palmares Cultural Foundation, a federal agency in the Ministry of Culture. Once the quilombola communities receive this certification, they must wait for INCRA to approve the land title, a lengthy bureaucratic process. Although the National Coordination of Quilombola Communities (CONAQ) estimates that 5000 quilombola communities exist throughout Brazil, the Palmares Cultural Foundation has only certified 2278, less than half. Furthermore, of these 2278 certified quilombola communities, only 207 of them have received land titles. Thus, less than five percent of all self-identified quilombola communities in Brazil have actually succeeded in receiving land rights. The map below visually illustrates this first set of numbers: the quilombola communities that have been certified in each state, and the total estimated number of quilombola communities (Figure 1).

The litigation process whereby groups receive status as quilombos leaves in the background a much broader set of issues that are considered by groups in making their livelihoods, e.g. history, culture and intra-group conflicts. Furthermore, the legal process is complicit in reducing the meaning of quilombos to land and, hence, to a potential commodity with a value within the current capitalist system. This could be understood as a form of 'symbolic violence', insofar as the legal reductionism erases the quilombo's complexity (Bourdieu 1989) This deprives us of understanding the other dimension, that there is something beyond materiality that involves people and their culture.

We see clearly today that the government dismisses claims of justice in the name of development and profits. Thus, it is evident that it is important to give land rights to people of African descent, on the same terms that the state has given them to European

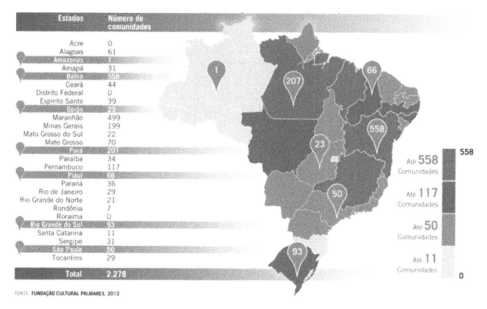

Estados	Número de comunidades
Acre	0
Alagoas	61
Amazonas	1
Amapá	31
Bahia	558
Ceará	44
Distrito Federal	0
Espírito Santo	39
Goiás	23
Maranhão	499
Minas Gerais	199
Mato Grosso do Sul	22
Mato Grosso	70
Pará	207
Paraíba	34
Pernambuco	117
Piauí	86
Paraná	36
Rio de Janeiro	29
Rio Grande do Norte	21
Rondônia	7
Roraima	0
Rio Grande do Sul	93
Santa Catarina	11
Sergipe	31
São Paulo	50
Tocantins	29
Total	**2.278**

FONTE FUNDAÇÃO CULTURAL PALMARES, 2013

Até 558 Comunidades

Até 117 Comunidades

Até 50 Comunidades

Até 11 Comunidades

Figure 1.   Quilombo communities with official certificates.
Note: The map was created by Palmares Cultural Foundation in 2013. Available from: http://www.pagina22.com.br/index.php/2013/11/movimento-de-resistencia/#sthash.OGTiPWie.dpuf (accessed 19 November 2013). The columns on the left indicate the numbers of quilombo certificates that have been issued in each state by the Palmares Cultural Foundation. The right column represents the approximate numbers of quilombos that exist throughout Brazil, estimated by the Coordenação Nacional dos Comunidades Quilombolas (National Coordination of Quilombola Communities, or CONAQ).

immigrant communities in the past until today. The quilombo denounces a history of injustice by introducing the African diaspora into the nationalist agenda. Once 'quilombo' becomes a right, it does not lose its historical relationship to the idea of African resistance to slavery. However, it does bring this resistance within a new order, or, in other words, it brings the quilombo from the outside – from the context of Diaspora – to the inside, to the context of a nation. The sense of rebellion becomes a social pact with the Brazilian state. From the point of view of State policies, the incorporation of the quilombo into Brazilian society inverts the idea of resistance as conflict to the idea of resistance as a new form of social order. The struggle of the quilombola communities is spreading. The idea of quilombo as resistance masks the many conflicts that continue to exist between these marginalized communities and the government, and the slow rhythm of actual social changes.

Alfredo Wagner Berno Almeida has developed several important arguments concerning the land market in Brazil (Almeida 2005, 15–44), which describe why it is difficult to obtain quilombo land rights. He argues that Brazilian agrarian politics is resistant to the use of land for social proposes, because international capital is not interested in shielding thousands of hectares of land from market speculation and economic exploitation. This creates a significant political barrier to the regulation of land for social interests.

There are also many big development projects organized by the Brazilian government itself, which come into direct conflict with social interests. One example is the coastal railroad that will link Brazil to the other countries in Mercosul, which is a key piece of the Brazilian government's developmental project. In Santa Catarina, this railroad will displace, according to recent estimates by INCRA, at least seven quilombola communities. These communities will all have to be relocated because of this railroad. There is also the case of the community Invernada dos Negros, in the municipality of Campos Novos in Santa Catarina, whose land was taken in order to plant pine trees for a paper factory. This latter action was supported by the government, and companies were never fined for their negative environmental impacts. These examples illustrate that at the same time as the Brazilian government now recognizes territorial, cultural and political rights for Afro-descendants, these rights are often not realized in practice.

The collective ownership of land that is required in Decree 4887 is complex. There are many difficulties and impediments to recognizing collective land ownership, principally due to other competing land claims. Therefore, the state must compensate these other land claimants in order to give the land to the communities that are demanding it. The land titling process faces countless administrative barriers in the government and the courts. Thus, these government land actions proceed slowly through the Brazilian courts and through INCRA.

This lengthy process sometimes results in the violent death of community leaders. This was the case in Brejo das Crioulos, located in the northern part of Minas Gerais, immediately following INCRA's initiation of the process of land titling of the quilombola community in 2003. This community has begun denouncing the various acts of violence that have occurred since 2003 by other claimants to the land. These examples illustrate tensions between the quilombola communities and other interest groups. The Brazilian press has accused the quilombola movement of 'racializing' agrarian reform, in order to spread or foment conflict and manipulate identities. The result is that the quilombola communities are often assumed to be the problem, and are themselves considered to be violent in the Brazilian national imaginary. In the discourse of elites and big companies, these communities are simply manipulating identity to obtain land. There are also critiques of the quilombo movement by some intellectuals, who accuse the quilombola association of being controlled by the State.

Finally, the case of quilombola territorial rights is often critiqued because the land is administrated collectively. These production models are on the periphery of the capitalist system. Despite the fact that descendants of Africans brought to Brazil complex social systems that contributed to the persistence of these communities, they are considered by critics as 'backwards' and an impediment to modernization. These misconceptions ignore the fact that collective organization and practices of solidarity were critical for the growth of this African-descendent population. These processes are part of the African legacy, and the creative self-administration of Afro-Brazilian communities.

Despite the increasing strength of quilombola social movements in every region of Brazil, every year new bureaucratic requirements are created that impede the administrative and juridical process of land titling. According to the head prosecutor of the Ministry of Public Defense, Debora Duprat, the quilombo land titling process faces prejudice, misunderstanding and resistance. In her analysis, 'if the current pace of land regularization continues, the 2007 quilombos would have to wait 175 years to have their land processes concluded'. She continues in the rest of the article to argue that this slowness of the quilombo process reveals the vestiges of hegemonic practices, principally those practices that create a classification and opposition between us and them (Duprat 2013). The process of quilombo formation, currently, pass through many economic, political and legal hurdles.

Metaphorical processes and the meaning of quilombo in historical context

Marshall Sahlins (1997) refers to the ways in which old words acquire new meanings in different historical contexts as 'metaphorical processes'. The metaphorical processes associated with the term quilombo, and the definitions I outline above, are relevant because they indicate how, in the construing of the concept and its different meanings, the term has always been utilized to question structures of domination in different contexts. As I have shown, the term has been associated with Africans and their descendants, as well as with Brazilian national identity. The term quilombola is a way actors identify with Afro-descendants in order to achieve political recognition.

Barth offers us a way to conceive of quilombo identity with respect to ethnicity. Ethnicity, for Barth (2000), is a condition used to explain the persistence of borders created by colonial racist orders. Social otherness recreated in relations between oppositional groups also emphasizes the positivity of black identity. An important aspect of the analysis highlighted by Barth is the symbolic dimension of social identity, marked by the notion of a presumed common origin and its effect on self-identification through kinship ties. Ethnic identity, as described by Barth, is especially important in order avoid certain culturalist analyses that reduce social groups to sets of objective traits. Barth highlights the relational context of creating otherness, especially with respect to those aspects recognized by the subjects themselves and others. Barth creates a powerful concept to describe the cultural dynamics of self-recognition.

However, the degree of generality and comprehensiveness leads to a certain impoverishment in some analyses, especially in the way that the criteria for membership are mechanically translated. A consequence of this reductionism is the recurrence in the way in which historical, cultural and structural contextual differences of groups (in my case, quilombos) are only understood in racial terms. However, Barth's theory of ethnicity, with its degree of generality, describes how quilombos emerge in local contexts and are constituted in particular spaces or places through historically situated familial relations, not necessarily reduced to race or to biology. The emergence of a collective consciousness does not appear spontaneously as a result of political interventions, but as a special and particular moment of historical development.

In addition to ethnicity and kinship, the metaphorical transformations that the term quilombo has undergone illustrate the direct effects at different historical moments of struggles for social and political recognition of territorial rights. Some scholars deny the validity of these definitions of quilombo, in favor of more 'developmentalist' and 'modernization' models. By questioning the force of the legal definitions of the term quilombo, large businesses and corporations attempt to invalidate the ongoing processes of recognition because they interfere with land acquisition by agrobusiness. The so-called 'counter-reports' intended to support the administrative and legal defense against the territorial rights of quilombos are motivated by companies that pay to defend their land interests. It is important to note that while many of these actions are unconstitutional, they are supported in the Brazilian Congress by the rural voting block (*bancada ruralista*), one of the main groups attempting to deny the Constitutional rights of quilombola communities.

Some of the more recent approaches towards analyzing the impact of political processes, such as the work of Mignolo (2000) and Quijano (2007), attempt to analyze territorial social movements as alternatives to inequality, unbridled accumulation and dehumanization. These approaches are based in the concept of 'decolonization', and they attempt to see beyond the dichotomies introduced by colonial processes in Latin America, in which the notion of race had a fundamental role in creating unequal power relations. These scholars, in introducing questions of inter-subjectivity, emphasize the web of micro and macro power relations and the clear articulation between social, political and economic factors. This literature seeks to identify more than just a contradiction between different political meanings, but rather alternative ways of conceiving and living that ensure the sustainability of marginalized groups. It is through this prioritizing of new voices that the very idea of the quilombo became a political project and rapidly gained prominence in the Brazilian political scene over the past 20 years.

The term quilombo is the product of a complex and tortuous history that has become part of the social and cultural fabric of Brazilian identity. Quilombos represent both memory and presence. It is not possible to understand them without rethinking the different translations and projects the term encompassed before its current usages. These historical moments were grounded in relations of submission, slavery and violence. Brazilian history is replete with documented cases of killings of poor families, the expulsion of people from their land, the burning down of houses and the unjustified imprisonment and torture of different members of quilombo communities. In this sense, the current use of quilombo by various actors carries with it these historical traumas. The form of collective agricultural practices contrasts with dominant plantation monocropping from colonial times to the present day. And while the term is often used to denote runaway slave communities and their descendants, contemporary uses highlight how the term extends beyond a simply racial determinism to include particular historical and social understandings of territory.

The quilombo represents a memory, which is grounded in having an African ancestry, and represents the search for a place to live, raise children and establish a new socio-cultural order. Additionally, it is a journey, if we can call it that, grounded in past struggles against the dominant relations of production in Portuguese society. The humanity of black populations was systematically deconstructed through 300 years of colonization. Nonetheless, Indians and Africans, while systematically treated as inferior, began to resignify their identity and to develop a self-identification rooted to a particular place. It is not possible to understand the 'quilombo' without this idea of a 'place'.

The quilombo has also become part of the re-imagined Brazilian nation. Article 68 of the Brazilian Federal Constitution recognizes the rights of the descendents of African diasporas that were produced through colonial enterprises. This term, 'African Diaspora', expresses what are commonly identified as subaltern places (Gilroy 2001). The cultures that become integrated in these places represent a range and diversity of practices and narratives. Furthermore, the 'right to the quilombo' that the Brazilian juridical system created in its application of Decree 4887 represents the right to land and to the means of production. Through the many examples that quilombola communities bring us, it is possible to perceive that these territories weave together the complex relationships among time, space, experiences, destinies and desires.

Slavery installed an economic and social mode of production based on abstract individuality as the maximum expression of the human condition. In the total suppression of collective identity, it aimed to destroy previous forms of African culture and history. The enslavement of Africans attempted to break all their former social ties. Slavery represented the denial of kinship, language, religious systems and aesthetics, or, in other words, all of the factors that allowed for the expression of the diversity of African cultures. Colonial slavery created a specific culture of individualism that attempted to expropriate previous forms of collective seeing and doing. Community was replaced by individual, laboring bodies that were to be stripped of any collectivity. The collective mapping of black territories registers, therefore, places and life expressions not found in everyday life. In order to perceive the form and shape of this map, it is necessary to remember that the notion of rights emerging among subaltern groups is still largely invisible in Brazil. They are demanding land to live on, and also recognition of modes of kinship and types of histories that have been marginalized.

Quilombo lands correspond to the territories identified by black populations as sites of specific experiences that were consolidated through social and historical connections, with notions of kinship and common origin, and the expression of their own modes of social organization and a collective patrimony. The quilombo territories, therefore, have gained a strategic dimension, both as a vital resource and as a symbolic space of collective reproduction that allows for members to organize for political recognition. Territorial disputes surrounding quilombos are affected by several factors, including external interests, the possible disappearance of their spaces and vulnerability to the diverse mechanisms of land exploitation. In the political imaginary associated with the quilombo, its members are viewed as slave descendants – that is, people who had been deprived of citizenship. The land rights the quilombo confers are thus positive, affirmative rights related to those who previously were rendered invisible. These rights now apply to a large portion of the Brazilian population, both rural and urban.[6] The diverse 'quilombo' legal devices that were created in the last two decades in Brazil highlight not only the notion of race, but also the incomplete citizenship of descendants of enslaved Africans. The theoretical basis for quilombo policies is different from the justification for black quotas in universities, which is, more than anything else, based on the color of an individual's skin. In contrast, the recognition of quilombos is based in the *collective use of the land*.

There is no quilombola territory in Brazil that comes in the form of individual property. This is due to a legal requirement that quilombo recognition can only come through networks of kinship and social organizations, where the collective use of land and housing predominate. This process of self-identification necessitates a collective fight of a community for a determined territory, with members who are not all necessarily *pretas*, *negras*, *morenos* or *mulatas* (all terms for black or brown skin color). The historical formation of these territories implodes the criteria of 'skin color', valorizing much more the collective processes of organization and marginalization.

Local understandings of quilombo: a focus on Casca

The community of Casca was the first community recognized as a quilombo community in southern Brazil. I focus on this community to show how current meanings of quilombo expose conflicts that had been concealed by almost two centuries of institutional and historical invisibility. I also use this case to illustrate how quilombos now are best understood in the context of Barth's understanding of ethnicity, and as a form of social organization intended to guarantee actors political recognition. Traveling back and forth to Casca for

---

[6]In 2005, a public action was taken by two federal agencies, INCRA and the Palmares Cultural Foundation, to support the establishment of the Silva Family Quilombola Association, in the capital city of Porto Alegre. This was the first officially recognized urban quilombo in Brazil. The judge made his decision by arguing for the need 'to assure the protection of those who have resisted for a long time and fought for their survival in the margins of the established order'. The open language of this court decision has allowed for the possibility of quilombola territories to be established in urban as well as rural areas. Thus, quilombo rights have now become central to the territorial politics in both rural and urban Brazil, re-shaping the struggle for land and the recognition of alternative social and cultural practices.

field research during the past 10 years has allowed me to follow the successive changes that have also occurred in people's understanding of quilombo.

The meanings and use of quilombo were just as much a point of contention among researchers as among quilombolas themselves. The relationships among the members of the community were discussed at the community's meeting concerning its own history being rewritten in light of the new term 'quilombo'. Previously, quilombo had acquired a pejorative meaning in the region. Thus, there was a need to provide more information about quilombo struggles and how the legislation allowed for the recognition of land titles. The criteria on which the social organization, the traditions of succession and the administrative political system were based had also to be revised and redefined.

In Brazilian law, quilombola land is based on collective ownership. This is not adequate for some situations where black families already own land as private property. In the case of Casca, much of the land was already individually owned. This was a product of what I call the 'Political Law of Casca' (Leite 2002), a decision to continue respecting the land clause introduced in the 1824 law, which stipulated that landlords could leave their land to slaves, but also prohibiting its sale. Due to the fact that these lands could not be sold, they had to be passed on indefinitely to one's children. Those who broke this rule were rejected by their own kinship groups. This presented challenges for the community of Casca in achieving land rights. Although the community struggled for over a decade, legal land rights were only given to the community of Casca in 2012.

When I started visiting the region, there existed a strong factionalism among those who were seeking recognition and land title. In fact, some carried firearms to meetings. The strongest argument from certain leaders against titling was that some inhabitants were not legitimate heirs of the land. It was claimed that some from outside the group had arrived, taken land and wanted to exercise power, to concentrate land for themselves and their children. It was also alleged that they would act like plantations; land owners would beat people and negotiate land deals with real estate companies. Despite differences, both sides agreed to the measurement and demarcation of the land, to be performed by INCRA. Given that the consensus of all was needed to make decisions, the most tense issues were shelved for later debate and discussion began on where housing and production areas would be located. Two years later, the external boundary line of the territory was again questioned by the quilombo community. After much debate between the leadership and INCRA, a new survey was conducted and published.

When trying to understand what was happening, I realized that the challenge was related to the rules of the group itself and its relations to their self-recognition and desire for social, political and legal recognition. I began by asking community members questions about what they had previously done, listening and participating in local life and investigating more thoroughly what was hindering the process of land titling. The discussion focused on what one of the elders called the 'Law of Casca Politics', which in my view constitutes the guiding principle of access to land and the operating principle belonging to the group of heirs and their ethnicity. This 'law', in the understanding of one of the elders, was deference to the will left by Dona Quitéria, the owner of the plantation who freed the slaves and left the land to them. According to her will, they cannot sell any part, even from father to son. They all must be able to live there. Any action to the contrary represents a break in the founding group rule. By breaking such rules of the community, the current leader demonstrated that he had disobeyed the principle that guided the formation of the group. Therefore, at the inquiry, he was accused of acting against the rules of the original will and, as a result, he lost his leadership position and his claim to land.

I realized that being grounded in the notion of being a relative was a fundamental principle and part of the moral rule of the group. This speaks to Barth's point concerning kinship ties and ethnicity: that the community had constructed its own form of relations over time that grounded its sense of identity. I found that the disputed leader was, in fact, descended from the heirs named in the will, and therefore was indeed a member of the community. It was evident that he had presented his argument to highlight the priority of kinship ties, which meant not only blood, but also respect for the rules established in the will. This sense of being related to the group assured access to land. However, he had appropriated more than his own share of land for himself and his descendants. He had even sold some to a real estate company which intended to build a resort there. This enraged other members of the community and led them to question both his right to leadership and his claim to membership. This happened even though he was black, which indicates that color in itself was not enough to guarantee his membership in the community. This is similar to other cases in which members can be brown or white as a result of matrimonial alliances. History demonstrates that phenotype is not a sufficient criterion for belonging. Recognition and respect for the rules that support this group are also central to defining community.

When, therefore, scholars refer to ethnic processes of territorialization involving Afro-descendants in Brazil, they must always consider the permeable nature of Brazilian identity and the relationship between that identity and the construction of the Brazilian nation. The system's racist roots are found in the ideology of miscegenation, whose whitening criterion was an important factor in establishing status and privilege. Quilombo identification challenges this, proposing that belonging to a quilombo community is based more on rules than solely on race. The most interesting fact here is that being considered black, by itself, does not guarantee one's legal membership to a quilombo. The constructed nature of quilombo identity, as well as its close connection to social conflict, provides a way to understand Brazilian ethnic politics that challenges the dominant narrative.

Conclusions

Throughout this contribution, I have sought to analyze the present situation of quilombos through examples, some of them historical and some based on my own field research carried out in quilombola communities in Southern Brazil. My goal in this paper has been to illustrate the meaning of quilombo at different historical moments, and to illustrate how these meanings operate in the construction and deconstruction of local identities, as well as of the Brazilian nation. I argue that the best understanding concerning quilombos requires analyzing them as social organizations, claims to African ancestry and fights over territoriality. None of these alone captures the quilombo in its entirety. My position is that the quilombo represents not only new meanings and understandings, but, more importantly, social actions, political organizations, administrative processes and transformations in public policies. The quilombo introduces new dynamics to the democratic system in Brazil. These pressures for social change are directly correlated with new social and political representation of groups, and the visibility of their specific demands.

The government's current slowness in the regularization of land titles and the recognizing of these social rights is a central characteristic of the quilombola struggle. In the present context, many local leaders find it difficult to have any kind of dialogue with the government about their communities' demands for social services and benefits. The most relevant current issues are related to access to loans for agricultural production. Even with their titles to land, small farmers face huge difficulties, especially in gaining access to agricultural technology. Rural assistance programs do not always end up producing satisfactory results, and when this happens small farmers are forced to work as day laborers in large agrobusinesses. All of these issues continue to threaten the viability of black small-farming communities across Brazil.

I emphasized the existence of a broad framework of definitions of the 'quilombo' from diverse sectors of society and diverse social actors over the past centuries. It is because of this diversity of meanings that metaphorical processes have taken a prominent place in my analysis. A multiplicity of political dynamics is currently being produced in response to the continuing resignifying of the term quilombo. The quilombo represents a struggle for rights and for a consciousness about the nature of the African diaspora and its descendants. The leaders of the quilombo movement and quilombola associations have been protagonists, and will continue to be critical to the politics of Brazilian land throughout the twenty-first century. The central importance of the quilombo in the Brazilan imaginary concerns what it teaches us about nation formation – instead of disavowing conflict and resistance, it places these features at the heart of national origins.

Acknowledgements

I would like to thank Rebecca Tarlau and Anthony Pahnke, for their excellent work organizing this collection. I also want to thank my colleagues Mauricio Pardo, Diana Brown and Mario Bick for our discussions about the ideas in this contribution, and for help making the necessary adjustments with my wording on concepts in English.

References

ADCT. n.d. *Ato Das Disposiçoes Constitucionais Transtórias* [Temporary Constitutional Provisional Acts]. Article 68. http://jurisite.com.br/codigosjuridicos/codigos_pdf/adct.pdf (accessed May 8 2015).

Almeida, A.W.B de. 2005. Nas bordas da política étnica: os quilombos e as políticas sociais. *Boletim Informativo do Neur* 2, no. 2: 15–57.

Almeida, A.W.B de. 2008. *Terras de quilombo, terras indígenas, babaçuais livres, castanhais do povo, faxinais e fundo de pastos: terras tradicionalmente ocupadas*. Manaus: Universidade Federal do Amazonas.

Arruti, J.M. 2006. *Mocambo: antropologia e história do processo de formação quilombola*. Bauru: Editora da Universidade do Sagrado Coração.

Associação Brasileira de Antropologia (ABA). 1996. Grupo de trabalho de ABA sobre comunidades negras rurais: documento dirigido à Fundação Cultural Palmares. *Boletim Informativo do NEUR* 1, no. 1: 81–82.

Barth, F. 2000. Os grupos étnicos e suas fronteiras. In *O guru, o iniciador e outras variações*, ed. F. Barth, 25–67. Rio de Janeiro: Contra Capa.

Bourdieu, P. 1989. A identidade e a representação, elementos para a reflexão crítica sobre a ideia de região. In *O poder simbólico*, 107–132. Lisboa: Difel.

Constitution of the Federative Republic of Brazil. n.d. Article 215 and 216. http://www.stf.jus.br/repositorio/cms/portalstfinternacional/portalstfsobrecorte_en_us/anexo/constituicao_ingles_3ed 2010.pdf (accessed 8 May 2015).

Duprat, D. 2013. Luta contra a lentidão. *Folha de São Paulo*, 20 Nov.

Foucault, M. 1999. *A verdade e as formas jurídicas*. Rio de Janeiro: Editora NAU, Pontifícia Universidade Católica do Rio de Janeiro.

Fundação Cultural Palmares (FCP). 2013. Perguntas Mais Frequentes. http://www.palmares.gov.br/quilombola/ (accessed November 10, 2013).

Gilroy, P. 2001. *O Atlântico Negro: modernidade e dupla consciência*. Rio de Janeiro: Universidade Cândido Mendes, Centro de Estudos Afro-Asiáticos.

Leite, I.B. 1995. Classificações étnicas e as terras de negros no sul do Brasil. In *Terra de quilombo*, ed. E.C. O'Dwyer, H1–H5. Rio de Janeiro: Associação Brasileira de Antropologia.

Leite, I.B. 2002. *O legado do testamento: a comunidade da Casca em Perícia*. Florianópolis: Universidade Federal de Santa Catarina.

Mendes, D. 2006. *Governo lança política para comunidades tradicionais*. Brasília: Ministério de Meio Ambiente. http://www.mma.gov.br/informma/item/3462-governo-realiza-consultas-p (accessed November 12, 2013)

Mignolo, W.D. 2000. *Local histories/global designs: coloniality, subaltern knowledges, and border thinkings*. New Jersey: Princeton University Press.

Nascimento, A. 1948. Nós. *Jornal Quilombo*, 1(1).

O'Dwyer, E.C. 1992. *Terra de Quilombos*. Rio de Janeiro: Associação Brasileira de Antropologia.

O'Dwyer, E.C. 2002. *Quilombos: identidade étnica e territorialidade*. Rio de Janeiro: Editora Fundação Getúlio Vargas.

O'Dwyer, E.C. 2012. *O fazer antropológico e o reconhecimento de direitos constitucionai:. o caso das terras de quilombo no Estado do Rio de Janeiro*. Rio de Janeiro: Editora E-papers. http://www.e-papers.com.br/produtos.asp?codigo_produto=2231 (accessed January 19, 2015)

Parreira, A. 1996. *Economia e sociedade em Angola na época da rainha Jinga*. Lisboa: Editora Estampa.

Presidência da Republica. 2003. Decreto n° 4.887, de 20 de Novembro de 2003. http://www.planalto.gov.br/ccivil_03/decreto/2003/d4887.htm (accessed 8 May 2015).

Presidência da Republica. 2004. Decreto 5051/04 | Decreto n° 5.051. http://presrepublica.jusbrasil.com.br/legislacao/97798/decreto-5051-04 (accessed November 12, 2013).

Presidência da Republica. 2007. Decreto n° 6.040, Article 3, II. http://www.planalto.gov.br/ccivil_03/_Ato2007-2010/2007/Decreto/D6040.htm (accessed 8 May 2015).

Quijano, A. 2007. Colonialidad del poder y classificación social. In *El giro decolonial: reflexiones para una diversidad epistèmica mas allá del capitalism global*, eds. S. Castro-Gómez and R. Grosfoguel, 98–124. Bogotá: Siglo del Hombre Editores.

Ratts, A. 2007. *Eu sou atlantica: sobre a trajetória de vida de Beatriz Nascimento*. São Paulo: Imprensa Oficial/Instituto Kuanza.

Sahlins, M. 1997. O 'pessimismo sentimental' e a experiência etnográfica: por que a cultura não é um 'objeto' em extinção. *Mana* 3, no. 2: 103–150.

Secretaria Especial de Politicas de Promoção da Igualdade Racial (SEPRIR). 2005. *Programa Brasil Quilombola*. Brasília: Governo Federal.

Silva, L. O. 1996. *Terras devolutas e latifúndio: efeitos da Lei de 1850*. Campinas: Editora da Universidade Estadual de Campinas.

Silva, E. 2003. *As Camélias do Leblon e a abolição da escravatura: uma investigação de história cultural*. São Paulo: Companhia das Letras.

Can urban migration contribute to rural resistance? Indigenous mobilization in the Middle Rio Negro, Amazonas, Brazil

Thaissa Sobreiro

Given the importance of land for indigenous peoples, rural out-migration is usually associated with a disruption of indigenous culture. This paper suggests that instead of being a disruptive process, migration can serve as the means for a 'scale shift' that transports mobilization capacity from one location to another. This contribution presents the case of Barcelos, in the Brazilian Amazon, where an indigenous movement first arose in an urban area, due to the migration of indigenous activists from other locations, and later spread to rural communities as a result of local migratory circulation. Through alliances with the regional indigenous movement, these rural communities became part of a broader mobilization network that supported the indigenous resurgence in Barcelos.

Introduction

All over Latin America, the rise of indigenous movements and other 'new social movements' is seen as a response to military dictatorships, re-democratization and the rise of neo-liberal policies (Jackson and Warren 2005; Yashar 2005; Stahler-Sholk, Vanden, and Kuecker 2007; Deere and Royce 2009). Many countries enacted constitutional reforms in the 1980s that recognized multiculturalism, including Brazil, Mexico, Paraguay, Guatemala, Nicaragua, Colombia, Ecuador, Argentina, Peru and Venezuela (Jackson and Warren 2005; Yashar 2005).

Indigenous political mobilization focuses on claims of indigenous identity as being distinct from other groups, both to maintain that identity and also as an attempt to assert rights over their territories (Occhipinti 2003; Jackson and Warren 2005). These claims were posed against state policies that were based on ideologies of assimilation into national societies (Jackson and Warren 2005). In many cases, the need to assimilate indigenous populations is the main argument used for governments and elites to dismiss the legal recognition of indigenous status and associated rights, especially territorial claims (Santos and Oliveira 2003).

The indigenous rights achievements in the 1988 Brazilian Constitution were the result of a continuous political mobilization among indigenous groups and non-indigenous allies during and after the military regime (Ramos 2011). This constitution was considered a benchmark in the Latin American indigenous rights movement (Ramos 1998). For the first time since European colonization, Brazilian policy for indigenous peoples ceased to be based on the goal of assimilation (Ramos 2011). Indigenous peoples were given rights to maintain their culture, and rights over their lands (with the exception of underground resources), as well as rights to use the judiciary without interference from the state, via the National Indian Foundation (FUNAI).

Indigenous political participation has been increasing on different scales and in various spheres (Brysk 1996), and many indigenous leaders have become well known nationally and internationally (Conklin and Graham 1995; Ramos 1998). Alliances and networks with national and international organizations were crucial for the visibility of indigenous causes (Brysk 1996). In Brazil, there was a proliferation of local, regional and national indigenous associations after 1988 (Albert 2005). Local associations created official leadership positions with clear responsibilities, including elected boards of directors. Most of these associations are based on a particular ethnicity, gender (e.g., there were many women's associations), or professional categories within ethnic groups (e.g., artisans, teachers, etc.; Albert 2005; Chernela 2012). Many indigenous associations have maintained continuous political mobilization, especially those struggling to have their constitutional rights respected.

The literature on Latin American indigenous movements offers several different explanations for mobilization. These perspectives have been driven by considerable theoretical work, and notable here are political process theory and identity theory. Political process theory focuses on a movement's organizational capacity, motives and political opportunities in order to explain levels of mobilization (Van Cott 2001; Yashar 2005). In parallel, identity theory highlights cultural and symbolic content of movements as the basis for mobilization and its outcomes (Alvarez et al. 1998; Warren 2001; Bolaños 2011). In this paper, I provided a case study for detailed analysis of the processes through which mobilization networks are built, contributing to the political process approach to social movement analysis. I suggest migration and mobility as factors accounting for new networks of local and regional mobilization of indigenous groups in the Brazilian Amazon.

One key priority among indigenous movements is to have indigenous territorial rights acknowledged (Almeida 2004; Schwartzman and Zimmerman 2005). According to the new constitution, all indigenous lands were to be demarcated in a period of 5 years. While many groups had land demarcated, there are still more than 100 awaiting the process.[1] There have also been requests for additional land demarcation. It is necessary to explain the difference between 'indigenous land' and 'indigenous territory'. Indigenous land is a legal category related to the political process conducted by the State to delimitate an area of exclusive use of an indigenous collectivity. 'Territory' refers to the relationship between a particular society and its territorial base, based in their construction and experience, which is culturally variable (Gallois 2004). When a territory (or part of it) is delimited as indigenous land, the material and symbolic relationships with the space transform into a new conception of property rights (Oliveira 1996).

The necessity to establish territorial limits demarcating indigenous land is a result of historical contact and conflict with the colonizers. Conflicts have resulted in 'territorialization' processes, where the political struggle for land is intertwined with the rearranging of social and ethnic boundaries (Oliveira 1996; Almeida 2004; Gallois 2004). Different groups have been reaffirming forgotten indigenous identities and assertions of rights over land to secure cultural continuity (Occhipinti 2003; Jackson and Warren 2005; Bolanos 2011). This indigenous 'resurgence' (Warren 2001) or ethnogenesis 'often denotes a gradual process through which older ethnic categories and boundaries are redefined. Sometimes it is also used to refer to the transformation or shifting salience of preexisting cultural identities as

---

[1]For more information about indigenous land in Brazil, see FUNAI (2014).

they become politicized in new contexts' (Bilby 1996, 119). In the territorialization struggle, it is common to invoke cultural ties to land and cosmologies, and the importance of land for social and cultural reproduction, as well as guaranteeing access to essential natural resources (Cunha 2012).

Given the importance of rural territory for indigenous peoples, rural out-migration may be associated with the disruption of indigenous culture. In the Amazon region, indigenous peoples tend to be labeled as attached to the forest and bounded by their territories (McSweeney and Jokisch 2007). However, the violent penetration of national fronts in their territories in each economic cycle produced strong displacements (Alexiades 2009). More recently, there are multiple examples of indigenous migration to urban areas, due to reasons ranging from search for jobs, education and healthcare to community conflicts (Brandhuber 1999; Baines 2001).

Both rural and urban spaces are important for indigenous peoples, and the literature on this topic recognizes the multi-local nature of indigenous communities (Brandhuber 1999; Alexiades 2009), with households participating in 'social and economic activities in several places' (Trager, 2005, 28), spreading their functions between rural and urban areas, which is only possible through sustaining extended kinship social networks (Pinedo-Vasquez and Padoch 2009; Eloy and Lasmar 2012).

The lack of economic opportunities for indigenous communities in the Amazon rural areas, combined with urbanization, increases the interdependency and mobility between rural and urban spaces. This can result in rural out-migration or the constitution of new multi-local livelihood strategies (Eloy and Lasmar 2012). The causes of migration go beyond economic reasons, varying from forced migration induced by missionaries (Chernela 2012), threats to traditional territory (Romano 1981) and internal conflicts (Brandhuber 1999), to voluntary desire for upward mobility (Lasmar 2005, 2008).

Mobility represents both challenges and opportunities for indigenous people's livelihoods. Even when land rights are guaranteed by demarcation, there are other challenges such as a lack of suitable health services, and educational and economic opportunities, in rural areas. As indigenous people try to procure these services, rural–urban mobility and rural out-migration have increased (Bernal and Mainbourg 2009).

Historical indigenous population movements in Rio Negro (RN) are well described: forced displacement during colonization (*descimentos*), mobility related to prophetic indigenous movements, displacement to work for merchants extracting forest products and movement toward missionary centers (Wright 1992; Wright 2005). More recently, there are various records of indigenous migration/mobility to urban areas. They tend to focus on Manaus, the largest urban center in the region (Romano 1981; Pereira da Silva 2001; Bernal 2003; Bernal and Mainbourg 2009; Melo 2009; Figoli and Fazito 2009), or Upper RN (Fígoli 1982; Brandhuber 1999; Lasmar 2005, 2008; Andrello 2006; Chernela 2012; Eloy and Lasmar 2012). In cities, indigenous people reorganize their social networks, creating an extended kinship network that includes more distant kin and friends, and is maintained by visits among them (Pereira da Silva 2001). This support network contributes to maintain their identity inside the urban context (Fígoli 1982) and to resist against prejudice (Romano 1981).

More recent studies suggest that there is a growing fluidity between the forest and the city, forming multilocal communities on a regional scale – structured by kinship networks and flows of goods and people between various nodes (Albert 2005; Eloy and Lasmar 2012). Based on studies with Tukanoan people, Brandhuber (1999) suggested that mobility is inherent to the socio-cultural system of RN.

Little is known about the influence of this culture of mobility on the political mobilization of indigenous people in RN. However, there is growing evidence that migration and urbanization can contribute to indigenous political mobilization (McSweeney and Jokisch 2007; Bernal and Mainbourg 2009). Formal indigenous organizations are commonly located in urban areas. Many migrants are linked to political networks related to urban indigenous organizations, and this relationship depends on the capacity of indigenous leaders and organizations to attract migrants (Pereira da Silva, 2001, 76). Urban indigenous organizations unite ethnic diversity into a specific place, creating new networks and forms of social mobilization (Pereira da Silva 2001). In these organizations, migrants are connected to members of their own group, as well as other ethnic groups and identities. These places can be gathering places for migrants, where they can receive valuable information about their rights, and where they learn strategies for mobilization (Bernal and Mainbourg 2009).

Mobility and increasing amounts of interactions with urban areas may also influence a community's social capital and political organization, which may in turn provide access to important resources and rights. Amazon rural–urban networks permit the circulation of cash, remittances, goods and gifts, but also new technologies, tools, consumer preferences and skills (Pinedo-Vasquez 2009). Sometimes these networks expand towards the 'white world', through marriages between indigenous women and non-indigenous men (Lasmar 2005, 2008). Networks are not only constituted and maintained by kinships, but they also create political relationships that serve to mobilize resources.

In this paper, I argue that migration and mobility permit the extension of the social networks of rural Amazonians into local and regional cities, and that these networks connect these urban associations to rural communities. Migration can serve as the means to shift the scale of indigenous mobilizations seeking state recognition of indigenous identities and territorial claims. This paper discusses how this migration contributed to indigenous mobilization by articulating rural communities with urban movement associations, and how such mobilization increasingly depends on ties to other organizations.

Specifically, I suggest that migration can support mobilization and territorialization by serving as the mechanism that permits a movement *scale shift* (McAdam, Tarrow, and Tilly 2001; McAdam 2003; Tarrow and McAdam 2005). Scale shift is 'a change in the number and level of coordinated contentious action leading to broader contention involving a wider range of actors and bridging their claims and identities' (McAdam, Tarrow, and Tilly 2001, 331). I argue that two key historical migration flows involving indigenous peoples in RN permitted the emergence of new mobilization: one flow involved migration of indigenous people with mobilization experience to a new location that did not have a history of indigenous mobilization, thus providing the conditions for local urban mobilization. This constituted a regional scale shift by extending the conditions for indigenous mobilization to new locations. A second migration process involved local rural–urban circulation, which connected rural indigenous communities to urban indigenous activists. This network process was key for urban mobilization to incorporate rural communities, and constituted a local scale shift by expanding participation in mobilization from local urban centers to rural communities. In addition to these two historical migration flows, the context of Brazilian indigenous politics since the 1988 Constitution is also considered.

The remainder of this paper is organized as follows: in the next section, I present a short history of recent Brazilian ethnic politics in order to situate the context in which indigenous groups in the Amazon have been organizing their movements. Second, I will present the current literature dealing with networks and indigenous mobilization. Third, I present a case study in the municipality of Barcelos in the Middle RN, where an indigenous movement emerged in an urban area and later spread to rural areas via rural-urban circulation.

This movement subsequently became part of a larger network of RN indigenous associations.

Brazilian ethnic politics and Amazon indigenous movements

The Upper RN indigenous movement is one of the most prominent in the Amazon. This movement has followed the broader national trends of indigenous struggles. To understand questions of indigenous identity and political mobilization in Brazil, it is important to highlight Brazilian ethnic politics in the last four decades, and how it is related to the RN movement's history. This will set the stage for the recent scale shift from Upper RN towards Barcelos.

Brazil has a long history of ambiguous policies towards indigenous peoples (Perrone-Moisés 1992; Cunha 1992; Lima 1992). The policies, until the Constitution of 1988, varied from the indigenous people's slavery in colonial times, to various attempts to assimilate tribal groups into Brazilian national society. Here, I will focus on the more recent history related to the rise of indigenous political mobilization beginning in the 1960s during Brazil's military regime. In this period, large-scale development programs were launched in the Amazon, causing intense competition for land and resources among different interest groups (Schmink and Wood 1992; Ramos 1998). Diverse ethnic groups were negatively affected by land grabbing, mining, and the construction of dams and roads (Davis 1977), which drew the attention of national and international media and generated criticism against the Brazilian government (Albert 2005).

The military government created new legal instruments in order to deal with the 'problem' of indigenous lands, which prevented the implementation of development policies. At the same time, this government was accused in national and international forums of violations of indigenous rights (Ramos 1998, 2011). In 1967, the government transformed the Indian Protection Service (SPI) into FUNAI as a response to international accusations of SPI's corruption, exploitation and coercion of indigenous peoples (Albert 2005).

In 1973, the military government promulgated the Indian Statute (IS; Law 6001), which classified indigenous peoples as being 'relatively capable' – a legal standing on par with that of minors – and thus under FUNAI wardship. The law's idea was to 'pacify' and assimilate indigenous groups into Brazilian society, and classified all ethnic groups under a generic category of *silvícolas* (forest dwellers) (Albert 2005). The statute also created new territorial categories which restricted indigenous groups to specific areas with defined boundaries determined by the government. Until the end of the military government, there were many political attempts to manipulate the IS in favor of various development interests. Among these were attempts to 'emancipate' indigenous peoples, which carried implications that they would then have no more rights to the land as descendants from original inhabitants. There were also efforts to create criteria for 'indigenousness' as a means of identifying some groups as authentically indigenous but not others, therefore excluding 'acculturated' indigenous groups (Ramos 1998).

In this context, the indigenous political movement emerged in the Amazon to engage in resistance against the Brazilian state and development interests, in order to maintain their traditional lands and their culture, and assert their citizenship rights (Albert 1996; Ramos 1998). With support from the Catholic Church, anthropologists, artists and volunteers, the first independent multi-ethnic Brazilian indigenous organization was founded in the 1980s: the Union of Indigenous Nations (UNI). UNI amounted to a national-level organization composed of several well-known indigenous leaders. However, UNI had difficulty in connecting effectively with local groups (Ramos 1998). UNI members were also persecuted by the military government because of their 'relatively capable' legal status. But despite these difficulties, UNI constituted an important social actor in the negotiations for the new Constitution in 1988 (Ramos 1998).

Article 231 of the 1988 Constitution (Brasil 1988, 181) recognized 'the indigenous social organization, customs, languages, beliefs and traditions and the rights to the lands they traditionally occupy, and is the responsibility of the Federal government to demarcate, protect, and respect this land'. It was the first time that indigenous groups were recognized as culturally distinct, which allowed indigenous groups to invoke cultural ties, cosmologies and ethnic reproduction arguments when claiming for territorial rights. The concept of land traditionally occupied is:

> land inhabited by them on a permanent basis, those used for their productive activities, those indispensable to the preservation of the environmental resources necessary for their well-being and for their physical and cultural reproduction, according to their uses, customs and traditions. (Brasil 1988, 181).

Once demarcated, these lands are registered in the name of the Federal government, but for the permanent possession and exclusive use of the indigenous communities.

In RN the struggle for recognition of indigenous territory preceded the new Constitution. At the end of the 1980s, despite the optimism of Brazil's re-democratization and the end of the military dictatorship, the military increased its presence in the Amazon region through a strategic plan of territorial control called Calha Norte (CN). CN was planned in 1985 and its logic was focused on military control of Brazilian borders in the Amazon for security reasons. Those borders, however, including the Upper RN, were also the territory of 63,000 indigenous people from more than 50 ethnic groups (Ramos 1998). The military transformed a border area, of 6500 km in length and 150 km in width, into a national security area, with military checkpoints and open areas for private mining companies. The military controlled FUNAI from 1989 to 1992, and thus was able to interfere in policy concerning indigenous peoples. Many indigenous groups consequently experienced threats to their lives and territory by the creation of different categories of indigenous 'colonies' with ambiguous criteria that fragmented large territories. Impacts varied among different regions and groups. The military also forced the creation of Indigenous Colonies surrounded by National Forests in the Upper Solimões River and in the Upper RN. Military posts were even created on these indigenous territories.

Since the 1970s, different Upper RN indigenous associations have made requests for official recognition of their traditional territories. In this context, leaders from different indigenous associations created the Federation of Indigenous Organizations of Rio Negro (FOIRN) in São Gabriel da Cachoeira in 1987. Despite early internal disagreements, indigenous peoples from diverse ethnicities united against the creation of indigenous colonies that would reduce 58% of their traditional territory. They were in favor of the demarcation of a continuous indigenous land that included multiethnic indigenous communities. FOIRN used diverse strategies, from destroying the physical marks delimiting the 'indigenous colonies', to legally claiming the recognition of their land by drawing arguments from the rights officially acquired in the 1988 Constitution (Cabalzar and Ricardo 2006). After years of litigation in the federal courts, 10.6 million hectares of continuous land were demarcated in 1998 by Brazilian President Fernando Cardoso.

Currently, FOIRN is one of the most recognized indigenous organizations in the Amazon. FOIRN is an umbrella organization composed of 89 local associations representing about 750 villages spread throughout both demarcated and non-demarcated indigenous lands (Soares 2012). Their coverage area comprises the municipalities of São Gabriel da Cachoeira, Santa Izabel do Rio Negro and, more recently, Barcelos, corresponding to a total area of 108 million km^2. Over 35,000 indigenous people, belonging to 23 ethnic groups, representative of Tukano, Arawak and Maku language families, live in this area (FOIRN 2013).

The importance of networks for indigenous mobilization

Several studies on indigenous movements stress the importance of alliances and networks between local groups and national and international organizations (Conklin and Graham 1995; Brysk 1996; Keck and Sikkink 1999; Mato 2000; Tilley 2002). Research has particularly focused on the role of advocacy networks formed by social movements and civil society supporting a common cause. The literature on transnational advocacy networks (TANs) (Keck and Sikkink 1998) emphasizes the transmission of strategic information from international to national to local indigenous organizations, to frame indigenous claims and pressure national governments for their rights. TANs create transnational public spaces outside of state control, where the local and global scales are interdependent (Mato 2000), contributing to a 'global civil society' (Moghadam 2012).

The emergence of indigenous movements on the international scene through TANs has helped to give indigenous groups a global visibility (Brysk 1996). The circulation of information about the marginalized position of indigenous identity was turned from a weakness into strength (Brysk 1996). Indigenous identity was therefore used strategically to call attention to issues of indigenous rights (Turner 1991; Conklin and Graham 1995; Gray 1997; Ramos 1998). International media attention to environmental issues, including Amazon deforestation, biodiversity loss and global warming, also helped to reinforce the image of indigenous peoples as reliable protectors of nature, thereby connecting human rights and environmentalist networks (Conklin and Graham 1995; Cunha and Almeida 2000). The creation of the International Alliance of the Indigenous Tribal Peoples of the Tropical Forests, and the participation of the indigenous movement in the Rio Summit in 1992, formalized the conservation and indigenous alliances (Cunha and Almeida 2000).

The ratification of International Labor Organization Convention 169 on the rights of Indigenous and Tribal Peoples by 14 Latin American countries was one the most significant results of this international mobilization (Van Cott 2010). The United Nations also approved the Declaration of the Rights of Indigenous Peoples in 2007. These international declarations are not only the result of indigenous mobilization, but also constitute resources that can be used to pressure states to protect indigenous rights (Brysk 1996; Gray 1997; Sieder 2002; Van Cott 2010). With these international norms and repercussions, a wide range of organizations such as the World Bank, environmental nongovernmental organizations (NGOs) and governments have included indigenous issues in their agendas and projects.

National networks were also crucial for indigenous mobilization. During the discussions for the new Brazilian constitution, indigenous leaders and allies in civil society organizations played a crucial role in lobbying for reforms in legislation (Ramos 2011). The Catholic Church was also an important ally of indigenous people in their struggles in the 1970s and 1980s (Ramos 1998; Yashar 2005; Albert 2005). More specifically, the development of indigenous alliances in the Brazilian Amazon is directly related to the participation of a progressive branch of the Catholic Church (Albert 2005), especially the Missionary Indigenist Council (Conselho Indigenista Missionário, or CIMI). CIMI promoted meetings and assemblies of indigenous chiefs from different ethnicities in the 1970s. CIMI's logistical support and political education of indigenous leaders helped to establish the base for Brazil's national indigenous movement. Many lay volunteers started to join the mobilization. At the end of 1970s, there were around 30 pro-indigenous NGOs in addition to these church organizations (Albert 2005).

These national and international networks provided numerous opportunities and resources, and political support, for indigenous groups, which started to constitute themselves as formal associations. The previous networks formed during these first alliances developed into a second generation of indigenous movement (Yashar 2005). While there is an extensive literature on this latter process, less attention has been given to the role of these local advocacy networks for indigenous mobilization, especially after these communities achieved victories concerning indigenous rights. In the next section, I present a case study that illustrates the important role of local networks for indigenous mobilization, with a particular focus on how migration contributes to the scale shift of the mobilization process by connecting rural communities with urban associations and support organizations.

Indigenous mobility and resistance: the case of Barcelos

In the remainder of this paper, I use the municipality of Barcelos, Amazonas, Brazil, as a case study to show the importance of indigenous migration as a means of rural–urban articulation for indigenous mobilization. This case study is based on interview data and participant observation collected over numerous research trips to Barcelos since 2006. This work started during my master's research from 2005 to 2009, which focused on the political ecology of fishing in Barcelos. My research in Barcelos has continued as part of my doctoral field research. In 2011 and 2013, I collected qualitative data about the role of indigenous movements in local development through semi-structured interviews with 20 key informants. Interviewees included representatives from indigenous movement associations, interest group associations and rural communities. During this period, I also participated in several activities and meetings organized by the indigenous movement in the municipalities of Barcelos, Santa Izabel do Rio Negro and São Gabriel da Cachoeira. Information about the indigenous movement's earlier history is primarily based on secondary data (Peres 2003, 2011), interviews, project reports and publications from the Instituto Socioambiental (ISA).

The municipality of Barcelos (Figure 1) is located in the Middle RN basin, in the northwestern Brazilian Amazon. Barcelos is still a relatively isolated region, maintaining most of its forested area with no road access to other municipalities. The municipality has a large area of 122,476 km^2 (IBGE 2013a), and most rural land has no official titling. Upstream from Barcelos are the municipalities of Santa Izabel do Rio Negro and São Gabriel da Cachoeira. In those municipalities, most rural land has been demarcated as indigenous lands, and the region currently has the largest indigenous population in Brazil (IBGE 2013b). Downstream from Barcelos is Novo Airão, where rural land is largely in protected areas. Beyond Novo Airão is Manaus, the capital of Amazonas State, with a population of 1.8 million inhabitants (IBGE 2013a). Transportation among these municipalities is mostly based on river navigation. The town of Barcelos has strong spatial, economic and cultural connections to the river and forest, which in turn translate into an urban spatial organization oriented towards the river (Trindade Jr. et al. 2008).

Barcelos was founded in 1728 as a Catholic mission, and 30 years later it became the first capital of the Amazonas State, which then represented the main colonial power in Western Amazon (Reis 1999). At that time, the urban population consisted of about 2000 indigenous peoples of various ethnicities (Ferreira 1959). The Middle and Upper RN are recognized as

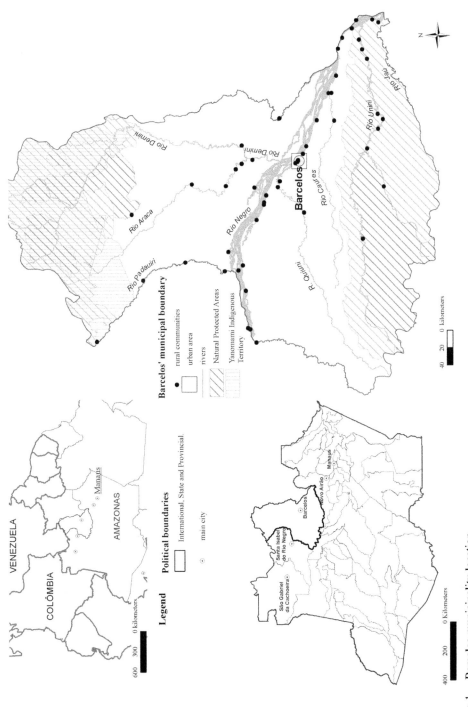

Figure 1.   Barcelos municipality location.
Source: Sobreiro (2015). Cartographic base from IBGE (2010), MMA (2011), FUNAI (2014), and FOIRN and ISA (2014).

having a large number of different ethnic groups from three linguistic branches: Arawak, Tukano and Maku. These groups had inter-ethnic relations predating Portuguese colonization (Guzman 2009). During the early colonial period (1700–1758), Barcelos was an important port for trade of enslaved indigenous people as well as the export of extractive products (Prang 2001). The regional inhabitants suffered a process of intermarriage with European and African peoples during the eighteenth and nineteenth centuries (Guzman 2009). As part of that process, the region exhibited intense population movement. During colonization, indigenous groups were forced to move downriver as a result of Catholic missionary influence, and to work on extraction of rubber and other products.

After the transfer of the state capital to Manaus in 1807, Barcelos experienced a period of economic stagnation and declined as a political and urban center (Prang 2001). The RN went through another period of economic growth during the rubber boom from 1880 to 1925. After the 1920s, rubber exports declined, except for a brief resurgence during World War II (Santos 1980). Rubber production ceased in the mid-1980s with the end of federal subsidies. Since then, piassava fiber (*Leopoldinia piassaba*) and ornamental fishing have become the main extractive products of the Middle RN region. Extraction of these resources operated under the traditional system of patronage that carried over from rubber production (Prang 2001). Between 1970 and the 1980s, many indigenous people moved from Upper to Middle RN tributaries to work on piassava extraction (Emperaire and Eloy 2008). Since 2000, agriculture, sport fishing, commercial fishing and tourism have increased in importance for the local economy.

A Salesian Catholic mission was founded in Barcelos at the beginning of the twentieth century (Machado 2001). The missionaries' centers in RN were a bridge between the indigenous rural life and the 'civilized' and urban world (Figoli 1982; Andrello 2006). Missionaries repressed the remaining cultural indigenous traditions and languages, and kept indigenous children in boarding schools. In the 1970s, Salesian missionaries organized dispersed indigenous and riverine people into formal rural settlements (*comunidades*, or communities), and this model of socio-spatial rural organization has prevailed until today.

Indigenous livelihoods and mobility

Most of the rural population of Barcelos lives in communities that are recognized by the municipal authorities. These communities are located adjacent to water bodies. Households were settled around a chapel, school and social center, an organization that was reproduced and adapted to the urban context when these populations moved to the city (Peres 2003, 2011). Almost every community has primary schools maintained by the local government, but few have secondary schools.

There are approximately 48 rural communities, 30 composed by a multiethnic indigenous population. Indigenous people in Barcelos are descendants of different ethnicities whose languages come from the Arawak and Tukano linguistic families. The vast majority of the population is of the Baré ethnicity, followed by Baniwa, Tukano and Desana (Sobral and Dias 2013). Baré culture was heavily impacted in the historical process of colonization of RN. These Baré peoples lost their native language and, until recently, were considered extinct (Gourevitch 2011).[2]

Indigenous communities are composed of households that have diversified livelihood strategies, generally based on a combination of shifting agriculture, hunting, fishing and

---

[2]More information about Barés can be found in Vidal (2000), Figueiredo (2009) and Melo (2009).

extraction of non-timber forest products such as piassava fiber and Brazil nuts. Spatial mobility is necessary to have access to seasonal resources that are dispersed throughout these territories. There are periods of the year in which the communities are more dedicated to agriculture, combined with expeditions to other locations for fishing, hunting and collection of piassava and nuts. It is also common for households to spend time with their distant relatives, helping with work or just visiting. In addition, people travel to other communities to participate in Catholic saint feasts. Rural mobility is therefore common and important for household reproduction.

Visits to the urban center have increased more recently as result of access to federal programs for the inclusion of minorities and poverty reduction (for example, cash transfer programs such as *Bolsa Família* and retirement benefits). People go monthly to town in order to receive their payments, and to buy industrialized goods. Urban centers are also spaces where rural households access services such as education and health care, and sell their agricultural and extractive products. Thus, while visits to town are not a novelty, they are occurring with more frequency than ever before.

Deficiencies in the educational and health care systems in rural areas have led some families to move more permanently to the Barcelos urban area. This process of rural out-migration tends to be gradual, with a few members of the family establishing economic activities in the town, complemented by resource extractivism from their rural community. To support themselves, many households pursue a livelihood strategy of multi-locality or, in other words, living in both rural and urban areas. Many families live in urban settings, but also have agricultural land (*roça*) in rural or peri-urban locations. Some families may return to fish or hunt in their rural community, in order to sell the catch in the urban center. In the process, the migrants maintain social networks with their rural origin community and territory (Peres 2011).

Although urban areas provide new opportunities and access to goods, migration to town also has some negative impacts. For example, finding jobs in the city is difficult, especially for indigenous people with no formal education. Another difficulty is the cost of living; as one informant said, 'in Barcelos town you have to buy everything' (Interview, 2013). The town is also perceived as a dangerous place, with violence and drugs. Parents say that in town they lose control of their children. However, even with these difficulties, many families insist on settling in the city for their children to be able to study. Education for children and adolescents is considered a priority for rural households as the best way to improve their lives (*melhorar de vida*). While some rural communities have schools managed by the local government with federal government resources, these communities do not have high schools, leading parents to move to town or to send their children to live with relatives in the city.

The growth of the indigenous population in both rural and urban areas makes Barcelos a useful study site for an evaluation of migration processes, indigenous mobilization and ethnic identity. In the last 30 years, there were two migration flows into Barcelos: (1) from São Gabriel da Cachoeira and Santa Izabel do Rio Negro municipalities to rural and urban areas in Barcelos, and (2) from rural communities and *sitios* (individual settlements) in Barcelos to Barcelos town. With regard to the first of the migration flows, Barcelos attracted migrants because, compared to the Upper RN, it was considered to be a region that was relatively rich in important natural resources such as fish and land for agriculture. The increase in tourism and the recent establishment of an army battalion in Barcelos town also attracted migrants searching for work opportunities. Barcelos is also closer to Manaus, the major urban center in the region. The first migration flow thus brought indigenous people with mobilization experience to Barcelos. This would later prove crucial as a basis for indigenous mobilization based in this city.

Table 1. Barcelos population according to 1991, 2000 and 2010 Brazilian censuses.

	Year	Total population		Non-indigenous population		Indigenous population	
		(total)	(index)	(total)	(index)	(total)	(index)
Urban	1991	4018	100	3888	100	130	100
	2000	7954	198	7785	200	169	130
	2010	11,157	278	9787	252	1370	1054
Rural	1991	7017	100	5544	100	1473	100
	2000	16,243	231	10,225	184	6018	409
	2010	14,561	208	7564	136	6997	475
Total	1991	11,035	100	9432	100	1603	100
	2000	24,197	219	18,010	191	6187	386
	2010	25,718	233	17,351	184	8367	522

Source: Sobreiro (2015). Data from IBGE (2013a).

With regard to the second migration flow, this rural out-migration was mainly related to the lack of adequate education and health services. The lack of schools in rural communities is pointed out as the main factor for rural out-migration. This second migration flow has connected rural communities to Barcelos. While this was initially for livelihood reasons, it became a means of linking rural communities to indigenous associations and thus became a mechanism to support indigenous mobilization via rural–urban networking.

Data from the Brazilian census indicate that the Barcelos indigenous population increased five-fold from 1991 to 2010 (Table 1).[3] While the non-indigenous population decreased 7 percent, the indigenous population increased 136 percent between 2000 and 2010. In the Barcelos urban area, the indigenous population increased by 934 percent while the non-indigenous population grew 52 percent. In rural areas, the total population declined by 23 percent; the non-indigenous population declined by 48 percent, while the indigenous population grew by 66 percent.

This steep increase in the indigenous compared with the non-indigenous population, especially in urban areas, suggests that ethnic resurgence is the main reason for the increase of indigenous population in Barcelos. This phenomenon is not restricted to Barcelos. Perz, Warren, and Kennedy (2008), analyzing Brazilian census data, found that the indigenous population doubled from 1991 to 2000. The 71 percent growth of the indigenous population in Brazil during this period was due to the reclassification of race. A significant proportion of indigenous reclassification occurred in urbanized areas. Rural out-migration also probably contributed to the growth of the indigenous population in Barcelos, but the available census data make it difficult to estimate the extent of this impact. In the next section, I describe how mobility and migration contributed to mobilization and the process that led to this ethnic resurgence.

Indigenous mobilization in Barcelos

Although the indigenous population was always present in Barcelos, rural communities and urban populations were until recently considered 'acculturated', and classified instead as

---

[3]Brazilian census methods have probably underestimated the indigenous population in 1991 (Santos and Teixeira 2011; IBGE 2013b).

caboclos[4] (Pereira 2007) or 'civilized Indians' (Peres 2003, 2011). Even among people maintaining indigenous cultural practices in their private life, the public manifestation of any indigenous ancestry was received negatively by the non-indigenous elite, as a sign of underdevelopment (Adrião 1991; Peres 2003). The presence of indigenous people in Barcelos was routinely denied, except for reference to Yanomami groups that live in distant and isolated areas in the municipality. Local discourses denying the presence of the indigenous people reflect the ambiguous strategies of the Brazilian State to 'pacify' and 'integrate' them into national society since colonization (Cunha 1992; Lima 1992; Ramos 1998).

Influenced by the national context of indigenous rights acquired in the 1988 Constitution and the demarcation of indigenous land in Upper RN the National Institute for Historical and Artistic Heritage (Instituto do Patrimônio Histórico e Artístico Nacional, or IPHAN) visited Barcelos in 1999 to investigate influences of the indigenous culture in the city. In the same year, indigenous leaders living in Manaus were hired to map the location of regional indigenous populations, as part of a study for the implementation of an Indigenous Special Health District (Distrito de Saúde Especial Indígena, or DSEI) in the Upper and Middle RN. Surveys and interviews with people in Barcelos stimulated local meetings to discuss indigenous issues, which initiated a process of redefinition of ethnic identities in the municipality.

In November 1999, 90 people were present in a first indigenous meeting. Most participants were descendents of indigenous people who migrated from the Upper RN to rural Barcelos to work in forest extractivism for local merchants (*patrões*), or indigenous households that lived and moved among different rural communities and *sitios* in RN until they settled in Barcelos. Some of them had previous experience participating in the indigenous movement in São Gabriel da Cachoeira via FOIRN, and had studied in Salesian boarding schools. The participants at the first meeting discussed the urgency of recognizing the presence of indigenous culture in Barcelos. At that meeting, they proposed the creation of an indigenous association.

The organizational model based in associations is characteristic of the Upper RN indigenous movement (Peres 2011). Albert (1996) argues that the spread of this model in the Amazon is due to both national and international processes: the legalization of indigenous associations after the 1988 constitution; national state decentralization from indigenous issues (except for land demarcation) combined with budget cuts in indigenous policies; and networks between indigenous and international agencies for financing local ethnodevelopment and conservation projects.

The Barcelos Indigenous Association (Associação Indígena de Barcelos, or ASIBA) was created in a second meeting in December 1999. Discussions underscored questions about structural and bureaucratic aspects of the new organization. The meeting included the participation of FOIRN and *Coordenação das Organizações Indígenas da Amazônia Brasileira* (COIAB), which represented respectively the RN and Amazon indigenous movements. Their presence meant an official recognition of ASIBA as part of the indigenous movement (Peres 2003, 2011). The NGOs CIMI and ISA were also present.

---

[4]As an academic concept, *caboclo* comprises groups from multiple origins, initially mixed descendants of indigenous and Europeans, and later mixed with Northeast Brazil migrants (Lima 1999). They are treated as the historical Amazon rural peasants, with ecological knowledge and customary practices influenced by indigenous cultures (Nugent 1993).

It is critical to note that the history of migration in Barcelos provided the context for the flashpoint of mobilization resulting from this study. The migration from the Upper RN of indigenous people with mobilization experience explains the presence of indigenous people in Barcelos who had an interest in discussing indigenous issues. Furthermore, the circulation of indigenous people among rural communities and the city of Barcelos connected rural communities with urban activists and organizations. Hence, the two migration flows constituted a local base of indigenous people with a shared interest in mobilization in support of indigenous claims to identity, territory and other resources.

From a political process perspective, the capacity for indigenous mobilization in Barcelos was built upon alliances between regional movements and NGOs, which helped to consolidate new urban–rural networks. With the political and financial support of FOIRN, ASIBA became part of a local network that constituted an indigenous resurgence. The institutional presence of actors such as IPHAN and DSEI was a political opportunity for highlighting the indigenous presence in Barcelos. With the help of this mobilizing structure (McAdam, Tarrow, and Tilly 2001), ASIBA was able to build support for their demands and struggles. This process of territorialization developed within a national context, where policies of affirmative action for ethnic groups and minorities were being developed and implemented (Almeida 2004). For indigenous peoples, these policies were a direct consequence of the recognition of their rights by the Constitution of 1988.

In 2001, ASIBA started to engage rural communities, making visits to survey indigenous populations, and discussing issues of identity and information about indigenous lands. Four rural assemblies were set up, and, as a result of the discussions, ASIBA sent a document to the local FUNAI office requesting the demarcation of indigenous lands in the sub-regions of the Aracá/Demeni basin, Padauiri/Rio Preto basin and the right bank of the RN and Caurés basin. Each of these sub-regions created their own rural indigenous associations. Each association elected their representative leaders in sub-regional assemblies, which in turn is connected with elected ASIBA leaders. In 2007, FUNAI established several technical groups (GTs) to conduct studies to evaluate proposals for the legal recognition of indigenous lands.

ASIBA had initial support from two NGOs, *Caldes Solidaria* (a Catalan NGO) and *Fundação Vitória Amazônica* (FVA), for economic projects involving agriculture and handicrafts. More recently, ASIBA strengthened its relations with FOIRN and ISA. In this partnership, ASIBA implemented a communication and transportation network by providing radios to rural communities and boats for sub-regional associations. Since 2008, ASIBA, FOIRN and ISA have held workshops on participatory mapping of natural resource use and conflicts, which have resulted in proposals for management plans to improve natural resource governance in rural areas (ISA 2010, 2011, 2012). In November 2008, ASIBA, FOIRN and ISA organized a workshop (*I Seminário de Ordenamento Territorial*) to discuss territorial planning in the municipalities of Barcelos and Santa Izabel do Rio Negro (ISA 2008, 2009a). Rural communities, local associations and NGOs discussed indigenous territories and state proposals for the creation of protected areas in Barcelos and Santa Izabel do Rio Negro. A preliminary mapping exercise was conducted to organize the demands of rural communities in relation to conflicts, as well as areas of traditional use and importance for conservation. They discussed legal mechanisms that could ensure the protection of important natural resources for rural community residents.

In July 2009, ASIBA organized a public meeting with their associations to discuss the infrastructure for education and healthcare, as well as the migration to urban areas (ISA 2009b). On the last day of the meeting, the participants marched to demand the demarcation of their indigenous lands. At the end of the protest, they went to the local government office to demand the legal recognition of indigenous peoples in the municipal law.

Early in October of the same year, ASIBA held a second seminar (ISA 2009c, 2010), in order to continue the discussion about territorial planning and participatory mapping of rural areas with regard to natural resource use and conflicts. The meeting was well attended by local urban and rural associations, NGOs and local government representatives. However, due to some misinformation that the seminar intended to 'close rivers' by delimiting areas through mapping exercises, there were protests in the city against the demarcation of indigenous land. These protests were promoted by the Piassava Extractors Cooperative (Cooperativa de Piaçabeiros do Médio e Alto Rio Negro, or COPIAÇA-MARIN) and the Fishermen's Association (COLPESCA Z-33), with the support of councilors of the local government. Carrying banners against the demarcation of indigenous lands, the protesters were invited to attend the indigenous territorial planning seminar. Those against the indigenous movement argued that land demarcation was being made in an obscure way, and disregarded the town's residents, fishermen and extractivists who were born and raised in the region. These protests created a hostile environment for rural communities living in Barcelos (Peres 2011).

Those who opposed to the demarcation of indigenous land in Barcelos questioned the authenticity of the indigenous population in Barcelos, since they consider this population to be 'mixed.' They accused the NGOs and their international financier of influencing *caboclos* to identify themselves as indigenous. Their discourse aligned with a dichotomous vision of traditional/rural versus acculturated/urban indians. At stake for this urban opposition was the loss of their rights to natural resource extraction as result of land demarcation. Natural resource extraction has been carried out without restrictions, which compromises the availability of resources for rural communities.

From an ethnic identity perspective, indigenous mobilization emerged as result of distinct ethnic groups demanding recognition of indigenous identity and culture (Jackson and Warren 2005). These arguments are posed against state policies based on the ideology of assimilation. In Barcelos, the presence of indigenous people and culture was historically denigrated and officially denied. Engaging in ASIBA, indigenous people and their descendants living in Barcelos were able to question the *caboclo* label and thereby revive their indigenous identity (Peres 2003, 2011; Pereira 2007).

Perceptions of ethnic identity were also influenced by the national and state context. The installation of infrastructure for indigenous health care, financed by the federal government, legitimized indigenous people's presence, in the face of denials by urban elites. Although indigenous authenticity is questioned in discourses against land demarcation, it is not questioned by local governments that are benefiting from the federal government's investment in indigenous healthcare. In addition, despite resistance, ASIBA has contributed to the strengthening of rural indigenous identity by providing political support for the process of demarcation of indigenous land (Pereira 2007; Peres 2011). This land demarcation is still being evaluated by FUNAI and the Brazilian government.

Historical migration flows were a key factor in the constitution of the local indigenous movement. Both migration flows – the regional flow from São Gabriel da Cachoeira and Santa Izabel do Rio Negro to Barcelos's rural and urban areas, and local circulation among rural communities and *sitios* and Barcelos town – were the key for the spread of the indigenous movement in Barcelos. Migration linked previously unconnected sites, which became a key mechanism for a 'scale shift', which McAdam, Tarrow, and Tilly (2001) refer to as 'brokerage'. This scale shift allowed for the transmission of information from one place to another. Migrants from the Upper RN River were brokers who contributed to the mobilizations by drawing on their previous experience in social move-ments, helping to maintain connections between the Upper and Middle RN. It is important to highlight that the first ASIBA president had just migrated from Upper RN with his family to Barcelos, looking for improved life conditions, at the time of ASIBA creation. He and his wife had previous experience working in the indigenous movement.

The spread of a movement between two places depends on actors in both locations having a mutual identification about the reasons for mobilization. This is the second mech-anism of a social movement scale shift: the 'attribution of similarity' (McAdam, Tarrow, and Tilly 2001; Tarrow and McAdam 2005) or 'frame bridging' (Snow and Benford 1988, 1992). Mobility and migration contributed to the linkage of people from different regions who held in common cultural practices as indigenous peoples. Interactions between them contributed to the (re)creation of common identities. Migration may hinder relationships with the origin community, but unites migrants around a common identification of their social position inside the urban world (Figoli and Fazito 2009).

The indigenous movement scale shift in Barcelos started first with ASIBA's emergence in the urban area, and later spread to rural areas. The movement leveraged rural–urban inter-actions established via local circulation and kinship networks, both by maintaining a radio communication system and by holding meetings in rural areas, all of which increased opportunities for interactions that supported a mobilization network in the service of rural communities. Rural communities thus engaged with ASIBA and created their own associations, incorporating the same model as employed previously in the Upper RN. The participation of rural communities in the indigenous regional network in turn built capacity for coordinated action and resistance against labor exploitation, paternalistic relations with the local government, and disputes over natural resources including land. Interactions through political networks thus strengthened the shared sense of indigenous identity among rural and urban indigenous peoples by increasing their participation in local and regional political spaces.

Conclusion

ASIBA's organizational model followed the larger process of indigenous 'associativism' and ethnic resurgence that had begun in the 1980s in the Upper RN. The alliance between ASIBA and FOIRN inserted these regional indigenous groups into a larger multiscale network of indigenous mobilization. Common identification between different sites served to justify pol-itical action in the form of 'associativism' (Peres 2003, 2011), leading to coordinated actions and tactics based on other experiences (Tarrow and McAdam 2005).

Contrary to the movement in São Gabriel da Cachoeira and Santa Izabel do Rio Negro, which was born under the flag of demarcation of indigenous land, the movement origins in Barcelos developed in a context of mobilization to recognize and appreciate indigenous culture, to demand better urban living conditions (such as access to education and health care services), and to open markets for handcrafted products. Now that the indigenous pres-ence is openly acknowledged in Barcelos, and ASIBA is institutionally recognized, the focus of indigenous mobilization has shifted towards the demarcation of indigenous land and rural ethno-development. Presently, the mobilization for land demarcation is the main reason for political contention in Barcelos. This demarcation would confer legal rights to rural indigenous communities not only over the land but also over natural resources, which are in dispute by the urban elites.

McSweeney and Jokisch (2007) suggest that urbanization and indigenous rural out-migration are factors contributing to indigenous mobilization in rural territories. This paper has corroborated this argument, presenting a case study of how indigenous mobiliz-ation spread in RN and specifically how rural-to-urban migration contributed to indigenous political mobilization in both urban and rural areas. I argued that migration and mobility can become key mechanisms for a 'scale shift' of social movements and indigenous territoria-lization. Mobility and migration do not necessarily result in the emptying of traditional ter-ritories; to the contrary, these migration flows can support regional multilocal networks for mobilization, with the potential to contribute to processes of territorialization. Future research is necessary to clarify whether this positive relationship between migration and social movement scale shifts is present in other Amazonian regions and indigenous groups.

Acknowledgements

The author thanks S. Perz, A. Pahnke, R. Tarlau and S. Schramski for comments and revisions. Thanks to Barcelos rural and urban communities, ASIBA, FOIRN and ISA for collaboration with the research. I also thank the International Foundation for Science, School of Natural Resources and Environment at the University of Florida, the Tropical Conservation and Development Program and the Gordon and Betty Moore Foundation for their support.

References

Adrião, D.G.S. 1991. *O processo de identificação étnica: a recriação da identidade indígena de Barcelos, AM*. Master thesis. Universidade Estadual de Campinas.
Albert, B. 1996. Associações indígenas e desenvolvimento sustentável na Amazônia brasileira. *Povos indígenas no Brasil, 2000*, 197–207.
Albert, B. 2005. Territorially, ethnopolitics, and development: the indian movement in the Brazilian Amazon. *The land within: Indigenous territory and the perception of the environment*, 200–228.
Alexiades, M.N. 2009. *Mobility and migration in indigenous Amazonia contemporary ethnoecologi-cal perspectives*. New York: Berghahn.
Almeida, A.W.B. 2004. Terras tradicionalmente ocupadas. Processos de territorialização e movimen-tos sociais. *Estudos Urbanos e Regionais* 6, no. 1: 9–32.
Alvarez, S.E. et al. 1998. *Cultures of politics/politics of cultures: re-visioning Latin American social movements*. Boulder, CO: Westview Press.
Andrello, G. 2006. *Cidade do índio: transformações e cotidiano em Iauaretê*. SciELO-Ed. São Paulo: UNESP.
Baines, S.G. 2001. As chamadas "aldeias urbanas" ou índios na cidade. *Revista Brasil Indígena* 2, no. 7: 15–7.
Bernal, R.J. 2003. *Indiens urbains – Processus de reconformation de lìdentité ethnique indienne à Manaus*. Université de Paris: Ecole des Hautes Etudes en Sciences Sociales, 2003. Thesis (PhD), Paris.
Bernal, R.J., and E.M.T. Mainbourg. 2009. *Índios urbanos: processo de reconformação das identi-dades étnicas indígenas em Manaus*. Manaus: EDUA, Editora da Universidade Federal do Amazonas, Manaus.
Bilby, K. 1996. *Ethnogenesis in the Guianas and Jamaica Two Maroon Cases. History, power, and identity : ethnogenesis in the Americas, 1492 - 1992*. Iowa City: Univ. of Iowa Press.
Bolaños, O. 2011. Redefining identities, redefining landscapes: indigenous identity and land rights struggles in the Brazilian Amazon. *Journal of Cultural Geography* 28: 45–72.
Brandhuber, G. 1999. Why Tukanoans migrate? Some remarks on conflict on the upper rio Negro (Brazil). *Journal de la Société des Américanistes* 85, no. 1: 261–80.
Brasil. 1988. Artigo 231. *Constituição da República Federativa do Brasil*. Imprensa Oficial. São Paulo. p.181.
Brysk, A. 1996. Turning weakness into strength: the internationalization of indian rights. *Latin American Perspectives* 23, no. 2: 38–57.
Cabalzar, A., and C.A. Ricardo. 2006. *Mapa-livro povos indígenas do Rio Negro*. São Paulo and São Gabriel da Cachoeira: Instituto Socioambiental (ISA) and Federação das Organizações Indígenas do Rio Negro (FOIRN).

Chernela, J.M. 2012. Indigenous rights and ethno-development: The life of an indigenous organization in the Rio Negro of Brazil. *Tipití: Journal of the Society for the Anthropology of Lowland South America* 9, no. 2: 93–120.

Conklin, B.A., and L.R. Graham. 1995. The shifting middle ground: amazonian indians and eco-politics. *American Anthropologist* 97, no. 4: 695–710.

Cunha, M.C. 1992. Política indigenista no século XIX. In: *História dos índios no Brasil*. Companhia das Letras: Secretaria Municipal de cultura: FAPESP.

Cunha, M.C. 2012. *índios no Brasil: história, direitos e cidadania*. São Paulo: Claro Enigma.

Cunha, M.C., and M. Almeida. 2000. Indigenous people, traditional people, and conservation in the Amazon. *Daedalus* 129, no. 2: 315–38.

Davis, S.H. 1977. *Victims of the miracle: development and the indians of Brazil*. Cambridge: Cambridge University Press.

Deere, C.D. and F.S. Royce. 2009. *Rural social movements in Latin America: Organizing for sustainable livelihoods*. Gainesville: University Press of Florida.

Eloy, L., and C. Lasmar. 2012. Urbanization and transformation of indigenous resource management: the case of upper Rio Negro (Brazil). *International Journal of Sustainable Society* 4: 372–88.

Emperaire, L., and L. Eloy. 2008. A cidade, um foco de diversidade agrícola no Rio Negro (Amazonas, Brasil)? *Boletim do Museu Paraense Emílio Goeldi* 3, no. 2: 195–211.

Federação das Organizações Indígenas do Rio Negro- FOIRN. 2013. Available from: http://www.foirn.org.br (acessed November 28, 2013).

Federação das Organizações Indígenas do Rio Negro- FOIRN, Instituto Sociambiental, ISA. 2014. Cartographic data on Barcelos communities.

Ferreira, J.P. 1959. *Enciclopédia dos municípios brasileiros*. Rio de Janeiro: IBGE.

Figoli, L. 1982. Identidad Étnica y Regional. Dissertation (MS). Universidade de Brasilia (UnB).

Fígoli, L., and Fazito, D. 2009. Redes sociales en una investigación de migración indígena: el caso de Manaus. *Encontro Nacional de Estudos Populacionais* 15: 77–95.

Figueiredo, P. 2009. Desequilibrando o convencional: estética e ritual com os Baré do Alto rio Negro. Thesis (PhD), PPgas/Mn/rJ.

Fundação Nacional do Índio, FUNAI. 2014. Demarcação de Terras Indígenas. http://www.funai.gov.br/index.php/todosdtp/154-demarcacao-de-terras-indigenas (acessed July 15 2014).

Gallois, D.T. 2004. Terras ocupadas? Territórios? Territorialidades. *Terras indígenas e unidades de conservação da natureza: o desafio das sobreposições*. São Paulo: Instituto Socioambiental, 37–41.

Gourevitch, A. 2011. Os Baré da Venezuela e do Brasil da área indígena e da cidade-ontem e hoje. *Cadernos CERU* 22, no. 1: 39–56.

Gray, A. 1997. *Indigenous rights and development: self-determination in an Amazonian community*. Vol. 3. New York and Oxford: Berghahn Books.

Guzmán, D.A. 2009. Mixed Indians, caboclos and curibocas: historical analysis of a process of miscegenation; Rio Negro (Brazil), 18th and 19th Centuries. In *Amazon peasant societies in a changing environment*, eds. C. Adams, R. Murrieta, W. Neves, and M. Harris, 55–68. Dordrecht: Springer Netherlands.

Instituto Brasileiro de Geografia e Estatística, IBGE. 2010. Geociências. http://downloads.ibge.gov.br/downloads_geociencias.htm (acessed 5 April 2010).

Instituto Brasileiro de Geografia e Estatística, IBGE. 2013a. Municipalities' data. http://www.cidades.ibge.gov.br/xtras/perfil.php?lang=&codmun=130040&search=amazonas|barcelos (acessed August 28, 2013).

Instituto Brasileiro de Geografia e Estatística, IBGE. 2013b. Census data. http://downloads.ibge.gov.br/downloads_estatisticas.htm (acessed August 28, 2013).

Instituto Sociambiental, ISA. 2008. Lideranças e associações indígenas debatem ordenamento territorial do Médio e Baixo Rio Negro (AM). http://www.socioambiental.org/nsa/detalhe?id=2821 (acessed April 29, 2012).

Instituto Sociambiental, ISA. 2009a. *Boletim socioambiental*. Número 1. Fevereiro.

Instituto Sociambiental, ISA. 2009b. Mobilização geral dos povos indígenas do médio e baixo Rio Negro reúne mais de 300 pessoas. http://www.socioambiental.org/nsa/detalhe?id=2921 (acessed April 28, 2012).

Instituto Sociambiental, ISA. 2009c. Seminário no médio Rio Negro (AM) reforça debate democrático sobre ordenamento territorial. http://www.socioambiental.org/nsa/detalhe?id=3006 (acessed April 28, 2012).

Instituto Sociambiental, ISA. 2010. *Boletim socioambiental*. Numero 3. Agosto.

Instituto Sociambiental, ISA. 2011. Oficinas discutem zoneamento das atividades de pesca no médio Rio Negro (AM). http://www.socioambiental.org/nsa/detalhe?id=3415 (acessed April 28, 2012).

Instituto Sociambiental, ISA. 2012. *Ordenamento pesqueiro do Médio Rio Negro é discutido com governo do Amazona*s. http://www.socioambiental.org/nsa/detalhe?id=3524 (acessed April 28, 2012).

Jackson, J.E., and K.B. Warren. 2005. Indigenous movements in Latin America, 1992–2004: controversies, ironies, new directions. *Annual Review of Anthropology* 34: 549–73.

Keck, M.E., and K. Sikkink. 1998. *Activists beyond borders: advocacy networks in international politics*. Ithaca, NY: Cornell University Press.

Keck, M.E., and K. Sikkink. 1999. Transnational advocacy networks in international and regional politics. *International Social Science Journal* 51, no. 159: 89–101.

Lasmar, C. 2005. *De Volta ao Lago de Leite: Gênero e transformação no Alto Rio Negro*. São Paulo: Edusp; ISA; Rio de Janeiro: NuTI.

Lasmar, C. 2008. Irmã de índio, mulher de branco: Perspectivas femininas no alto rio Negro. *Mana* 14, no. 2: 429–54.

Lima, A.C. de S. 1992. O governo dos índios sob a gestão do SPI. In *História dos Índios no Brasil*, ed. Manuela Carneiro Cunha, 155–72. São Paulo: Companhia das Letras.

Lima, D. M. 1999. A construção histórica do termo caboclo: sobre estruturas e representações sociais no meio rural amazônico. *Novos Cadernos NAEA*, 2, no. 2.

Machado, R. 2001. Life and culture on the Rio Negro, Brazil. In *Conservation and management of ornamental fish resources of the Rio Negro Basin Amazonia, Brazil – Project Piaba*, eds. N.L. Chao, P. Petry, G. Prang, L. Sonneschien, and M. Tlusty, 27–36. Manaus: Editora Universidade do Amazonas.

Mato, D. 2000. Transnational networking and the social production of representations of identities by indigenous peoples' organizations of Latin America. *International Sociology* 15, no. 2: 343–60.

Melo, J.G. 2009. Identidades fluidas: ser e perceber-se como Baré (Aruak) na Manaus Contemporânea. PhD thesis. UnB. PPGAS. Brasília.

McAdam, D. 2003. Beyond structural analysis: toward a more dynamic understanding of social movements. In *Social movements and networks: relational approaches to collective action*, eds. M. Diani, and D. McAdam, 281–98. Oxford University Press.

McAdam, D., S. Tarrow, and C. Tilly. 2001. *Dynamics of contention*. Cambridge and New York: Cambridge University Press.

McSweeney, K., and B. Jokisch. 2007. Beyond rainforests: urbanization and emigration among lowland Indigenous societies in Latin America. *Bulletin of Latin American Research* 26, no. 2: 159–80.

Ministério do Meio Ambiente, MMA. 2011. *Protected Areas*. http://mapas.mma.gov.br/i3geo/datadownload.htm (accessed 12 October 2011).

Moghadam, V.M. 2012. Global social movements and transnational advocacy. In *The Wiley-Blackwell companion to political sociology*, eds. E. Amenta, K. Nash, and A. Scott, 408–20.

Nugent, S. 1993. *Amazonian caboclo society: an essay on invisibility and peasant economy*. Berg Publishers Ltd. Providence. 278pp.

Occhipinti, L. 2003. Claiming a place: land and identity in two communities in Northwestern Argentina. *Journal of Latin American Anthropology* 8, no. 3: 155–74.

Oliveira Filho, J.D. 1996. Viagens de ida, de volta e outras viagens: os movimentos migratórios e as sociedades indígenas. *Revista do Migrante*, 5–10.

Pereira, R. 2007. *Comunidade Canafé: história indígena and etnogênese no rio Negro*. Doctoral thesis. Brasilia: Universidade de Brasília.

Pereira da Silva, R.N. 2001. O universo social dos indígenas no espaço urbano: identidade étnica na cidade de Manaus. Thesis (MS). Porto Alegre:Universidade Federal do Rio Grande do Sul.

Peres, S. 2011. Associativismo, etnicidade e conflito: sociogênese do movimento indígena no Rio Negro. *Paper presented at the Congreso Internacional da Sociedade latino Americana de Sociologia*.

Peres, S.C. 2003. *Cultura, política e identidade na Amazônia: o associativismo indígena no Baixo Rio Negro*. Doctoral thesis. Universidade Estadual de Campinas, Instituto de Filosofia e Ciências Humanas. Campinas.

Perrone-Moisés, B. 1992. Índios livres e índios escravos: os princípios da legislação indigenista do período colonial (séculos XVI a XVIII). In *História dos índios no Brasil*, ed. M. C. Cunha. Companhia das Letras, 116–132. São Paulo.

Perz, S.G., Warren, J., and Kennedy, D.P. 2008. Contributions of racial-ethnic reclassification and demographic processes to indigenous population resurgence: the case of Brazil. *Latin American Research Review* 43: 7–33.

Pinedo-Vasquez, M., and C. Padoch. 2009. Urban, rural and in-between: multi-sited households mobility and resource management in the Amazon flood plain. In *Mobility and migration in indigenous Amazonia: contemporary ethnoecological perspectives. Studies in environmental anthropology and ethnobiology*, eds. M.N. Alexiades, 11, 86. New York; Oxford: Berghahn.

Pinedo-Vasquez, M. 2009. Urbano e rural: famílias multi-instaladas, mobilidade e manejo dos recursos de várzea na Amazônia. *Novos Cadernos NAEA* 11, no. 2.

Prang, G. 2001. *A caboclo society in the Middle Rio Negro basin: ecology, economy and history of an ornamental fishery in the state of Amazonas, Brazil.* Doctoral thesis. Wayne State University, Detroit, Michigan.

Ramos, A.R. 1998. *Indigenism: ethnic politics in Brazil.* Madison: University of Wisconsin Press.

Ramos, A.R. 2011. Os direitos humanos dos povos indígenas no Brasil. *Desafios aos direitos humanos no Brasil contemporâneo*, 65–87.

Reis, A.C.F. 1999. *Manaós e outras villas.* 2ª Ed. Rev. EDUA. Governo do Estado do Amazonas.

Romano, J. 1981. *Índios Proletários en Manaus.* Thesis (PhD). Brasilia: Universidade de Brasilia (UnB).

Santos, R. 1980. História econômica da Amazônia (1800-1920) T.A. São Paulo.

Santos, A.F.M., and J.P. Oliveira Filho. 2003. *Reconhecimento étnico em exame: dois estudos sobre os Caxixó* (Vol. 9). Rio de Janeiro: LACED.

Santos, R., and P. Teixeira. 2011. Indígena que Emerge do Censo Demográfico de 2010. *Cadernos de Saúde Pública* 1048–1049.

Schmink, M., and C.H. Wood. 1992. *Contested frontiers in Amazonia.* New York: Columbia University Press.

Sieder, R. 2002. *Multiculturalism in Latin America: indigenous rights, diversity and democracy.* London: Institute of Latin American Studies.

Snow, D., and R. Benford. 1988. Ideology, frame resonance, and participant mobilisation. In *From structure to action: comparing social movement research across cultures*, eds. B. Klandermans, H. Kriesi, and S. Tarrow, 197–217. Greenwich, CT: JAI.

Snow, D., and R. Benford. 1992. Master frames and cycles of protest. In *Frontiers in social movement theory*, eds. A. Morris, C. Mueller, 133–55. New Haven, CT: Yale University Press.

Soares, R.M. 2012. *Das comunidades à federação: associações indígenas do alto Rio Negro.* São Paulo: Dissertação de Mestrado, Faculdade de Filosofia, Letras e Ciências Humanas, Universidade de São Paulo.

Sobral, C., and C. Dias. 2013. *Barcelos indígena e ribeirinha: um perfil socioambiental.* C. Sobral and C. Dias, eds. ISA/ASIBA/FOIRN Press.

Sobreiro, T. 2015. Urban-rural livelihoods, Fishing conflicts and indigenous movements in the middle Rio Negro Region of the Brazilian Amazon. *Bulletin of Latin American Research*, 34: 53–69. doi:10.1111/blar.12259.

Stahler-Sholk, R., H.E. Vanden, and G.D. Kuecker. 2007. Introduction: Globalizing resistance: The new politics of social movements in Latin America. *Latin American Perspectives* 34, no. 2: 5–16.

Schwartzman, S., and B. Zimmerman. 2005. Conservation alliances with indigenous peoples of the Amazon. *Conservation Biology* 19: 721–27.

Tarrow, S., and D. McAdam. 2005. Scale shift in transnational contention. In *Transnational protest and global activism*, eds. D. Della Porta and S.G. Tarrow, 121–50.

Tilley, V.Q. 2002. New help or new hegemony? The transnational indigenous peoples' movement and being indian in El Salvador. *Journal of Latin American Studies* 34: 525–54.

Trager, L. 2005. *Migration and economy: Global and local dynamics.* Walnut Creek, CA: AltaMira Press.

Trindade, S.C.Jr. et al. 2008. Das" janelas" às" portas" para os rios: compreendendo as cidades ribeirinhas na Amazônia. *Cidades ribeirinhas na Amazônia: mudanças e permanências.* Editora Universitaria UFPA.

Turner, T. 1991. Representing, resisting, rethinking. *Colonial Situations: Essays on the Contextualization of Ethnographic Knowledge*: 285–313.

Warren, J.W. 2001. *Racial revolutions: antiracism and Indian resurgence in Brazil*. Durham, NC: Duke University Press.

Wright, R.M. 1992. História indígena do noroeste da Amazônia: hipóteses, questões e perspectivas. *História dos índios no Brasil* 2: 253–66. São Paulo: Claro Enigma.

Wright, R.M. 2005. *História indígena e do indigenismo no Alto Rio Negro*. Campinas: Unicamp.

Van Cott, D.L. 2001. Explaining ethnic autonomy regimes in Latin America. *Studies in Comparative International Development* 35, no. 4: 30–58.

Van Cott, D.L. 2010. Indigenous peoples' politics in Latin America. *Annual Review of Political Science* 13: 385–405.

Vidal, S.M. 2000. El rol de los líderes Barés en el surgimiento y desaparición de confederaciones multiétnicas en el noroeste amazónico (siglo XVIII). In *Historia y etnicidad en el noroeste amazónico*, eds. A. Zucchi, and S. Vidal, 83–94. Universidad Los Andes.

Yashar, D. 2005. *Contesting citizenship in Latin America: the rise of indigenous movements and the postliberal challenge*. New York: Cambridge Univ. Press.

Lula's assault on rural patronage: Zero Hunger, ethnic mobilization and the deployment of pilgrimage

Aaron Ansell

This paper explores the Workers' Party government's attempt to use anti-poverty policy to disrupt rural patronage, and the implications of this effort for theories of patronage. I argue that state officials and race-based activists implementing President Lula's flagship 'Zero Hunger Program' (2003–2005) turned mundane program exercises into pilgrimage rites in an effort to build lateral solidarities among Afro-Brazilians and undermine their vertical patronage alliances. The partial success of such efforts suggests that there are circumstances in which vertical and horizontal alliances are compatible, and that investigating this compatibility entails consideration of local categories of exchange.

In 2000, Luis Inácio 'Lula' da Silva, the leader of Brazil's left-wing *Partido dos Trabalhadores* (Workers' Party – PT), went on television and told the nation,

> Unfortunately, in Brazil, the vote is not ideological. Unfortunately, people don't vote for a political party. And unfortunately, you have a part of society that, because of its great poverty, is led [*conduzida*] to think with its stomach and not with its head. That's why there are so many food baskets and milk packs distributed at election time – because this is a kind of trade. And in this way, you depoliticize the electoral process .... You have the logic of maintaining a politics of domination ... [1]

In this brief polemic, Lula summarized the Brazilian Left's conviction that grave material deprivation had long buttressed the patron–client exchanges ('patronage' or 'clientelism') that pervert Brazil's representative institutions. These perverse exchanges short-circuited Brazilian democracy by depoliticizing the vote, depriving the electorate of an 'ideology' of governance grounded in claims about the public good. Accordingly, the elector could not construct an ideological perspective if she did not reckon herself as a member of a collective (the nation, the working class, etc.). Thus, hunger was not just a human

---

[1] Lula released this televised statement as part of a public critique of President Fernando Henrique Cardoso's redistribution measures as mere palliatives tantamount to state patronage. In 2008, the video took on a wide Internet circulation, when Lula's adversaries cited it as evidence of his hypocrisy (titonio2000 2010). The premise of those critiques was that Lula's *Bolsa Família* program was no different from the 'food basket and milk distribution' programs of President Cardoso.

tragedy; it was a political tragedy that atomized the electorate, foreclosing those solidarities among workers, neighborhood residents or marginalized ethnic groups that allowed for proper democratic engagement.

These comments foreshadowed President Lula's flagship social program, 'Zero Hunger' (*Fome Zero*), which sought to ameliorate extreme poverty, originally focusing on the rural residents of the Northeast. But that was not its sole purpose. Here I draw on two years of fieldwork accompanying the implementation of Zero Hunger in Piauí State[2] (2003–2005) to argue that the state officials who implemented this program (and, later, *Bolsa Família*) infused their administrative activities with techniques aimed at recalibrating the political culture of the rural underclass, the program's primary beneficiaries. By enlisting the participation of regional social movements that championed the struggles of disenfranchised groups, Zero Hunger's assault on patronage gained added rhetorical depth and cultural specificity. Still, the joint efforts of the state and its allied social movements were only partially successful.

It is not surprising that a left-wing Latin American government would generally oppose patron–client relationships. The historical association between the region's large plantation (hacienda or latifúndio) economies and its 'semi-sacred cultural mechanisms such as ritual kinship and other ceremonial ties' has made patronage in the countryside a particular concern throughout the continent (Wolf and Mintz 1957, 391). In Brazil, Communists, unionists and Catholic organizers have long struggled to organize rural laborers along class lines, only to be frustrated by their 'lamentable state of ignorance' regarding government policies, national and state-level party politics and political ideologies (Leal 1948, 137; see also Graham 1990).[3] Yet scholars studying patronage are divided regarding the basic assumption that personalistic vertical alliances impede the poor from forging empowering horizontal ties with one another. At stake in this debate is the issue of whether patronage alliances are necessarily inimical to democratic participation. Analysis of the Lula administration's attempt to fortify lateral ties and diminish vertical reciprocities offers new insights into this unresolved question. It also builds towards a more historically specific understanding of the social engineering project that underlay President Lula's 'market-friendly Brazilian state activism' (Arbix and Martin 2011, 64).

Here, I offer an ethnographic analysis of one component of President Lula's Zero Hunger Program in order to make two general points about rural patronage, state-led efforts to dismantle patronage and the role of regional social movements in such efforts. First, subsistence cultivators in Northeast Brazil (and likely elsewhere) distinguish between modes of political exchange (patronage) that fragment intra-village solidarities and those that facilitate them. In the absence of clear evidence to refute their reflections, scholars should assume that such distinctions have real (albeit only local)

---

[2]The choice to launch Zero Hunger in Piauí State made sense in light of two factors: First, the same year Lula of the PT was elected president of Brazil, Piauí's citizens elected a PT governor, Wellington Dias. This meant that Lula had a natural ally and partner to help him implement programs in that state, whereas other Northeastern states were largely dominated by conservative governors who might try to sabotage Zero Hunger or claim credit for it. Second, Piauí is a notoriously poor state, and one that has a large subsistence cultivator population with a relatively small rural proletariat and landless workers' movement. For that reason, the face of poverty in Piauí state was less militant and controversial.

[3]Leal (1948) and Graham (1990) are concerned with a historically specific (early twentieth-century) configuration of rural patronage in Brazil, one in which local bosses compensated for their declining economic power by channeling state and federal resources to the poor in exchange for their votes.

merit. As a result, state-led initiatives that assume that all patronage undermines lateral solidarity may confuse policy beneficiaries, as well as ignore or degrade the forms of agency that they hope to exercise. The second point pertains to the means by which left-wing governments such as President Lula's actually go about subverting patronage and the outcome of such efforts. Zero Hunger officials undertook a subtle project of culture engineering that sought to build class- and race-based horizontal solidarities among beneficiaries that would ultimately change their understanding of the state and reorient their affective posture towards local authorities. Here, I claim that their primary means of doing so was to turn mundane programmatic encounters into symbolically dense rituals, such as pilgrimage. To help them ritualize program activities, state officials relied on local social movements to supply programmatic encounters with evocative narratives of past heroes, villains and courageous acts, along with imagined futures of achievement and redemption. These narratives could then be rhetorically linked to the assault on patronage in such a way as to provide mythic legitimacy and increased multivocality to that social engineering effort. Having helped to solidify these collective identities, Zero Hunger officials then linked them to an ideology that positioned the state as an indebted party to an exchange with these groups, rather than a creditor whose programmatic gifts required electoral repayment. In this way, Zero Hunger officials (and the activists with whom they worked) linked the sociological project of organizing lateral collectivities to an ideological project of creating citizens who pursued state resources as a political right. But this does not mean that the cultural changes they intended occurred as planned.

It is a predictable irony of such efforts that their proponents often become unwitting patrons themselves. This is what occurred in Piauí State, where individual state officials and social activists alike became resource brokers for beneficiary communities even after the Zero Hunger project discussed here ended. A cynical reader might assume that the PT strategists had this in mind all along, that they themselves were no more than would-be patrons using social policy to lure clients from rivals. To my knowledge, no Zero Hunger official ever distributed any public resource on the condition of beneficiary political support. (Whether officials hoped that beneficiaries would reward the PT for Zero Hunger at the polls is another matter.) Here, I am less concerned with defending the PT against such allegations than I am with showing how the patronage dynamics between Zero Hunger officials (including activists) and beneficiaries emerged as a consequence of the beneficiaries' habituated modes of exercising agency within the patronage framework. They mobilized the strategic forms of etiquette associated with moral forms of vertical exchange that had long worked to their advantage.

I proceed by elaborating scholarly debates pertaining to patronage's compatibility with horizontal modes of organization, suggesting that such compatibility depends on local categories of exchange. I show how rural residents of 'Passarinho' Municipality (one of Zero Hunger's two pilot municipalities) have traditionally differentiated between two different modalities of vertical exchange. This local distinction is key for understanding how representatives of Passarinho's villages engaged the state officials and social activists implementing Zero Hunger. Following some general comments about the Zero Hunger Program, I explore one of its many component projects, a regional initiative that involved Afro-Brazilian villagers in a pilgrimage experience designed to build their sense of race-based community and replace their putative deference towards government authority with an assertive political attitude grounded in an ideology of rights (a là Amartya Sen (1988) on entitlement to food).

The compatibility of vertical and horizontal alliances

The 'patron–client relationship' emerged as an anthropological concept to explain the importance of kin relations in peasant societies that lacked the unilineal descent groups that organized many African political systems (Mintz and Wolf 1950; Foster 1963; Forman 1975).[4] Early ethnographic studies did not condemn patronage, but the spread of electoral institutions throughout the global south eventually led to scholarly concerns that the vote had become a currency of patronage exchange. Political scientists especially began to worry that electors' rational evaluation of their collective interests was subverted by patron buy-offs, and, relatedly, that electors' propensity to organize into horizontal collectivities (to demand good policies) was subverted by the atomizing effects of private, illicit patronage deals (see, *inter alia*, O'Donnell 1992; Putnam, Leonardi, and Nanetti 1993; Stokes 1995). Scholars studying Brazil have been particularly sympathetic to this perspective; many attribute the shortcomings of the nation's democracy to the divisive effects of patronage on the rural and urban masses (e.g. Leal 1948; Schmitter 1971; Martinez-Alier and Boito Júnior 1977; Weyland 1996; Holston 2008; Nelson and Finan 2009).

More recent scholarly work challenges the assumption that horizontal associations and vertical patronage ties are mutually antithetical (Burgwal 1995; Marques 1999; Goldman 2000; Auyero 2001; Villela 2004; Collins 2008; Auyero, Lapegna, and Poma 2009). Javier Auyero's (2001, 16) investigation of 'poor people's politics' in Argentina suggests that intimate, horizontal 'problem-solving networks can be envisioned as concentric circles surrounding [patron] brokers'. Far from implying fragmentation of the client population, this image evokes a kind of centripetal force: patron–brokers pull clients together. He and others (Auyero, Lapegna, and Poma 2009) have gone on to explore those circumstances in which such patron brokers promote militant collective action among their clients (e.g. mobilizing protests). Jorge Mattar Villela's (2004) historical ethnography of the interior region of Pernambuco (Piauí's neighboring state) goes so far as to jettison the concept of patron-clientelism altogether, arguing that its emphasis on centralized control occludes the diffusion of power across rural households and the segmentary quality of conflict (pacé Foucault and Evans-Pritchard). Villela's account suggests that not only do vertical and horizontal alliances coexist, the rural poor do not make sharp distinctions between them. I draw from these analyses two findings: first, patronage can, but does not necessarily, spur collective democratic action, and, second, investigating this possibility entails an ethnographic analysis of any local population's thinking about power and exchange.

Anthropological studies of gift exchange point to the internal complexity inherent to most societies' classifications of power, wealth and exchange-based agency (e.g. Bloch and Parry 1989; Graeber 2001; Elyachar 2005). This cross-cultural literature suggests that many societies construe exchange in bimodal terms. In particular, Bloch and Parry's (1989) generalization, that long-term 'transaction orders' tend to be considered moral and community-enhancing, while short-term reciprocities are often construed as selfish and self-destructive, provides a useful starting point for thinking about the complexity of local exchange modalities in rural Piauí. A brief sketch of the ethnographic context bears this out.

In the municipality of Passarinho, Piauí (population about 5000 in 2003), most families live on small plots of land (less than 50 hectares) where they plant corn, beans and manioc, and raise goats and sheep. About 80 percent of the municipality's inhabitants live in villages

---

[4]I am indebted to anthropologist Sidney Greenfield who conveyed this insight to me in a personal conversation in 2011.

surrounding the town center, which is located 12 km from a small city. Labor-intensive farming on their small lands yields a very little surplus for sale in local markets, and chronic hunger affects many households; at least, it did prior to the PT's federal antipoverty policies (CESACF 2003). In that regard, Passarinho is fairly typical of Brazil's vast, semi-arid backlands where some 15 million people reside.

Villagers and politicians alike distinguish between moral and amoral exchanges between municipal politicians (usually from the rancher or small-business class) and 'the masses' (*o povão*). Local talk about the merits of political candidates posits that a good politician is 'always there for us'. Such talk valorizes the long-term character of exchange relations between subsistence cultivators and local elites (ranchers, business owners, etc.). These moral ties are stereotypically realized through material gifts (farm inputs, medical care, village reservoirs, etc.) aimed at ensuring the cyclical perpetuation of crops, families and village neighborhoods. Such material gifts not only bear fruit over long periods of time, they are often repeated across successive electoral cycles, creating lasting sentimental bonds. Rural people who commit their votes to such benefactors claim that their support 'comes from the heart'. Politicians pursue this heartfelt support by enacting what Villela (2012) calls an 'etiquette' of politics, a utilization of personal memory, kin connections, geographic familiarity, etc. to engender fellow feeling. They also make themselves vulnerable to these villagers by conveying to them their deficits of political support and their consequent need for help in their campaigns (Ansell 2014). Conversely, villagers court long-term alliances with politicians by enacting a reciprocal etiquette and vulnerability, reminding politicians of the gift objects that each has given the other and the happiness and gratitude that these exchanges have brought them, tracing kin ties and conveying some of the (potentially shameful) needs that plague their households. Crucially, both politicians and electors claim that the 'unity' of electors' families (nuclear and extended) and interfamily networks improve the moral standing of those who approach politicians, and that this unity makes their allied politicians more attentive to their needs, more generous in their gifts, and more likely to fulfill those promises they make during the campaigns. Indeed, from a practical perspective, it is easy to see why a politician would seek to appease the members of tightly knit groups, who influence one another's votes.

In contrast to these moral exchanges, both cultivators and elites lament the prevalence of short-term 'vote-buying' (*compra de voto*) that usually occurs immediately before an election through the medium of cash or booze. These buy-offs are (stereotypically) devoid of affect and mutual concern, and are condemned as symptoms of an increasing selfishness and desire for cash that afflicts all municipal inhabitants (Ansell 2014). Villagers claim that this second kind of exchange is easier to come by, and that politicians' perceptions of intra-village or intra-family strife make them more likely to solicit votes through buy-offs. This condemnation of short-term exchanges is diminishing as cash becomes a more predominant feature of the local economy, and as young adults increasingly see cash exchange with politicians as a means to start their independent lives. Still, the traditional distinction between long- and short-term exchange carries moral and practical weight, and, thus, the question of whether Passarinho's rural poor perceive horizontal solidarities to be compatible with vertical reciprocities depends on the kind of vertical reciprocities in play.

This dichotomous exchange system is not the only feature of rural Brazilian political culture that shaped the Zero Hunger beneficiaries' engagement with the program. Typically, rural families want to forge alliances with politicians who are likely to win elections, and thus control key resources (e.g. municipal jobs). The same logic leads them to build alliances with local brokers who have good connections with stable higher-level authorities (e.g. state and federal legislators and bureaucrats). Beneficiaries' perception that the political tides in Piauí and Brazil were turning in favor of the PT likely motivated them to establish alliances with officials and activists who could broker their relationship to PT-controlled governments.

Appreciating these features of Passarinho's political culture helps us to understand why Zero Hunger's rural beneficiaries embraced the program's effort to expand their lateral solidarities, while they rejected its assault on their vertical affiliations. It also explains how the program ultimately improved the political standing of its rural beneficiaries; Zero Hunger unintentionally gave these rural beneficiaries the means to upgrade their patronage relations, expanding the moral, long-term relations that affirmed egalitarian group organization.

The Zero Hunger Program

Lula's plans for a national 'food security' program emerged during the 1990s, following his loss of the 1989 presidential election. The PT and its sympathizers needed a bold platform that would rally the nation but not scare off the middle class that generally feared Lula's association with the socialist Left. In 1993, they formed the National Food Security Council that would draw together a number of civil society experts (nutritionists, agronomists, sustainable developers), social movements (land reform, race-based movements, trash-pickers), and multilateral development agencies (e.g. the United Nations Food and Agriculture Organization, FAO) to discuss the prospects of a national food security[5] policy. The complex structure of the Zero Hunger Program was drafted by 2001.[6] By the 2002 presidential election, Lula's association with the fight against hunger had eclipsed his earlier, more radical socialist persona, making him electable (Hunter 2010, Ch. 5). He launched the much-anticipated Zero Hunger (*Fome Zero*) Program in the Northeast, the symbolic epicenter of Brazilian poverty and the locus of the nation's moral tradition (Alburquerque Jr. 1999). It quickly became apparent that Zero Hunger was too complex for its own good; its structure tried to reconcile too many competing interests, and many of its planned components were never realized. By the end of 2003, the Lula administration shifted its focus to a new program, the now-famous *Bolsa Família* cash grant. Zero Hunger continued officially, but petered out in practice. Its short lifespan notwithstanding, the program embodied the deeper aspiration of the Lula administration 'to radicalize democracy' through an attack on the cultural complex of patronage (Maluf et al. 1996, 88).

---

[5] 'Food security' was the more technical organizing term for the Zero Hunger Program. According to the1996 World Food Summit held in Rome in which many Brazilian activists participated, food security refers to a situation in which 'all people at all times have access to sufficient, safe, nutritious food to maintain a healthy and active life' (World Food Summit 1996).

[6] Zero Hunger was technically comprised of three pillars: structural, local and emergency initiatives. The key structural initiative was to be a national land reform policy, but the administration never implemented this. The emergency initiatives revolved around food distribution and the temporary cash stipend, Food Card (*Cartão Alimentação*), which was the only Zero Hunger policy that spread throughout most of the Northeast. The local initiatives included the race-based *Quilombola* Project discussed in this text, along with a host of small-scale development actions, community gardens, cooperative enterprises, and water infrastructure projects. These were implemented in select locales in tandem with social movements and nongovernmental organizations (Fome Zero 2003).

This aspiration did not disappear with the advent of *Bolsa Família*. It continues to define the culture of the PT administration.

Zero Hunger's leadership in Brasília adopted a defensive posture towards patron–client relations in the countryside. The PT administration, afraid that its federal resources would be co-opted by municipal officials, consciously reflected on how to 'depersonalize [the program's] administration ... to reduce the clientelistic use' of its component initiatives (Maluf et al. 1996, 83). Even if destroying existing clientelist structures seemed beyond their capacity, they hoped to at least 'break with the logic, and to overcome the identifications ... with the fragmenting power of clientelismo' (Yasbek 2004, 112). In the eyes of the administration, the 'fragility of social mobilization' was both a cause and a consequence of clientelist reciprocities (Burlandy et al. 2007, 92, also see Vaitsman, Andrade, and Farias 2009; Bichir 2010).

Zero Hunger propaganda framed the rural poor's deficit of lateral solidarity as cultural loss: the once-fraternal countryside had become individualistic. Cultivators no longer sustained their traditional collective labor forms, the so-called *mutirão*, which one Zero Hunger brochure defined as 'a practice of working together that has its roots in indigenous and peasant cultures' (Fome Zero 2003, 12). Zero Hunger's nostalgic rhetoric invited myriad interpretations and renditions among allied social movements whose own discourses also thematized loss (rural people whose lands had been taken by agribusinesses, fishing communities displaced by commercial development, indigenous culture forgotten due to assimilation, etc.). Zero Hunger's general call to recover a bygone age of solidarity could thus be translated into and out of these movement-specific discourses, as could its opposition to patronage authority. In this regard, Zero Hunger rhetoric exemplified a general motif of Brazilian redemocratization discourse. In the wake of the military dictatorship (1964–1985), various social movements, especially race-based movements, have called for the recovery (*resgate*) and valorization of heritage, memory and culture in an effort to reimagine the foundations of the nation and to enlist the state in carving out a dignified place for its historically marginalized citizens (Sansi 2007, 110–15; Araujo 2010, 255; Collins 2011).

Zero Hunger documents extended the metaphor of the *mutirão* to encompass social movements and other organized civil society entities in northeastern municipalities. It was as if these entities embodied both the extension of the PT administration in the countryside and the future trajectory of rural beneficiaries who would hopefully join these organizational entities in opposition to conservative power. The administration assumed that these entities operated outside the circuits of clientelistic alliance, and even created local councils dominated by civil society organizations to select the program's beneficiaries, deliberately bypassing distrusted municipal authorities (Betto 2003, 18).

Most Zero Hunger officials who had direct contact with the rural beneficiaries were not federal employees but the state-level personnel who actually traveled to the remote rural municipalities to implement the program's various components. In Piauí State, where the program's two pilot towns were located (Passarinho being one of them), about 20 state officials worked as part of an ad hoc group called the *Coordenadoria de Segurança Alimentar e Erradicação da Fome/Programa Fome Zero* (the Food Security and the Eradication of Hunger/Zero Hunger Coordination Office, hereafter the 'Zero Hunger Coordination Team'). Most of them lived in the lower middle-class outskirts of the state capital, Teresina, and had activist backgrounds in various social movements. Their perspective on Zero Hunger's confrontation with rural patronage differed in an important respect from that of their superiors in Brasília: for the state-level officials, the program would ideally do more than evade the grasp of clientelist mayors; it would subvert clientelism itself.

Whereas the federal-level administrative efforts to shield Zero Hunger from patronage were conscious and preplanned, the more aggressive, state-level efforts to dismantle patronage emerged within 'the embodied work that [was] done' during the real-time flow of policy implementation (see John French 2009, 367, for comparison). Only occasionally did state-level officials articulate their intent to subvert patronage, and they never spoke of these efforts as if they had any forethought or technique. Still, there was a discernable coherence of meaning in the way the officials spoke to the beneficiaries and in how they arranged the latter's activities. In the context of the initiative I examine below, state officials and their allied activists enacted a sequence of activities resembling pilgrimage, including the temporary geographic dislocation of beneficiaries to stimulate a powerful empathetic solidarity among beneficiaries (what Victor Turner called 'communitas'), and to instill within the beneficiaries an indignant and confrontational attitude towards government authorities (Turner and Turner 1978, also see Daniel 1984, 237–39).

The particular Zero Hunger project in question was fairly local in scope (it applied to eight or so contiguous municipalities, Passarinho included), and took as its 'target population' Afro-Brazilian villages that identified as descendants of slave communities, known as *quilombos*.[7] To implement this particular, race-based initiative, the Zero Hunger Coordination Team partnered with the regional Black Movement (also called the *Quilombola* Movement) headquartered in a small city adjacent to Passarinho. The relationship between the activists and the Coordination Team was tense at first, but ultimately symbiotic. The tension derived from the Movement's distrust of most institutional authority. The activists were also keenly aware of the class differences between themselves and the Coordination Team members. None of the activists had stable incomes or much accumulated wealth. A couple owned old motorcycles but they could not afford to fuel them. Furthermore, while they supported the PT, most were not convinced that the Zero Hunger Program really cared about their cause. Relations between the Coordination Team members and the activists softened as the latter began receiving consultant fees that allowed them to pay their bills and to increase their visibility to the villages they wanted to mobilize. The two groups converged in two other regards: first, the *quilombola* activists were critical of local clientelism, and they too collapsed the distinction between moral exchange and 'vote-buying'. Second, the state officials shared the activists' suspicions of government authority.[8] Because of these convergences, it proved easy for both groups to translate Zero Hunger's generic anti-patronage project into the *quilombola* idiom. This framing would help the activists secure ongoing relevance in the beneficiaries' engagements with the state (even after Zero Hunger ended). But this is not to say that the *quilombola* idiom was more vernacular or familiar to the beneficiaries than other Zero Hunger discourse. In fact, before this project, the Afro-Brazilian villagers who I had been living with in Passarinho's interior had never heard of a *quilombo/la*. Many were reticent to identify themselves as 'black' (*negro* or *preto*), and even fewer were prone to make their racial identity relevant to interactions with political authorities.

---

[7]The term *quilombo* refers to a community of escaped African slaves or their descendants, while *quilombola* refers to the people (singular or plural) living in such a community. The term *quilombola* can also function as the adjectival form of *quilombo*, as in *comunidade quilombola*.

[8]Zero Hunger discourse, influenced by the culture of the PT, generally distrusted large institutions, believing that these often tried to control workers and other social movements to the latter's detriment. With regard to social policy, Zero Hunger discourse labeled the initiatives of prior administrations *assistencialismo*, roughly translatable as 'welfare statism', though the Portuguese term critiques the recipients' subordination to the state rather than state waste (Ansell 2014, 29–31).

The Quilombola/FAO project

The substance of the Quilombola project (launched officially in 2004) was not predetermined; it would be decided through consultation among the Zero Hunger Coordination Team, the Quilombola activists and representatives of the beneficiary communities at a 3-day training/planning seminar in the capital city, Teresina. The Coordination Team relied on the activists to identify these beneficiary communities for the state and to convince them to participate.

As a North American anthropologist, I became involved in the project in March 2004. I had traveled from Passarinho to the state capital to attend the Second National Conference on Food Security, which is where I met Maria, the leader of the regional *Quilombola* Movement. She told me of the incipient *Quilombola* project and of her intent to make contact with the Afro-Brazilian village where she heard I'd been living. She asked if I would broker a meeting between her and the village's development association. I agreed.

Like many other state and regional branches of Brazil's loosely organized *Movimento Quilombola*, Maria's group had been trying to mobilize Passarinho's Afro-descended villages as *quilombos* since 1988, when Brazil's post-dictatorship constitution obliged the federal government to grant collective land titles to the 'descendants (*remanescentes*[9]) of *quilombola* communities' (Article 68; see Arruti 1997, 27). The criteria for the state's recognition of a *quilombo* had been hotly debated for several years prior to Zero Hunger (Jan French, 2009). Some in the government had maintained that state recognition required reports from outside experts (e.g. anthropologists), but in 2003, Lula issued a presidential decree making a community's 'self-identification' as a *quilombo* sufficient for state recognition, and this remains the case (O'Dwyer 2007, 51). Still, community 'self-identification' remained problematic because many rural Afro-Brazilians had little historical memory of slavery and slave resistance.[10]

Knowing that Passarinho's communities had little familiarity with her movement and the *quilombola* concept, Maria came to the village's association meeting one Sunday in early April ready to impart a comprehensive lesson in history and contemporary politics. She began by telling the villagers that the federal government was finally fulfilling its obligation to right the wrongs of 500 years, and continued:

> We've all heard the story of how the little white Princess Isabel freed the black slaves so many years ago (in reference to the Golden Law of 1888), but do you think your *negro* ancestors simply walked off the fields of the white masters and started their own farms? No, they had nowhere to go so they had to keep working. They still took the whip. The beauty of *negro* women was still exploited. They continued living under an abominably racist and exploitative

---

[9]José Maurício Andion Arruti (1997) explains the contemporary usage of the term *remanescente* (lit. 'remainder') to supplement *quilombo/ola,* claiming that it 'arises to reconcile the continuity and discontinuity with the historic past (in a context) in which descent does not seem to offer a sufficient tie' (21). Its usage, in short, affirms that direct descent is not a necessary criterion to warrant a community's recognition as a *quilombo.*

[10]Many scholars have noted that contemporary Afro-Brazilian communities lack a strong historical memory of slavery or historical *quilombola* resistance, and thus often do not identify with the contemporary *quilombola* ethnic category (e.g. Price 1999, 23; O'Dwyer 2007, 45). Scholars and activists have thus questioned the features of social life that motivate a community to 'assume' the identity despite this unfamiliarity. These features might include social memory of racial persecution (Oliveira 2010), autonomous (boss-less) labor (Carvalho et. al. in Price 1999, 15), collective land usage bespeaking an 'intimate ... relation with their territories' (Souza 2008, 7), and historical confrontation with the state and private capital (O'Dwyer 2002).

capitalist system. They were still slaves. So no matter when this village of yours was founded it was founded by escaped slaves. You are escaped slaves. Will you assume your identity as *quilombolas*? Will you fight for your rights?

'Yes, yes. Certainly', several villagers responded. 'We identify'.

Maria's charismatic and well-practiced speech aligned race and class in a way that figuratively extended Brazil's slave period into recent history. Whether she meant to imply that all workers, or only all *black* workers, who labored under capitalism were slaves is unclear. Certainly, Maria was more concerned with building solidarity among Afro-Brazilians than between them and economically marginal whites.

Only marginally aware of the history of slavery, the villagers who had gathered in the chapel to listen to Maria had some difficulty understanding the historical referent of the term *quilombola*, and no way to comprehend the contemporary extension of this historical category to people like them. They had responded affirmatively to her appeal for their self-identification. Maria was energetic, intelligent and sympathetic towards their daily struggles. Their ebullient affirmations of her presentation reflected, if nothing else, the habituated etiquette by which they kept doors open to many potential benefactors and champions. Yet while many were no doubt also convinced by her substantive claims, others voiced uncertainty (later on after the visit) about whether or not they really were who or what she said they were.

Even more basic than the *quilombola* issue was the question of ethnic identity itself. Maria's visit sparked a fundamental debate among the villagers about whether they were, in fact, *negros*. Most villagers typically avoided calling themselves 'black' (*negro*) even though, phenotypically, all but two showed strong traces of African descent. Racial classification in Brazil is graded, rather than binary, and it is not uncommon for Afro-descended Brazilians to describe themselves using terms that locate them in the middle of a continuum ranging from black to white.[11] During the evening following Maria's visit, I witnessed a debate as to whether they were *morenos* or *negros*; the conclusion was that they were both – that is, technically *morenos* but 'in reality' *negros* (see Sheriff 2003, 30–31). I summarize the prevailing argument as follows: even if their physical traits suggested racial mixture, they experienced the stark 'reality' of racist judgments from whites who glossed them all as 'drunken blacks'.

The leap from *negro* to *quilombola* identity was more challenging for the villagers, in part because the act of 'self-identification' required a new way of mobilizing their ethnicity. As John Collins (2011, 692–93) observes in his ethnography of UNESCO's world heritage site in Salvador, Bahia, the embodiment of *negro* ethnicity relies on a learned practice of

---

[11]The English term 'black' is often translated into (Brazilian) Portuguese as *negro* or *preto*. *Preto* is often used to describe the appearance of a dark-skinned person, whereas *negro* is often used to identify that person with the legacy of African slavery. Both are often used in pejorative, racist remarks. It is not so much that either term is a racial slur, but rather that 'polite talk' about race entails a kind of discursive lightening of a person being described through the use of terms such as *mulato* or *moreno* (brown, brunette). Recently, Brazil's social movements have reappropriated *negro* as a term of race-based solidarity. Still, both terms are subject to considerable semantic flux. Of course, the nature and dynamics of Brazilian ethno-racial classification is the subject of considerable academic debate (Harris 1956; Wagley 1963; Skidmore 1993; Hanchard 1994; Sheriff 2003; Sansone 2003; Mitchell 2013). Suffice it to say that early twentieth-century celebrations of Brazil's 'racial democracy' (usually attributed to the anthropologist Gilberto Freyre) have been debunked, and racial discrimination and structural violence have increasingly been recognized as very real social maladies.

'relating sings and evidence ... an ability to link, specific types of facts to interpretations that posit a direct and continuous relationship ... to changing perceptions of personhood and racial belonging'. The villagers rightly surmised that, at some point in the future, they would be called upon to justify their ethnic self-identification by citing specific facts about their lives as evidence of *quilombola* identity in order to mobilize that identity as a political resource. But they were unsure what kinds of facts counted as evidence of a *quilombola* ethnicity, and this led them to doubt the validity of their own self-identification. Yet they trusted Maria as an ally whose strong connection with the PT would help her smooth out any problems associated with these concerns. The villagers were impressed by Maria's fluent articulation of the prejudice they had suffered, the sense of inferiority they had internalized, the narrative of slavery that seemed to explain these features of their lives and her display of sheer *força* ('life-force', 'inner strength' or 'charisma'), which pointed to her raw capacity to get things done. Inspired by whatever possibilities this new way of figuring blackness might hold in store, the villagers elected one young man, a small shop-owner, to go to Teresina, learn what a *quilombola* really was, and, above all, to 'see what goods he could bring back for the community'.

A few days after Maria's visit, I found myself waiting with this youth, and with two other representatives from other Passarinho villages (and several from villages within adjacent municipalities), for a chartered bus to collect us and take us to a planning/training workshop Teresina. The bus arrived with several regional *quilombola* activists and one Zero Hunger Coordination official already on board, and we all got in.

Project excursion: a programmatic pilgrimage against racism and clientelism

The bus ride to Teresina and the 3-day project planning/training seminar that occurred there gave way to a series of engagements in which the state officials (including the *quilombola* activists) enlisted the village representatives in rituals and mundane exercises geared towards recalibrating their political consciousness. While their success was questionable, their efforts to replace vertical ties with lateral ethnic solidarities were compelling and internally coherent. They consisted of three logically coordinated themes that reemerged throughout the event. The first pertained to the unification of various villages' representatives into a single racially conscious, trans-village *quilombola* community, one that stood ready to 'fight for its rights' in ways reminiscent of the archetypal figure of the militant, slave-era *quilombolas*. The second was the alignment of racial injustice with patronage exploitation, making the 'ethnic' dimension of this project jive with Zero Hunger's anti-clientelist agenda. The third theme was the inversion of the stereotypic clientelist asymmetry through ritual affirmation of state indebtedness to the beneficiary population. Yet the intended transformations to the representatives' subjectivity did not always go as planned.

One reason pilgrimage rituals work to reorient their participants' identities is that they literally remove their bodies from the everyday contexts regulated by social hierarchies (Sangren 1993). Pilgrimages are also psychologically disorienting and often physically grueling. They grind down a person's identity by depriving them of the habituated practices that support that identity, making way for new practices that affirm new subjectivities (see Daniel 1984, 233–88). It is a process that invites people to experiment with desired, but normally suppressed, forms of agency that are otherwise forbidden to them. But the village representatives did not harbor desires for an insurgent citizenship (Holston 2008); they wanted to expand their habituated practices of diplomatic agency, extending these to more powerful allies in the state government and social movements.

The bus ride to Teresina lasted about 10 hours, and most people slept during some part of it. As I nodded off myself, I heard project officials speaking to villagers in ways that worked to endow the trip with identity-changing capacity. One Zero Hunger Coordination Team representative pointed through the window to the small dirt roads leading off from the paved highway into the vast underbrush, roads that presumably led to villages. He recounted,

> The Portuguese took all of this land from the Indians who were here and they killed or enslaved the Indians. There were no municipalities. Here it was all one big ranch. Then they put your grandfathers and grandmothers to work. Now we have municipalities and mayors but look, it's the same thing. Do you know what they plant? Corn and beans. They work in the hot sun just the way you do.

Such talk emphasized the ubiquity of agricultural toil across the landscape, which the official contrasted with the recent emergence of municipal boundaries and their local potentates ('mayors'). He depicted a scenario in which the passengers' solidarity with the imagined inhabitants of the landscape was real and natural. This solidarity was grounded in their collective descent from those who experienced literal slave labor. The official's words rendered the beneficiary representatives' contemporary life conditions equivalent to slavery, and implied that municipal authorities were culpable for this continuity. It would be an exaggeration to claim that program officials likened mayors to slaveholders in an explicit and sustained way, but their talk tended to treat both figures as embodiments of exploitative white power. Moreover, both figures stood for personalistic and intimate forms of oppression. In effect, the state official from the Zero Hunger Coordination Team slipped into the rhetoric of the *Quilombola* Movement. He used that rhetoric to elaborate Zero Hunger's more generic critiques of clientelism and exhortations to horizontal solidarity among disenfranchised people. This suggests that the relationship between the Lula administration and its allied subnational social movements was not merely one of mutual convenience; it was a part–whole relation in which the *Quilombola* activists were deputized into the state itself. In fact, the Zero Hunger Program was in some sense an exercise in imagining the state as a network of diverse struggles, less a government program than a throng of small initiatives. As one Coordination Team official described, 'Zero Hunger is not a state effort; it's a civil society effort that the state is merely instigating'.

The *Quilombola* Movement's narrative of race-based struggle translated Zero Hunger's anti-patronage project into legendary terms that addressed the beneficiary population in its specificity and justified its prioritization in this and future policies. As the bus passed through a municipality very close to Passarinho Municipality, one of the *Quilombola* activists from the Movement explained, 'Here we have already identified three *quilombola* communities. Only two accepted the challenge. The third has not found the courage to fight for its freedom. They're afraid of the mayor'. Her comment posited an antagonism between the *quilombola* villages and a mayor who presumably wanted to maintain both racial subordination and social fragmentation. Her use of the term 'fight'[12] evoked the iconic *quilombola* figures of the slave era, (e.g. the community stronghold, 'Palmares'

---

[12]John Cunha Comerford's (1999) ethnography of rural sociality in western Bahia shows that small farmers use the term *luta* ('fight') in reference to everyday efforts to make a living against all odds. The term already has a kind of revolutionary potential in Northeastern Brazil; it links everyday poverty to underlying social antagonisms.

and it hero king, 'Zumbi'),[13] pitting them in conflict with contemporary patrons. Through such language, the community representatives' journey acquired a sacred character; it invoked a transcendent community whose present actions assumed an allegorical quality. Officials' talk of historical *quilombos* aligned the mundane bus ride and training seminar unfolding in 2004 with the legendary acts of slave resistance undertaken in ancestral time. By aligning present persons with mythic heroes, the officials gave these mundane trips an epic gravitas (Eisenlohr 2004). Even if the founding legends of the slave resistance remained unfamiliar to the representatives, the prophetic quality of this rhetoric reassured them that future gains would result from alliances with the activists and state officials.

The bus ride ended at a government training center, a lush, somewhat dilapidated structure, on Teresina's outskirts. The locale itself exerted symbolic force on the event. While hardly a significant locus of *quilombola* identity, the training center was a symbolic epicenter of participatory development, which stood in contrast to the presumed passivity of everyday political deference. It was a secluded setting where exercises in community diagnostics, community-building, leadership and project critique tested the beneficiaries and taught them the values associated with cooperative participation, *quilombola* identity and political rebelliousness. Such was the sacred knowledge of the pilgrimage, the inculcation of which was more important (to both the state officials and the activists) than the material goods the project might ultimately funnel to the participating communities. Throughout the three-day affair, officials and activists engaged in consciousness-raising exercises aimed at inculcating left-wing ideological perspectives and enhanced lateral solidarity within and across rural communities. The officials and activists adapt 'rights talk' (Holston 2008, 300–12) to rural worldviews: 'Somebody can take away your goods, but only God can take away knowledge'. In such utterances, 'goods' (*bens*) seemed to function as a shorthand reference to currencies of clientelist exchange, while 'knowledge' (*conhecimento*) indexed ideals of awakened political consciousness. They uttered these words with the embodied protocols of political stump speech – upright stance, audience-projected voices, head movement punctuating key words, a pointed finger wagging in gestural emphasis – that worked together to produce tropes of righteous indignation. The state and its activist allies interwove these demonstrations into planned pedagogical exercises and improvised performances that involved remodeling beneficiary behaviors and attitudes.

The fact that the project was still undefined in this moment gave the leadership (officials and activists) a two-fold opportunity: they could invite the community representatives to practice the art of demanding (*reivindicar*, literally 'to take back') what they wanted from the government, as well as shape their understanding of their wants by leading discussions about beneficiary communities' interests. The leadership's hope was that their structured activities would prompt the representatives to articulate their demands based on an emergent sense of trans-village solidarity with their fellow *quilombolas*, and that this would manifest as indignant calls for project activities that would somehow salvage (*resgatar*) or valorize (*valorizar*) their African heritage.[14] The leadership's ritual activities focused on cajoling the beneficiaries to demand such a project from the state, in part

---

[13]Palmares, located in the present state of Alagoas, was a seventeenth-century *quilombo*, the most famous in Brazilian history, which was comprised of thousands of escaped slaves and indigenous communities united under a ruler. Under the famous leadership of Ganga Zumba and his nephew, Zumbi, Palmares existed as a state within a state, resisting colonial military incursions for nearly 100 years until 1710.

[14]The program leadership was not certain about the details of such projects, and at one point asked me, the cultural anthropologist, to help them isolate those African cultural traits that could be fortified by

because this aggressive posture portended their future assertiveness in confronting municipal (and other) authorities.

On several occasions during the days' workshops, the officials engaged in a pedagogical ritual aimed at encouraging the representatives to demand training projects (instead of
material goods). After glossing the general difference between material and knowledge-
based resources, the leadership asked each village representative to stand up and tell the
group why they had come to Teresina. The representatives were bashful but brave: 'I'm
here to see what good stuff I can bring back for my community, and I'm very grateful to
this government', one said, similar to others, believing such an answer testified to his
virtue. They spoke loudly enough to fill the room, and their gestures seemed to wrap
their words in a trope of righteous sermonizing. But the community representatives' acqui-
sitive parochialism, combined with the warm demeanor they demonstrated during training
exercises, drew frustration from the leadership. The village representatives' habituated etiquette of appreciation and tranquility was apparent to the officials, who concluded the need
for further consciousness-raising to imbue a more militant attitude in the representatives.
'It's not just about getting stuff; it's about running after and claiming [*reinvindicando*]
your rights', one official responded in frustration. According to the officials' and activists'
logic, these calls for material benefits signaled the representatives' ongoing attachment to
the conditions of servitude (as servile blacks and deferential clients). The representatives'
polite demeanor solidified this interpretation, which one state official glossed later on as
an indicator of the representatives' 'low self-esteem'.

Because the village representatives did not regard patronage relations as humiliating or
inimical to egalitarian solidarity, the project leaders' overall message was hard for them to
comprehend. Back in their barracks, the village representatives expressed their confusion to
one another (and to me). Their conversations moved across several topics: first, they
worried that the state might not send them back to their villages with any definitive
promise for future resources. Clearly the project had placed the representatives in a position
of some risk: if they brought back valued things, they would be gain personal status as
effective brokers of material resources and immaterial alliances with state officials. Conversely, their neighbors would blame them if they came back empty-handed. Second, while
anything that they could 'bring back to their community' would be good, material goods
(wells, farm inputs, clothing, food) would be ideal. Whereas the officials had framed the
knowledge offered by farming and culture-enhancing courses as a permanent good (in contrast to objects that elites could take away), the village representatives felt the opposite.
They worried that their fellow villagers would see training courses as ephemeral lessons
geared more towards educated people than towards themselves. Third, the representatives
were accustomed to bureaucrats who informed them of what sort of benefits would be
headed their way. They found it implausible that experts on policy would ask *them* to determine the nature of such benefits. They surmised that the officials were hiding something,
which was only to be expected given that these were new relationships in which mutual
trust had not been established.

The representatives did not understand that the officials wanted them to incorporate
aggressive styles of indignation and demandingness into their dealings with public authorities. For them, such an attitude probably seemed counterproductive. From their perspective, a gracious etiquette towards patrons was perfectly synergistic with dignified

---

paying for courses (e.g. in the Afro-Brazilian martial art, *capoeira*) in each village. I disappointed
them in this regard.

expressions of 'unity' with one's fellow villagers: 'I'm here to see what good stuff I can bring back for my community, and I'm very grateful to this government'. Performances of gratitude to patron-figures (here 'this government') were part of a diplomatic agency habitually used to evince a client's moral character, which directly reflected their embedd-edness in extended family groups and neighborhood communities. These were also asser-tions of their incorruptible fidelity to their villages. In fact, some representatives assumed that the entire training process was really an elaborate test, as if the project leaders needed to make sure that they were not out to enrich themselves somehow ('eat money') at the expense of their fellows. They continued to project politeness and loyalty, thinking that this would afford them good standing with the officials.

The state officials and *quilombola* activists resolved after the first day of the training to counteract the representatives' putative deference by orchestrating a spectacle of reverse indebtedness. The officials understood patronage as a relationship of perpetual debt in which the client owed the patron gratitude (and votes) for gifts whose monetary value they could never repay. To counteract that framework, the leaders began the next day's activities with a round of pledges that depicted the state as the indebted party, and the com-munity representatives as its creditors. One after another, each institutional representative (Zero Hunger Coordination Team, *Quilombola* Movement and other allied government agencies at the training) stood up and spoke of their organization's commitments to the *Quilombola* Project and to the representatives. They claimed they had an ongoing obligation to 'give back the people's money'. According to one Zero Hunger official, Zero Hunger was 'not a gift from Lula'; it was the administration's returning of poor people's taxes to them. 'You don't owe anybody anything for it, not Lula, not us, not the mayors, not the *Quilombola* Movement, not anybody. So you need to tell us what you want us to do with your money … '.

Commenting later on about these ritual pronunciations of state debt, the village repre-sentatives expressed appreciation that 'this government is on our side', but they did not internalize the message that the state was indebted to them. These were new relationships, after all (and several mused that they never paid taxes). Key for the representatives was not whether the state officials were their debtors or creditors, but that the officials clearly desired an ongoing relationship with them, one that would naturally cycle through various reversals of debt and credit.

In sum, the 3-day workshop frustrated all parties involved, and as the representatives' return trip approached, the leadership realized that it needed to make some decisions about what the project would actually entail in order to give the representatives something clear to convey to their fellow villagers. Before the village representatives boarded the bus to go home, the Zero Hunger officials told them that they had discussed the issue and concluded that each *quilombola* community would receive both an experimental agricultural initiative (a low-cost fertilization technique) and a participatory culture-enhancing initiative. The latter would include coursework for village women that would train them to bottle and sell local herbal remedies. The bottles would bear the *quilombola* village name and index the savvy survival skills of a people descended from slaves who had come to under-stand their natural environment.

Zero Hunger's partial transformation of political culture

A proper assessment of the material outcome of this project's agricultural techniques and herbal remedies lies beyond the scope of this contribution. Suffice it to say, the material gains from both were very modest. Here I reflect on whether the symbolic and ritual machinery that I've captioned as 'pilgrimage' successfully undermined patron–client relations. Along those lines, Zero Hunger's *Quilombola* project did indeed succeed in strengthening lateral solidarities among the region's villages by prompting their residents to make explicit their shared struggles against racism. That change certainly affected their political participation, though not entirely in the way that the officials hoped. It empowered villagers within the framework of patronage-based resource brokering, rather than divesting them of the matrix of practices and convictions surrounding patronage. It may have also led them to feel more entitled to government resources and thus more receptive to left-wing ideological rhetoric.

In the years following the Zero Hunger Program, the Coordination Team remained in sporadic contact with the beneficiary villages as they brought other projects to completion. Maria and the *Quilombola* Movement remained in close contact with the villagers. Stationed more locally, and having forged a working relationship with state officials as a result of the Zero Hunger project, *Quilombola* activists enjoyed enhanced cachet among local civil society organizations, agronomists and politicians. Over the course of several years, they successfully brokered additional resources from the municipal and state governments, including a course in *capoeira,* an artisanal perfume-making class and a headquarters for one village association. When I asked this village's residents how they came by these things, they responded 'because we're *negros* and *negros* have suffered a lot and so the state is prioritizing us'. Their relationship to the *quilombola* identity remains more ambivalent. They assert their *quilombola* identity in the public sphere (e.g. changing the formal name of their associations so that they contain this term), while admitting (to one another and to me) that they don't feel confident to explain what a *quilombo* is. Nevertheless, they recognize that a *quilombola* is 'a kind of *negro*', and more community members are gaining fluency in discourses of *quilombola* history. In this way, the officials', activists' and beneficiaries' different ways of talking about race began to 'bleed into and inform one another in ways that support sanctioned histories and that promise to alter the conditions of possibility' in rural Piauí (Collins 2011, 689–90).

Maria, the Zero Hunger Coordination Team officials and the government agronomists who helped the villages implement the project are now part of the support system that Passarinho's Afro-descended villagers turn to when they need to consult about *Bolsa Família* and also about candidates for higher office. They also approach regional *quilombola* activists and government agronomists when they need to sell raffle tickets to sponsor events or when they need advocates to help channel municipal resources to their communities. In return, the Movement has asked that villagers show up to regional *quilombola* rallies, encourage their youth to participate in *capoeira* training and gain fluency in explaining their *quilombola* ethnicity. In 2008, Maria succeeded in getting elected alderwoman in a neighboring municipality, no doubt due in part to her increased support from the Afro-descended villages there that participated in this project.

This expansion of the villagers' patronage network had a salutary effect on their existing relations with municipal politicians. On a subsequent visit during the 2012 municipal elections, some of the villagers claimed that local politicians were 'valuing us more' than before. They cited the increasing frequency with which candidates visited their village to promise support for the community associations (now named '*quilombola* associations') and to present private 'proposals' (*propostas*) to individual families behind closed doors (the stuff of long-term patronage negotiations). A few village consultants noted that politicians were increasingly delivering on their private campaign promises (though complaints to the contrary could be heard as well). The improvement of these 'patronage' relations (according to local criteria) was hardly due to a shift in the villagers' attitude towards defiance or indignation. Instead, politicians perceived the villagers as being more 'united' than ever before. As a result, the villagers discovered that their deployment of etiquette, courtliness and hospitality (too easily mistaken for submissiveness) proved increasingly effective at forging moral political reciprocities.

Conclusion

The *Quilombola* Project, like other Zero Hunger initiatives, mediated a cultural engineering mission aimed at the political empowerment of marginal populations. It sought to replace the beneficiaries' individualistic concerns and quid pro quo relations with politicians with an ideology of rights based on collective identity and the presumption of state indebtedness. Project officials and local, race-based activists rendered their respective discourses commensurate, such that Zero Hunger's assault on patronage became equivalent to, and consubstantial with, the Movement's rhetoric of *quilombola* legacy and collective insurrection. Together they pursued an assault on rural patronage through pragmatic and symbolic action, using a combination of carefully planned administrative techniques and ad hoc ritual exercises that felt right in context. The initiative fomented new horizontal ties among the beneficiaries along ethnic lines, and helped them to link a *negro* identity to both the history of racial injustice and to the general deprivations of the rural underclass. Yet these techniques failed to divest the beneficiaries of their attachment to moral vertical alliances with local politicians.

This outcome makes sense when we consider that the rural population in question routinely mobilizes horizontal solidarities to secure alliance with local elites and to improve the terms of these alliances. Stronger lateral communities promise more votes for friendly politicians, and they affirm the honorable character of individuals and groups, allowing a politician to feel virtuous and to boast of these moral ties. In this cultural framework, rural people's political agency manifests in three kinds of actions: the solidification of their lateral community ties, the extension of their patronage networks to new actors and the upgrading of episodic patronage exchanges (glossed as 'vote-buying') into long-term reciprocities that provide communities with a measure of security. Zero Hunger achieved all three during the course of the project examined here, despite the leadership's effort to divest them of those political habits. The outcome of the project suggests that a local people's reflexive thinking about their own diverse modes of vertical exchange determines the compatibility between patronage and horizontal alliances, and shapes new strategies for making ethnic belonging relevant to relations with the state. The villagers' moral distinctions and their embodied strategies for pursing distinct forms of exchange contributed to the realignment of ethnic identity and the creation of new alliances with local activists and state officials. This suggests that attention to such local distinctions should be a key point of departure for scholars and activists who worry about the corrosive effects of patronage. I am left wondering whether it is possible for social programs to harness certain vertical relations of trust and reciprocity, rather than seeking to dismantle them.

Acknowledgements
I am deeply grateful to Rebecca Tarlau and Anthony Pahnke for organizing this special collection of the *Journal of Peasant Studies*, and to Gregory Morton for brokering my participation in it. I also thank Rebecca and Anthony for the excellent suggestions they provided in response to a prior draft of this contribution.

References

Albuquerque, D.M. 1999. *A invenção do nordeste e outras artes*. São Paulo: Cortez.

Ansell, A. 2014. Zero Hunger: *Political Culture and Antipoverty Policy in Northeast Brazil*. Chapel Hill: University of North Carolina Press.

Arbix, G., and S. Martin. 2011. New dimensions in public policy and state-society relations. In *The Brazilian State: Debate and Agenda*, eds. M. Font and L. Randall, 59–83. Lanham: Lexington Books.

Araujo, A.L. 2010. *Public Memory of Slavery: Victims and Perpetrators in the South Atlantic*. Amherst, NY: Cambria.

Arruti, J.M.A. 1997. A Emergência dos 'remanescentes': notas para o diálogo entre indígenas e qui-lombolas. *MANA* 3, no. 2: 7–38.

Auyero, J. 2001. *Poor people's politics: Peronist survival networks and the legacy of Evita*. Durham: Duke University Press.

Auyero, J., P. Lapegna, and F. Page Poma. 2009. Patronage politics and contentious collective action: a recursive relationship. *Latin American Politics and Society* 51, no. 3: 1–31.

Betto, F. 2003. *Programa Fome Zero: como participar*. Rio de Janeiro: CECIP.

Bichir, R.M. 2010. O Bolsa Família na Berlinda?. *os desafios atuais dos programas de transferência de renda. Novos Estudos* 87, July: 115–29.

Bloch, M., and J. Parry. 1989. Introduction: Money and the morality of exchange. In *Money and the Morality of Exchange*, eds. J. Perry and M. Bloch, 1–33. Cambridge: Cambridge University Press.

Burgwal, G. 1995. *Struggle of the poor: Neighborhood organization and clientelist practice in a quito squatter settlement*. Amsterdam: Centre for Latin American Research and Documentation.

Burlandy, M., V. Schottz, G. Monnerat, and R. Magalhães. 2007. Programa Bolsa Família: nova insti-tucionalidade no campo da política social brasileira?. *Revista Kátál* 10, no. 1: 86–94.

CESACF. 2003. *Diagnóstico alimentar e nutricional, Programa Fome Zero—estudo piloto*. Teresina: Governo do Estado do Piauí.

Collins, J. 2008. Public health, patronage, and national culture: The resuscitation and commo-dification of community origins in neoliberal Brazil. *Critique of Anthropology* 28, no. 2: 237–55.

Collins, J. 2011. Melted gold and national bodies: The hermeneutics of depth and the value of history. *American Ethnologist* 38, no. 4: 683–700.

Comerford, J.C. 1999. *Fazendo a lula: sociabilidade, falas e rituais na construçao de organizações camponeses*. Rio de Janeiro: Relumé Dumará.

Daniel, V. 1984. *Fluid signs: Being a person the Tamil way*. Berkeley: University of California Press.

Eisenlohr, P. 2004. Temporalities of community: Ancestral language, pilgrimage, and diasporic belonging in mauritius. *Journal of Linguistic Anthropology* 14, no. 1: 81–98.

Elyachar, J. 2005. *Markets of dispossession: NGOs, Economic development, and the state in Cairo*. Durham: Duke University Press.

Fome Zero. 2003. *Mobilização e educação cidadã*. Brasília: Fome Zero: Setor de Mobilização.

Forman, S. 1975. *The Brazilian peasantry*. New York: Columbia University Press.

Foster, G. 1963. The dyadic contract in Tzintzuntzan II: Patron-client relationships. *American Anthropologist* 65: 1280–94.

French, J. 2009. Understanding the politics of Latin America's plural lefts (Chávez/Lula): Social democracy, populism and convergence on the path to a post-neoliberal world. *Third World Quarterly* 30, no. 2: 349–70.

French, J.H. 2009. *Legalizing identities: Becoming black or Indian in Brazil's Northeast*. Chapel Hill: University of North Carolina Press.

Goldman, M. 2000. Uma teoria etnográfica da democracia: a política do ponto de vista do movimento negro de Ilhéus, Bahia, Brasil. *Etnográfica* 4, no. 2: 311–32.

Graeber, D. 2001. *Toward an anthropological theory of value: The false coin of our own dreams*. New York: Palgrave.

Graham, R. 1990. *Patronage and Politics in Nineteenth-Century Brazil*. Stanford: Stanford University Press.

Hanchard, G. 1994. *Orpheus and power: The Movimento Negro of Rio de Janeiro and São Paulo, Brazil, 1945–1988*. Princeton, NJ: Princeton University Press.

Holston, J. 2008. *Insurgent democracy: Disjunctions of democracy and modernity in Brazil*. Princeton, NJ: Princeton University Press.

Harris, M. 1956. *Town and country in Brazil*. New York: Colombia University Press.

Hunter, W. 2010. *The transformation of the workers' party in Brazil, 1989–2009*. Cambridge: Cambridge University Press.

Leal, V.N. 1948. *Coronelismo, enxata e o voto: o municícpio e o regime reprensentativo no Brasil*. Rio de Janeiro: Revista Forense.

Maluf, Renato, Francisco Menezes, et al. 1996. Contribuição ao Tema da Segurança Alimentar no Brasil. *Revista do Núcleo de Estudos e Pesquisas em Alimentação NEPA/UNICAMP Cadernos de Debate* 4: 66–88.

Marques, A. C. 1999. Algumas faces de outros eus. honra e patronagem na antropologia do Mediterrâneo. *MANA* 5, no. 1: 131–47.

Martinez-Alier, V., and A. Boito Júnior. 1977. The hoe and the vote: Rural Labourers and the National Election in Brazilian 1974. *Journal of Peasant Studies* 4, no. 3: 147–70.

Mintz, S., and E. Wolf. 1950. An Analysis of Ritual Coparenthood (compadrazgo). *Southwestern Journal of Anthropology* 6, no. 4: 341–68.

Mitchell, S. T. 2013. Space sovereignty, inequality: Interpreting the explosion of Brazil's VLS rocket. *The Journal of Latin American and Caribbean Anthropology* 18, no. 3: 395–412.

Nelson, D., and T. Finan. 2009. Praying for drought: Persistant vulnerability and the politics of drought in ceará, Northeast Brazil. *American Anthropologist* 111, no. 3: 302–16.

O'Donnell, G. 1992. Transitions, continuities and paradoxes. In *Issues in democratic consolidation: The New South American democracies in contemporary perspective*, eds. S. Mainwaring, G. O'Donnell, and S. Valenzuela, 17–56. Notre Dame: University of Notre Dame Press.

O'Dwyer, E.C. 2002. Remanescentes de quilombos na fronteira Amazônica: a etnicidade como instrumento de luta pela terra. *Boletim Rede Amazônia* 1, no. 1: 77–86.

O'Dwyer, E.C. 2007. Terras de quilombo: identidade étnica e os caminhos do reconhecimento. *TOMO* 11, July-Dec.: 43–58.

Oliveira, O.M. 2010. Quilombos: memória social e metáforas dos conflitos comunidades do Sapê do Norte, Espírito Santo. In *Territórios quilombolas e conflitos*, eds. A.W.B. de Almeida, R.E.A. Marin, R. Cid, C.B. Muller, E. de Almeida Farias Júnior, 63–9. Manaus: Projeto Nova Cartografia Social da Amazônia/UEA Edições.

Price, R. 1999. Reinventando a história dos Quilombos: rasuras e confabulações. *Afro-Ásia* 23: 1–25.

Putnam, R., R. Leonardi, and R.Y. Nanetti. 1993. *Making democracy work: civil traditions in modern Italy*. Princeton: Princeton University Press.

Sangren, S. 1993. Power and transcendence in the Ma Tsu Pilgrimages of Taiwan. *American Ethnologist* 20, no. 30: 564–82.

Sansi, R. 2007. *Fetishes and monuments: Brazilian art and culture in the 20th Century*. New York: Berghahn Books.

Sansone, L. 2003. *Blackness without ethnicity: Constructing race in Brazil*. New York: Palgrave Macmillan.

Schmitter, P. 1971. *Interest, conflict and political chance in Brazil*. Stanford, CA: Stanford University Press.

Sen, A. 1988. Food entitlements and economic chains. In *Science, ethics and food*, ed. B.W.J. Le May, 58–70. Washington, DC: Smithsonian Institute.

Sheriff, R. 2003. *Dreaming equality: color, race, and racism in urban Brazil*. New Brunswick, NJ: Rutgers University Press.

Skidmore, T. 1993. *Black into White: Race and Nationality in Brazilian Thought*. Durham: Duke University Press.

Souza, B.O. 2008. Movimento quilombola: reflexões sobre seus aspectos político-organizativos e identitários. From the conference proceeding of the 26th *Reunião Brasileira de Antropologia*. June 1–4 in Porto Seguro, Bahia.

Stokes, S. 1995. *Cultures in conflict: Social movements and the state in Peru*. Berkeley: University of California Press.

titonio2000. 2010, 10/20. Lula falando dos programas de assestencia do governo.wmv. http://www.youtube.com/watch?v=-oXRjEZ3Mes

Turner, V., and E. Turner. 1978. *Image and pilgrimage in Christian Culture*. New York: Columbia University Press.

Vaitsman, J., G.R. Andrade, and L. Farias. 2009. Proteção social no Brasil: o que mudou na assistência social após a Constituição de 1988. *Ciência e Saúde Coletiva* 14, no. 3: 731–41.

Villela, J.M. 2004. *O Povo em armas: violência e política no Sertão do Pernambuco*. Rio de Janeiro: Relume Dumará.

Villela, J.M. 2012. Confiança, autonomia e dependência na política eleitoral no Sertão de Pernambuco. In *Cultura, percepção e ambiente: diálogos com Tim Ingold*, eds. C.A. Steli and I.C. de Morura Carvalho, 211–27. São Paulo: Terceiro Nome.

Wagley, C. 1963. *An introduction to Brazil*. New York: Columbia University Press.

Weyland, K. 1996. *Democracy without equity: Failures of reform in Brazil*. Pittsburgh: University of Pittsburgh Press.

Wolf, E., and S. Mintz. 1957. Haciendas and plantations in middle America and the Antilles. *Social and Economic Studies* 6, no. 3: 380–412.

World Food Summit. 1996. Rome declaration on world food security. Available at the UN Food and Agriculture website: http://www.fao.org/docrep/003/w3613e/w3613e00.htm

Yasbek, M.C. 2004. Fome Zero: Uma Política Social em Questão. *Saúde e Sociedade* 12, no. 1: 43–51.

Managing transience: Bolsa Família and its subjects in an MST landless settlement

Gregory Duff Morton

Bolsa Família, the world's largest conditional cash transfer, provides welfare payments to 13 million Brazilian households – and creates dilemmas for Brazil's rural landless movement, the MST. Through ethnographic analysis in two villages, this paper explores the daily practices and political conceptions of the program's beneficiaries. Bolsa Família does not, as is often believed, create a de-radicalizing sense of contentment. Instead, the program generates a temporality that makes the benefit feel unreliable to beneficiaries. These beneficiaries must mediate the tension between 'citizen' and 'manager' identities, the latter being a salient subject-position produced by Bolsa Família. The precarity of this position helps explain why Bolsa Família has *not* inspired significant mobilization by social movements.

In the ruins of the great wooden farmhouse, overlooking a dry valley that the government had redistributed, the movement's next generation was camped out for the weekend. It was a cloudless afternoon late in 2011. The leaders of the MST, as Brazil's landless movement is called, had convened a 3-day training, hoping to turn some new recruits from Bahia into *militantes*, or organizers. Clear skies notwithstanding, this training shivered with the feel of frank anxiety.

The anxiety was about Dilma Rousseff. Dilma, Brazil's Workers Party president, had been elected with support from MST activists, but as her first year in office came to a close, she was refusing to redistribute land at even a modest pace. Instead she had prioritized the continuation of a set of massive development projects and social service programs. The most important of these was Bolsa Família,[1] a conditional cash transfer that uses electronic debit cards to deliver modest monthly payouts, preferentially to low-income mothers. Bolsa Família, in the 10 years since its creation, has proven both extraordinarily popular among its recipients and crucially effective at relieving hunger, school dropout and child mortality (De Brauw et al. 2010; Rasella et al. 2013).

---

[1] 'Bolsa Família' is roughly translatable as 'Family Grant', 'Family Scholarship' or 'Family Purse'.

But, like other conditional cash transfers around the world, Bolsa Família is not a right; it is a social program. The money is delivered to families that comply with certain human-capital mandates for children, including school attendance and vaccination. Bolsa Família, I will argue, operates inside a deep politics of transience and management: transience as the time-system for bureaucrats and beneficiaries to think with, and management as the subject-position for them to occupy. Bolsa Família remains always temporary; it must be managed, never counted upon, never demanded. To the trainers in the wooden farmhouse, this politics seemed like a threat to the very logic of the MST.

That afternoon the anxiety was getting channeled through Otilo.[2] Tall, prone to smiling, disposed to both pessimism and friendliness, Otilo was in his late twenties and he had braved land occupations and marches for a decade. Now it was time for him to help train a new crop of organizers. Pacing near the edge of the valley, he restlessly asked the assembled crowd of about 50 landless farmers:

Otilo: Folks, who gives Bolsa Família to you? Answer me here–
Man 1: The working person.
Otilo: –it's going to have to be like that.
Man 1: The workers.
Otilo: No. Who gives Bolsa Família to you?–
Natan: It's Dilma!
Otilo: –Speak up, for real!
Natan: It's little Dilma.
Otilo: […] Who gives benefits to you? Retirement pensions, um, Bolsa Família, investment from Incra [the federal land reform agency] – who gives all of that to you?
Natan: Our government.
Otilo: The government. Right? Because we ourselves with everything – […]
Franciely: Taxes.
Otilo: Hm?
Franciely: The taxes that we pay on everything–
Otilo: Yes, but there's someone who has to take that money and handle it, worry about all of you. The way it works is that the money gets passed on.
Franciely: The people who manage it are the government officials.
Otilo: Who manage it.
Franciely: But the money is mine.
[general laughter]

Otilo: Pessoal, quem dá Bolsa Família pra vocês? Me respondem aí–
Man 1: O trabalhador.
Otilo: –vai ter que ser assim.
Man 1: Os trabalhadores.
Otilo: Não. Quem dá o Bolsa Família a vocês?–
Natan: É Dilma!
Otilo: –Fala de verdade!
Natan: É Dilminha.
Otilo: […] Quem dá benefícios a vocês? Aposentadoria, hm, Bolsa Família, investimento do Incra – quem é que dá isso pra vocês?
Natan: O nosso governo.
Otilo: O governo. Não é? Por que nós mesmos que tudo – […]
Franciely: Os impostos.
Otilo: Hein?
Franciely: Os impostos que a gente paga em todo–
Otilo: Sim, mas tem alguém que tem que pegar isso e manejar, preocupar com vocês. Trabalha em forma de recurso passado.

Franciely: Quem administra são os governantes.

Otilo: Quem administra.
Franciely: Mas o dinheiro é meu.
[general laughter]

Later in his presentation, Otilo outlined his view in more detail.

---

[2] People's names, and the names of villages, have been altered.

Because if we keep thinking, 'Oh, it was always that way, that's how God wanted it' – that's the secret to why the rich are always in power. As long as we continue to think that, they're like, 'Amen. That's exactly right'. Because it's better for us to keep thinking like that and get all docile. And if we start to get riled up – a Fried-Snack Bolsa, a Chicken-Snack Bolsa, a Gas Voucher Bolsa, a Child Bolsa, whatever. And then things calm down even more.

Porque se a gente ficar pensando assim, 'Ô, sempre foi assim, Deus quis assim' – esse é o segredo que o rico está sempre no poder. Enquanto a gente continuar pensando assim, eles, lá, Ô! 'Amém. É isso mesmo'. Porque é melhor a gente ficar pensando assim, e ficar todo manso. E se começou a enfezar – Bolsa Pastel, Bolsa Coxinha, Bolsa Vale Gás, Bolsa Menino, não sei o que lá. Aí sossega mais ainda.

Otilo's skepticism stood in contrast to the opinion that a farmer named Marco had expressed in front of the same crowd the day prior:

I felt a little – um – victorious, because I've seen a lot of people become better off with this. You know? Under Lula's government [i.e. Dilma's predecessor, who created Bolsa Família], I saw a lot of people become better off. Now we've got Bolsa Família, right? Which, thank God, for those who receive it, was a great privilege. And it continues to be a great privilege. You know? And things went forward, advanced. But what makes me most conf – sort of, moved is the fact that things are advancing.
And me? And you? Hm? What are we going to work with? Right? We lost! We're los – every day that we stay inside the land settlement, or the land occupation, we're getting demotivated. Because there's someone in front [i.e. the president], you know, whom we put there.

Eu me senti um pouco – ah – vitorioso, porque eu vi muitas pessoas melhorar com isso. Né? No governo de Lula [Dilma's predecessor, who created Bolsa Família], eu vi muitas pessoas melhorar. Hoje nos temos aí a Bolsa Família, né? Que graças a Deus, para quem recebe, foi um grande privilégio. E está sendo um grande privilégio. Né? E saiu, avançou. Mas o que eu fico mais acom – assim, comovente, é que avançassem as coisas.

E eu? E você? Hein? Vamos trabalhar com o que? Né? Nós perdeu! Nós estamos perd – cada dia que nós ficamos dentro do assentamento, ou do acampamento, nós estamos ficando desmotivados. Porque tem alguém lá na frente [i.e. the president], né, que nós colocou.

At 30 years of age, the Movimento dos Trabalhadores Rurais Sem Terra (Movement of Rural Landless Workers, or, more simply, 'MST') stands as Brazil's quintessential post-dictatorship social movement, an agglomeration of perhaps 1.5 million people who proudly proclaim themselves to be (or, often, avidly seek to become) small peasants (Raney and Heeter 2005). The movement has led thousands of plantation occupations, through which landless farmers and poor urbanites demand that the federal government expropriate land, compensate its owners, and redistribute it. A successful occupation leads to the creation of an *assentamento*, a community of small farmers cooperatively governed through the MST.

During their disagreement at the training, Marcos and Otilo both expressed an unease that, for the MST, serves as a symptom of times that are once again changing. In the early 1980s, the movement captured the spirit of an anti-dictatorial moment, successfully reviving the demand for radical land reform that the military government had repressed for 20 years. Then, over the course of the 1990s, the MST responded to the Washington Consensus by adroitly pivoting its message and methods. Movement strategists decided to begin targeting agro-business as 'the new *latifúndio* [plantation system]' and to propose an alternative agriculture based on peasant farming.[3] The movement thus became, by the

---

[3]On the Washington Consensus and neoliberalism, see Fortes and French (2012).

start of the twenty-first century, a leading force in favor of alter-globalization and partici-patory democracy, a key figure in Via Campesina and the World Social Forum – and a sometimes electoral ally of Brazil's Workers Party (Branford and Rocha 2002; Ondetti 2008; Wolford 2010).

It is this latter alliance that has so vexed the MST. The Washington Consensus has now faded into a new developmentalism (Ban 2013), and the Workers Party uses state coordi-nation to guide this development, wielding two tools: first, the promotion of major ventures by selected corporations, and, second, the implementation of outcome-oriented redistribu-tive social programs. Harnessing a spurt of growth driven by the export of primary com-modities, especially to China, the Workers Party has overseen a decade of expansion led by the poorest Brazilians (Fortes and French 2012; Mansor de Mattos and Carcanholo 2012). In a moment of growth that is simultaneously corporate and redistributionist, it is not clear what posture can be adopted by a participatory peasant movement.

Bolsa Família functions as a sign of this problem, and it provokes ambivalence among MST leaders. The movement's National Coordinators have kind words for the program, as Marco does; local-level *militantes* in southwest Bahia typically express a pessimism similar to Otilo's. But between Marco's 'victorious' view and Otilo's dark suspicions, a single thread connects. That thread is demotivation, which, everyone seems to agree, goes hand in hand with Bolsa Família. The resulting dilemma – what to do with a program that appar-ently demotivates your base by offering them a much-needed resource – causes the move-ment to display an uncomfortable passivity in regard to Bolsa Família. At the Southwest Bahia MST Secretariat, one will find *militantes* busily planning protests or mounting col-laborative ventures related to all sorts of government benefits, from water supply and health care to telephone service and farm loans. All sorts of benefits, that is, except for Bolsa Família. The Secretariat even had activists who helped landless farmers sign up for retirement pensions – but not for Bolsa Família. Bolsa Família, the targeted welfare program, the non-right, is not good to organize with.

Why? The question, indeed, reaches farther than Brazil. As the biggest conditional cash transfer (CCT) in the world, Bolsa Família, with its more than 13 million beneficiary families (MDS 2013), serves as exemplar for the CCTs that have become prominent devel-opment interventions in 45 nations (Ballard 2013). These programs are usually identified with national-level parties,[4] not with social movements. Why are CCTs so good at building parties, and so bad at building movements?

It is important to note that this paper's question is not the same as Otilo's or Marco's. We are not asking whether or not Bolsa Família actually demobilizes the MST. Federal cash benefits might well promote political quiescence among recipients – but Bolsa Família might also *encourage* land occupations, since it provides beneficiaries with an income that helps them sustain themselves in the countryside during extended periods of

---

[4]CCTs seem to be producing a form of 'policy lock', in which political parties have difficulty disown-ing the transfers because these transfers assemble sizable constituencies. (Thanks go to a wonderfully astute anonymous reviewer for this point.) For a good summary of relationships between parties and CCTs outside of Brazil, see de la O (2013). In the Brazilian case, a massive political science debate surrounds the question of whether Bolsa Família did or did not sway the 2006 presidential election. For an overview of arguments in favor of a Bolsa Família effect, see Zucco (2008) and Canêdo-Pin-heiro (2009); for arguments against, see Shikida et al. (2009) and Bohn (2011). Rennó and Hoepers (2010) offer an interesting model, and Kerbauy (2011) and Peixoto and Rennó (2011) examine the 2010 election.

protest.[5] This paper addresses a question that is the inverse of Otilo's: why has the MST, itself, found it so difficult to articulate a position on Bolsa Família? We are thus thinking with the framing tradition in social movement studies (Snow 2004). This inquiry acquires a special relevance when we consider that, despite the program's profound importance to a quarter of Brazil's population, *no* major social movement has taken up Bolsa Família as a central concern, and this pattern, in broad strokes, repeats itself with CCTs throughout Latin America.

An analysis of macro political-economic forces can help up appreciate the quandaries faced by peasant movement leaders, but a somewhat different diagnosis emerges if one focuses on the MST's base, on farmers and the political sensibilities that they form about Bolsa Família when they receive it. I have carried out this inquiry through an extended period of field work with MST land settlements and their non-MST neighbors in rural Bahia, near the city of Vitória da Conquista. The first focal village is a land settlement, Maracujá, that the MST established through a land occupation in the 1990s. In December 2011, Maracujá had 62 households with 205 people in all; 50 percent of the households received Bolsa Família, and I identified another 13 percent that were likely eligible but not receiving the benefit. Strikingly, Bolsa Família access was better in the second focal village, Rio Branco, despite the fact the Rio Branco lacked an affiliation with the MST. Rio Branco had 35 households with 103 people in all. 74 percent of the households received Bolsa Família, and 6 percent were likely eligible but not receiving.[6]

Ethnographic study revealed three factors that make Bolsa Família incompatible with the MST's organizing frames. First, the program does not fit inside the movement's productivist rhetoric, which glorifies small farming as a form of work capable of sustaining a space of semi-autonomy inside the capitalist world system. Second (and here enters a hint of political-opportunity theory), Bolsa Família federalizes: it directly links individual recipients to the federal bureaucracy, which leaves little room for the movement to affect program implementation through local or even state-level organizing. Finally, the program offers the sort of gradualist redistribution that distracts from the MST's utopian vision of radical change to the class structure.

But none of these arguments explains enough. All of them could be applied, for example, to old-age pensions, an area in which the movement has been active. Something else is also helping to make Bolsa Família unavailable as an issue for organizing. We can catch a glimpse of this something else if we pay attention to the term that Marco uses: Bolsa Família is not a right, but 'a great privilege'. And it is a form of privilege that is, above all, impermanent.

Impermanence

Down in the dry valley, Miguel, 54, sat with me in his white-painted house at Maracujá land settlement. As night grew deeper in the fields around us, he explained why he had joined the occupation that led to the creation of Maracujá 16 years prior.

---

[5] Indeed, landless farmers frequently report that it is Bolsa Família that permits them to stay in the movement and out of the city. In the words of one of my interlocutors, 'A opção aqui, se não tiver Bolsa Família, é todo o mundo ir para a cidade trabalhar.' ('What we'd have to do here, if there were no Bolsa Família, is everyone go to the city to work.') Thanks go to Gabriel Ondetti for making me realize the importance of this effect.

[6] Data comes from a house-to-house census that I conducted between October 2011 and February 2012. I estimated annual household income for 2011 by making use of a survey instrument based primarily on the PNAD (Pesquisa Nacional por Amostra de Domicílios, or National Sampled Survey of Households) with some modifications for the rural context. Households were judged 'likely eligible' for Bolsa Família if they had an annual income below (or within $R5 of) $R140 per person/ month. For details, see Morton (2013a).

Miguel: Because my dream was always to get land to work. There in Abelhas [a nearby town], I was a contract worker, a day laborer, and a skilled wellmaker. And then, I fenced off, man, a little area on the side of the road, so that I could plant a field. But since the spot belonged, man, to the state, the police came and prohibited it. The military police. So that I would take out my fence.

Duff: Mmmm.

M: You know? Because in that area – the ditches, wherever they make roads there are those big ditches for – along the side, you know?

D: Mmmm.

M: I cleared the area there, I fenced it all in nicely, and they came there and ordered me to take down my fence. And that was what I had to do: I took down the fence, I sold the fence parts to the farmer there, and I stayed working for the landowner [...] And then [with the land occupation at Maracujá] this opportunity came up, to make my dream real, and get a little bit of land to work, and I came here, to this place. And still today I don't – I don't regret it. And there's no question for me of ever being able to leave here, Duff.

Miguel: Porque meu sonho sempre era conseguir uma terra para trabalhar. Lá nas Abelhas [a nearby town], eu era empreiteiro, diarista, e cisterneiro profissional. Aí, eu fechei, moço, uma beira de pista lá, para mim poder fazer roça. Mas como a área pertencia, moço, ao estado, a polícia chegou lá e barrou. A militar. Pra mim tirar minha cerca.

Duff: Mmmm.

M: Né? Que naquela área – nas valetas, aonde que fizer a pista tem as valetonas pra – no lateral, né?

D: Mmmm.

M: Aí eu limpei, fechei todinho, aí chegou e mandou tirar minha cerca. Aí a opção: eu tirei a cerca, vendi para o fazendeiro lá, e fiquei trabalhando para o proprietário [...] e aí [with the land occupation at Maracujá] surgiu essa oportunidade, de fazer meu sonho, e conseguir uma terrinha para trabalhar, e vim embora para aqui. E até hoje não – não estou arrependido. E não tem questão para mim poder ir embora daqui, Duff.

Miguel summed up the themes that recur in so many stories that people tell about entering the landless movement. He came to the movement out of dream and desire, a dream that felt specifically like it belonged to him, and a desire that had been frustrated during earlier, more conventional attempts to make a place for himself. As he told the story, he condensed his motivation into the key orectic symbol of land. And land made it possible for Miguel to speak in the terms of permanence. Miguel began by noting that he *always* wanted land; that was *his own* dream. He concluded by indexing the land around him ('here'), reiterating the personal quality of the declaration ('for me') and asserting that there is no question of him *ever* leaving. By speaking of land, Miguel could also speak of himself – as a person who had wanted something always, and who would stay somewhere forever. In other words, he could speak of himself as a lasting subject.

It is no coincidence that Miguel both opened and closed his story with the trope of permanence, *always* and *ever*. Landless people typically narrate their arrival in the MST as a major life change and, specifically, as a move towards something permanent. The conversion-and-permanence motif can be heard even more clearly in another man's recollection about the moment he arrived on a land occupation:

I got on my knees and said, 'I'm not leaving here'.

Botei os joelhos no chão e falei, 'De aqui não vou'.

The directionality in such stories feels permanent because the narrators are heading through an unstable world towards plots of land that become, in some durable way, *theirs*. But the narratives are about permanence in another sense as well. To speak of one's arrival in the movement is to speak in a register that facilitates talk about that which is permanently true, a register that makes it possible to address questions on dreams, selfhood and one's life trajectory viewed synoptically. *There's no question for me of ever being able to leave here.* As is the case with many social movements, people believe that they make a major break in their lives when they enter the MST, and this break renders their entire lives visible and articulable in narrative form, with a before and an after; the decision to make a break this significant can only be motivated by ultimate questions. Movements like the MST do not attract adherents by holding out the possibility of marginal improvements. Such movements require an object that is lasting and major, capable of explaining why someone would change everything in order to durably become what she, in potentia, already was. For the MST, this object is land.

These are the kinds of objects that can help form subjects. We are considering, here, the phenomenon that Gecas (2000) refers to as 'value-based identities'. 'Value-based identities', Gecas argues, 'are more transcendent than are identities based on roles or even on most group memberships. That is, they are less situation-bound, typically relevant across a range of diverse situations' (94–95). A master symbol like land helps to consolidate an identity that operates portably across the full range of a life, casting earlier actions as prefigurative efforts to achieve a fundamental goal.

Gecas claims that 'value identities referring to general goals and end states are the basis of much of our feeling of authenticity', and a mobilizing strategy that focuses on these 'end states' can 'increase member loyalty and commitment to the social movement' (102). Gecas suggests that 'being true to one's values and principles is being true to oneself in a fundamental way' (102). We can say more: the symbol of land allows farmers not just to be true to an existing self, but to construct themselves as the kind of people who, despite social exclusion, *can have* fundamental values and principles. The MST, in other words, gives people the capacity to hold values publicly. It is through these public values – like land – that activists come to speak of themselves as lasting subjects.

Permanent, public values are capable of motivating not just the initial occupation, but subsequent mobilizations as well. Clara and Francisco, long-time Maracujá residents, invoked the many 'conquests' garnered through activism – a fence, a state-funded beehive, two wells – and explained them in terms of a permanent goal:

Francisco: We don't stop. The MST doesn't stop. And so, for us to get these – patrimonies that we have here today in Maracujá –
Clara: It wasn't easy, right? It wasn't easy –
Francisco: It wasn't easy. We struggled a lot. We suffered a lot of need.
Clara: A lot of government agencies that we had to be occupying.
Francisco: […] Through struggle. And we went after – after our goal. That's to live well, you know, with our family.

Francisco: Nós não pára. O MST não pára. E aí, para nós conseguir esses – patrimônios que nós temos aqui hoje no Assentamento Maracujá –
Clara: Não foi fácil, né? Não foi fácil –
Francisco: Não foi fácil. Lutou muito. Nós passou muita necessidade.
Clara: Muitos órgãos que a gente tinha que estar ocupando.
Francisco: […] Lutando. E corremos atrás de – de nosso objetivo. É viver bem, né, com nossa família.

Francisco indexed the fruits of activism in his surroundings ('here today') in order to pronounce them 'a patrimony', a permanent possession. Continuing the kinship idiom, he cast the total result of his activism as a contribution to 'our goal', which was 'living well with our family'. Activism made sense, in Francisco's discourse, because its conquests built towards the enduring core value.

But what if an issue is not speakable in the rhetoric of permanence? In the words of six different interlocutors (all women receiving Bolsa Família), on five different occasions:

(Woman, age 29): Because it's like this. Bolsa Família isn't a sure thing that you – yo – Today you have it, tomorrow – you don't know any more if you have it.

Porque assim. Bolsa Família não é uma coisa segura que vo – o – Hoje você tem, amanhã – já não sabe mais se você tem.

(Woman, age 26): The day when they cut it off, it's going to be a general cutoff for everyone […] This money is, is, isn't for your whole life.

O dia que cortar, vai ser geral […] Esse dinheiro, e, e, não é para toda a vida.

(Woman, age 18): I think you have to be ready, because it's not a thing that's going to last forever […] and you don't even know if you're going to get it your whole life […] The government cuts it off.

Acho que tem que estar preparado, que não é uma coisa que vai ser pra sempre […] e você nem sabe se você vai ficar recebendo a vida toda […] o governo corta.

(Woman, age 27): I'm scared that this thingy will get screwed up and they'll cut it off.

Tou com medo desse trem dar um pepino e eles cortar.

(Woman, age 19): Get it until the day when they want to stop [… .] 'Cause no one knows if it's for your whole life.

Receber até o dia que eles querem […] que ninguém sabe se é para toda a vida.

(Woman, age 30): It's not something – reliable.

Não é algo – confiável.

Bolsa Família's recipients, here, were speaking the same language of unknowability as the program's administrators. At the municipal Secretariat for Social Development, I asked the city's Bolsa Família coordinator whether the program would be permanent. He answered:

No one can express an opinion, because […] it's a political decision.

Ninguém pode opinar, porque […] é uma determinação política.

A different Bolsa Família administrator provided me with a similarly enigmatic statement:

The program isn't permanent, but it ends up being lasting.

Ele não é permanente, mas acaba sendo duradouro.

Suárez and Libardoni (2007, 141), talking to a municipal program administrator in a different part of the country, received the same kind of message:

So we don't know how long it is going to last, since it's a program that had a beginning and that can have an end. Therefore, the families have to get themselves ready to disconnect themselves from it.

Que a gente não sabe até quando vai durar, que é um programa que teve início e que pode ter fim. Portanto, as famílias têm que se preparar para se desligar disso.

This instability could be heard at higher levels as well. In the 2004 presidential decree that established Bolsa Família, the program is described as an effort to

stimulate the sustained emancipation of families who live in poverty and extreme poverty.

estimular a emancipação sustentada das famílias que vivem em situação de pobreza e extrema pobreza (Casa Civil 2004).

Bolsa Família is designed to 'stimulate' emancipation, not to enact or guarantee it. And this is the key legal truth about Bolsa Família: it, like CCTs worldwide, is a *social program*, not a *right*. It aims to 'promote' and to 'combat' – these are the dynamic verbs that its administrators use – and this fluid rhetoric, much of it borrowed from the social movement lexicon, defines Bolsa Família as a flexible tool to achieve a greater end. The program thus bears some resemblance to the workfare-style initiatives that have attempted, over the last 20 years, to turn recipients into self-activating subjects inside liberal states around the globe.[7] More specifically, Bolsa Família marks a break with the earlier trajectory of the Brazilian welfare state, which had progressively expanded access to cash benefits as a right, first for formal-sector retirees (in the 1930s), then for rural retirees (in the 1960s) and then for low-income people facing disabilities or advanced age (in the 1988 Constitution; implemented in 1993). Indeed, it is instructive to compare the phrasing of the Bolsa Família decree with that of the 1993 law, which spoke in terms of 'protection', 'vigilance' and 'universalization of rights', while specifically offering low-income retirement benefits as a 'guarantee' (Casa Civil 1993). By contrast, Bolsa Família is intended – in the words of the program's local coordinator – to 'quebrar o ciclo intergeracional da pobreza' ('break the intergenerational cycle of poverty'), to achieve a temporally specific objective, not to secure a permanent entitlement (Lavinas 2007, 1468).

Unlike most of the earlier and more permanent cash benefit programs in Brazil, Bolsa Família is directed specifically towards women. Program planners hope that Bolsa Família's targeted money will allow women to increase their autonomy by better controlling their households' finances, and to direct more of these finances towards their children.[8] It is as if women were imagined as the most appropriate subjects for a new, more flexible family payment. At the federal level, this mode of payment gets described as an investment in human capital (de Paula 2009); such rhetoric, however, largely drops out of the speech of municipal program administrators, and is utterly missing among program recipients. What carries through, from Brasília to the beneficiaries, is not the vocabulary of investment, but rather the orientation towards impermanence.

Dona Marlene, a farmer at Rio Branco, cogently described the difference between retirement benefits (guaranteed by the 1993 law) and Bolsa Família (established 10 years later, in 2003). This difference becomes clear when farmers go to the city each month to pick up their money:

Retirement benefits, you get there [to the bank in the city each month] – it's guaranteed money, so that you get there, you just have to arrive and pick it up. And Bolsa Família isn't. Because I myself have left here to go get mine, I arrived there, and it wasn't there.	Aposentadoria, você chega – um dinheiro certinho, que você chega lá, é só chegar e pegar. E Bolsa Família não é. Que eu mesma saí daqui para ir pegar o meu, cheguei lá, e não estava.

---

[7]Bolsa Família does not include work requirements. But its temporariness makes it resemble workfare in certain regards. It is instructive to listen to the explanation offered by a municipal Bolsa Família administrator whom I interviewed. For him, the program had a dual aim. First, it was designed 'to alleviate immediately the hunger of thousands of people who are going hungry' ('para aliviar de imediato a fome de milhares de pessoas que estão passando fome'). In the long run, he thought, Bolsa Família should be part of a broader effort towards 'emancipation [...] training courses, put women into the labor market' ('emancipação [...] cursos de capacitação, colocar elas no mercado de trabalho'). This administrator was a devoted PT (Workers' Party) activist, and his use of 'emancipation' is striking. To emancipate did not mean to free people from labor, but rather to put them in the labor market.

[8]The debate over how CCTs affect autonomy, and what 'autonomy' means, is too extensive to detail here. For an optimistic assessment, see Thomas (1990), Hoddinott and Haddad (1995), Suárez and Libardoni (2007) and Rego and Pinzani (2013). For more critical views, see Morton (2013a) (and especially the works cited therein) and Molyneux (2009).

Bolsa Família is not a right, like retirement benefits. It is not 'guaranteed money'; it constantly threatens to disappear and, frequently, because of bureaucratic obstacles, it actually does. This transience is the theme that connects the verbs used in federal pronouncements, the careful rhetoric of local officials, and the opinions of the program's rural beneficiaries, and it was not automatically that these very different people came to share a discourse about Bolsa Família. The feeling of impermanence got transmitted through quite specific practices. These practices are particularly evident at three key moments: program enrollment, program cutoff and change in the benefit amount.

1 Enrollment

Margarita, a single mother at Maracujá, was widely reputed to be one of the poorest people in the village. She recalled with happiness the day when her enrollment efforts finally paid off:

Margarita: Ah, it was a blessing. You know? I lost hope for a long time. So I did this, this Bolsa Família registration, but – I didn't have any hope, none. But then all of a sudden – a call came in for me, telling me to go – all of the way to, to the Bolsa Família program, you know? For me to see if the card was already there. I got there at the bank, and the card was already there. It was a happiness for me, you know?
[…] D: Yeah. You, you waited how long to ge – to, um get the benefit?
V: Because, from, from eight months – no, not eight months. Since when I was eight months pregnant, I started trying to apply for the Bolsa Família card. And I started receiving just recently after he [her only child] was already four years old. And today he turned five, you know? So like – I just got it recently, so it was like four years, it took four years.
D: Really!
V: Yep. It took a very long time, right? And I was like, 'We were – Am I wor – am I worse than the rest, and I'll never get a card?' But it was – I had no hope, you know? Until it arrived. The Bolsa Família card.
[…]
D: Did they explain why it took so long?
V: Yeah. I would go there, it wouldn't work out right. 'Oh, now, it hasn't arrived yet'. 'You can't reserve a space in the line yet'. I would always go there, and it would be a waste of a trip, you know?

Margarita: Ah, foi uma benção. Viu? Eu desesperei por muito tempo. Igual eu fiz esse, esse cadastro do Bolsa Família, mas – eu não tinha esperança não. Mas quando de repente – surgiu uma ligação para mim, que eu fosse – até lá no, no programa do Bolsa Família, né? Pra mim ver se já estava lá o cartão. Cheguei lá no, na Caixa, já estava lá o cartão. Foi uma felicidade para mim, viu?
[…] D: É. Vo, você demorou quanto tempo para re – para, umm, receber?
V: Porque desde, de, de oito meses – de oito meses, não. Desde quando eu estava grávida desde oito meses que eu mexi no cartão de Bolsa Família. Aí eu vim receber agora depois que ele já estava com quatro anos. Que hoje ele fez cinco anos, né? Então como – eu vim receber agora, parece que teve quatro anos, fez quatro anos.

D: É mesmo!
V: Foi. Demorou um tempão, né? Que eu falei assim, 'A gente era – Será que eu sa – que eu sou mais ruim que os outros, nunca vou receber cartão?' Mas foi – Não tive esperança, né? Até que chegou. O cartão do Bolsa Família.
[…]
D: Eles explicaram porque é que demorou tanto?
V: É. Eu ia lá, não dava certo. 'Ah, agora, ainda não chegou ainda não'. 'Ainda não tem a senha ainda não'. Eu sempre ia lá, perdia a viagem, né?

When villagers apply for Bolsa Família at the municipal Secretariat of Social Development – a building universally and unironically referred to as 'a Muvuca' ('the Havoc') – virtually everyone has to wait for months. Applications may simply have no result; or, like Margarita's, they may lead to a multi-year period of uncertainty.

Margarita and her neighbors face long wait times because of the structure of federal policy. The federal government allots a limited number of Bolsa Família slots to each municipality, based on the local incidence of poverty in annual statistical surveys. This practice is intended to prevent the corrupt overuse of the program by people above the poverty line, but it also creates unpredictably long waiting lists (Lindert et al. 2007). When slots become available, it is managers *in Brasília* who decide which applicants to accept, a decision for which the criteria are opaque. Applicants thus have little way of knowing how long they will stay on the waiting list. For low-income single men, the wait can extend for an excruciatingly long time, and, during my field work, 3 months seemed to be the minimum in Vitória da Conquista for even the most highly prioritized applicants, women with children.

Margarita understood her wait to be a consequence of the federal system, and, in particular, of the vertiginous distances and enormous size of the nation-state:

D: And how was it that – why did it take so long?

M: It's because it's a lot of people, you know? 'Cause – It's all of Brazil, you know? So – so many people. 'Cause it goes to Brasília, you know, the name, 'cause they take the person's name here in, in Conquista. Right? The papers go to, to Brasília, then they return to Conquista again.

D: E como foi que – porque foi que demorou tanto?

M: É porque é muitas pessoas, né? Que – É o Brasil todo, né? Tan, tanta gente. Que vai para Brasília, né, o nome, que eles pegam o nome da pessoa aqui no, no, em Conquista. Né? Os papel vai para, para Brasília, para depois voltar para Conquista de novo.

Jamaira, considerably younger than Margarita, expressed a similar idea:

Hey. Let me talk here. It – on the day that I did an interview there [to apply for Bolsa Família at the municipal office], she – the woman said this. That – everything that I was saying, it was all going to – what's it called? Where Dil – mm – to Brasília, 'cause Dilma was going to read everything. She was going to see if she would stamp it with her stamp, and then send it back, to here, to Conquista, to see if I would receive it or not. 'Cause everything goes there. All of that stack of paper all goes there. [...] So I hope she stamped it there, because I need it! [laughs]

Hein. Deixa eu falar aqui. E – no dia que eu fazer entrevista lá [to apply for Bolsa Família at the municipal office], ela – a mulher falou assim. Que – tudo que eu estava falando, ia tudo para o – como é o nome? Onde é que Dil – em – para Brasília, que Dilma ia ler tudo. Ia ver se ela ia bater seu carimbo, e aí retornar, para cá, para Conquista, para ver se eu ia receber ou não. Que vai tudo para lá. Aquele tanto de papel vai tudo para lá. [...] Então tomara que ela carimbou lá, que eu estou precisando! [laughs]

Bolsa Família, as other scholars predicted (de la O 2013), unquestionably federalizes one's consciousness. Margarita and Jamaira felt acutely aware that they had solicited a relationship with the nation-state. But this is a nationalism on pockmarked terrain. The federal system has not rendered outcomes homogenous; instead, its separations have added a layer of unknowability.[9] With programs whose administration is visible at the municipal and state levels – like most of the programs around which the MST mobilizes – decisions can be understood, sometimes darkly, as the result of personal favoritism by officials whom one could potentially pressure through social movement mobilization. With Bolsa Família, results become simply unpredictable.

---

[9]On uncertainty as a form of power, see Bauman (2000, 119).

2 Program cutoffs

Bolsa Família's unreliability gets reinforced by a second practice: program cutoffs. Almost all of the farmers in the dry valley had a story about canceled or blocked benefits. Interestingly, these cutoffs came almost entirely as a result of the program's administrative practices, not its 'conditionalities'. Conditionalities are human-capital requirements that families must fulfill in order to continue receiving the money; most prominently, these include regular vaccination and perinatal care (for young children) and 85 percent school attendance (for older children up to 16; 16–17 year olds must have 75% attendance.) Families at Maracujá and Rio Branco rarely had their benefits interrupted because of an actual failure to comply with these conditionalities. Beneficiaries invariably told me that they *wanted* their children to receive education and health services; they had been asking the government to provide just such services for years, and they did not need Bolsa Família to encourage them to make use of the school or the clinic. In the dry valley, teachers and health workers confirmed that only rarely did they have to report families for falling behind on the conditionalities.

For the most part, then, Bolsa Família cutoffs had to do with purely bureaucratic obstacles, often those related to biannual program re-registration. It was re-registration that made trouble for Saralinda. In September 2012, Saralinda, one of Maracujá's brightest students, recounted to me that she had finally figured out why her Bolsa Família was frozen. The local office had misprocessed her form. Now, fortunately, the mistake was solved, but she wouldn't receive benefits again until November, and, since Bolsa Família was not a right, she would get no back pay to cover the months when she had been erroneously cut off.

Blocked benefits without back pay were extremely frequent. But Saralinda was lucky. Tamara had four children, and her benefits were cut off for 11 months:

Duff: And they – did you end up getting the – ah – the back pay?
Tamara: No. I just got the money from that month itself. For the back pay, there was just no way. Not – a cent. […]
D: Wow. And – ah – they cut off the card, and you didn't find out the reason or anything?
T: There wasn't a reason. I didn't find out.
D: Wow.
T: In – in the firs – in the second month, you know, they said that it was because they had, um, just written everything down on the papers there and hadn't put it in the computer. Then in the third month, they didn't know what it was, what it was, what it was, and it kept getting delayed – delayed, and there never really was an explanation why – why it was. Because if it was just because it hadn't entered the system yet, after, if, after three months it should already have, mm, been worked out, right? And then – it stayed like that. No one knows why it was. The kids were all in school, the vaccines all up to date. There was no why.

Duff: E eles – você acabou ganhando o – ah – o atrasado?
Tamara: Não. Peguei só o dinheiro mesmo do mês. Para o atrasado não teve jeito não. Nenhum – centavo. […]
D: Uau. E – ah – cortou o cartão, e você não soube o motivo nem nada?
T: Não teve motivo. Não soube não.
D: Uau.
T: No – no prim – no segundo mês, assim, eles falou que foi porque tinha, eh, só anotado nas folhas lá e não tinha jogado para o computador. Aí no terceiro mês, já não sabia o que é que foi, o que é que foi, o que é que foi, e foi enrolando, enrolando e não teve um justificativo mesmo porque – porque foi. Porque se foi só porque não tinha ainda caído no sistema, om, se, com três meses já era para ter, mm, resultado, né? E aí – ficou nessa. Não se sabe porque é que foi não. Os meninos estavam todos na escola, as vacinas todas em dias. Não teve um porquê não.

Resolving such cutoffs proved especially challenging in rural areas. In the dry valley, each trip to the Bolsa Família office required 6 hours of travel and a bus ticket that cost a day's wages – one quarter the value of a month's Bolsa Família Basic Benefit. This expense was difficult to stomach even during enrollment, which at its simplest usually required three trips: one to get an appointment date, one to conduct the interview, and one to pick up the card. When benefits stopped coming unpredictably, the cost of travel could make it very difficult to resolve the problem.

Interestingly, when recipients tried to explain the causes behind Bolsa Família cutoffs, their explanations foregrounded the tools of bureaucracy. Saralinda's misprocessed form, Tamara's unentered paperwork and many similarly wayward pieces of office ephemera were *the reasons why* (or everyone's best guess as to the reasons why) benefits had stopped coming. I never heard anyone in the dry valley blame a specific government functionary, or even the government in general, for problems with Bolsa Família cutoffs. Recipients seemed to have accepted the bureaucratic, object-focused explanations that they had been told by the staff at The Havoc. More than that: bureaucratic tools had in some cases become a fixation shared by staff and recipients alike, and a number of women in the villages outlined to me their own plans for improving the process of program administration.

In over 2 years of fieldwork, I also never heard recipients attribute Bolsa Família problems (or, for that matter, successes) to a mayor, movement leader or other locally visible power broker. Bolsa Família here stands in instructive contrast to certain other government benefits, like, for example, Cesta Básica food baskets that were distributed ('by the mayor's office', as people explained) during a drought, or a massive irrigation project for which credit was avidly claimed by a deputy, by officials at the land reform agency, and by several activist leaders.

Bolsa Família had indeed achieved the objective, common to CCTs, of teaching beneficiaries to attribute power to federal forces rather than municipal authorities. In recipients' accounts, cutoffs occurred not because of local staff but because of the mysterious object-agency of misplaced papers and forms. And these agentive office supplies ultimately channeled responsibility upwards towards Brasília, the place where such problems got solved, and towards a single human person, the enigmatic figure of Dilma herself.

3 Changes in benefit amount

Unpredictability was also conveyed by a third feature of Bolsa Família: the frequent changes in the benefit structure. Lara, a devoted farmer and the mother of four children, told me one day that she never understood why Bolsa Família had so many increases. Her words were inflected with a mix of bemusement and frustration. Some of Lara's neighbors had begun to receive higher Bolsa Família payments if they had children under six – this was through a new initiative called 'Brasil Carinhoso', or 'Caring Brazil' – but Lara, with her 2-year-old daughter, had gotten no such thing. Brasil Carinhoso was only one of several modifications to Bolsa Família that appeared over the course of my time in the villages. Villagers heard about – but only sometimes saw – Bolsa Estiagem (Dry Spell Bolsa), ProJovem (a federal youth project), Bolsa Gestante (Pregnant-Mother Bolsa) and Bolsa Nutriz (Nursing-Mother Bolsa.)

Bolsa Família allowed for the implementation of new programs that could respond to new needs or political priorities – programs, also, that particular federal politicians could mark as their own projects. This policy churn produced a dizzying effect. Understandably, recipients had difficulty in ascertaining which monies they should be receiving, or even in predicting how much they would get from one month to the next. The rapid changes underscored Bolsa Família's unstable nature. Bolsa Família was unpredictable because it was, above all, a social program – that is, a tool that powerful people could use to shape behavior in accordance with the exigencies of the moment.

Speaking about Bolsa Família over her kitchen table, Dona Marlene summed up the cumulative effect of the enrollment process, the cutoffs and the changeable benefit structure. It was a worry that haunted people when they went to pick up their monthly benefits:

Bolsa Família, sometimes you can't leave here trusting that you'll be able to go to Conquista [the nearby city], get there, and do something with the Bolsa Família money. You know? Because you can get there and the money's not there [...] Because how many fathers of a family, how many mothers, leave here to go there and pick it up. All they've got is that little money. And they get there, that's even going to be the money for the bus ticket. They get there and don't receive it – and they end up there, looking around, without being able to even – eat something, not even bring something home, and not even really pay for the bus ticket.	Bolsa Família, às vezes você não pode sair daqui confiado de ir a Conquista, chegar lá, fazer alguma coisa com o dinheiro do Bolsa Família. Né? Que você pode chegar lá e não está [...] Que quantos pais de família, quantas mães saem daqui para ir lá pegar. Só tem aquele dinheirinho. E chega lá, vai ser o dinheiro até da passagem. Chega lá e não recebe – aí fica com a cara arriba sem poder nem – comer uma coisa, nem trazer nada para casa e nem bem pagar a passagem.

Bolsa Família's mundane practices conveyed to recipients the same principles embodied in high-level federal rhetoric: the need for constant change, the importance of achieving particular objectives, and the unfixity of the future. Bolsa Família felt always like 'a great privilege', a monthly indulgence from another, and never like 'patrimony', an object that one has rendered permanent and made a part of the world of one's kin.[10] The program was not thinkable inside the logic of the MST organizing story, which required a moment of transformation through which a landless farmer achieved a lasting place for herself. As a non-right, Bolsa Família was a difficult object for MST organizing.

In order to frame Bolsa Família as an MST organizing issue, movement leaders would have to speak about it as a resource that could become permanent in the lives of farmers – and Bolsa Família is designed precisely to evade such a frame. The point here is not that federal officials conspiratorially planned to produce uncertainty in the lives of Bolsa Família's recipients. Enrollment difficulties and accidental cutoffs happen even in the most well-intentioned social program. Rather, the point is that Bolsa Família's architects quite carefully set up the transfer *as* a social program – and not as a right. Consequently, recipients, when faced with the inevitable bureaucratic obstacles, have no remedy other than supplication. Brazil's retirement pensions, universalized in 1993, are a right. When pensioners face difficulty in accessing their monthly cash, they can retain a lawyer and enter into lengthy judicial proceedings, even receiving back pay for the months when they were unjustly denied. With Bolsa Família, recipients have no lawyers and no guaranteed back pay, and thus the program's administration, at a structural level, has no obligation to face the consequences of individual benefit errors. The result is apparent. In the words of Dona Marlene, retirement pensions are 'guaranteed money [....] And Bolsa Família isn't'.

A whole ideology emerges out of Bolsa Família's status as a non-right. As we have seen, administrators explicitly describe temporariness as a form of consciousness that recipients must acquire ('the families have to get themselves ready to disconnect themselves

---

[10]On women's efforts to make Bolsa Família permanent inside households, see Morton (2013a).

from it'). But things did not need to work this way. In 2004, Senator Eduardo Suplicy successfully championed a federal law calling for a guaranteed, permanent minimum income for all Brazilians (Lei 10.835 de 2004), and Bolsa Família was envisioned as the first step towards this project. MST organizers did not take up the Suplicy plan as a major mobilizing issue, perhaps because it fit poorly with another movement priority: the dramaturgy of production. Ultimately, the law did not find a constituency among either social movements or government planners, and, never implemented, it faded from public view. Its disappearance marked the limits of social-democratic governance in Brazil, which would henceforth approach redistribution by creating programs rather than rights.

If Bolsa Família has a demotivating effect, the effect operates through the principle of impermanence. Bolsa Família does not necessarily demotivate because it satiates recipients or makes them happy enough not to fight. It demotivates for the opposite reason: because its irregularities encourage recipients to forget how to dream for something durably good. In Margarita's words, when she was applying, she steeled herself so that 'I didn't have any hope – none'. Bolsa Família's vicissitudes teach its recipients to work around, plan for and expect bureaucratic uncertainty.[11] The program's very structure encourages people *not* to dream of Bolsa Família in the way that Miguel dreamed about land, because Bolsa Família is designed to feel unreliable; that structure thus encourages people *not* to fight for it, in the way that Miguel fought when he fenced off the side of the road and then joined a land occupation, because Bolsa Família seems like it will never become permanently a part of one's life.

But if it does not build personal power or family patrimony, Bolsa Família does contribute towards the construction of a different mode of durability: the nation. The program evokes for its recipients a sense of the national sublime. The nation appears unknowably big. It looms in one's awareness of who else is applying for Bolsa Família ('all of Brazil, you know? So – so many people'). It materializes in the vast distance between the nearby city and the national capital ('the papers go to, to Brasília, then they return to Conquista again'), and in the fact that one's own name will make that journey ('it goes to Brasília, you know, the name, 'cause they take the person's name here in, in Conquista'). The nation is visible in the size of the pile of papers one sees at The Havoc ('all of that stack of paper'). And it is humanized in the figure of Dilma, who has the gigantic task of supervising the entire operation.

The national sublime here operates through tropes of massive size and massive distance. In recipients' interactions with the line staff, this sublime gets constructed, very specifically, through a dual process of depersonalization and repersonalization. First, the personal incapacity of both the recipients and the line staff is displayed. Neither one can affect policy or do anything more than convey information. In fact, bad outcomes, to the extent that they are understandable at all, are due to malfunctioning objects like missing forms rather than decisions by people one can see. Second, the incapacity of the recipients and line staff gets inverted and repersonalized through the super-capacity of Dilma, who controls everything and has feelings for all of Brazil.

---

[11] Indeed, perhaps the most important behavior change associated with the program is not that recipients learn to send their children to get vaccines and schooling, but rather that recipients acquire the skills needed to engage with an unpredictable administrative system. This involves its own type of 'struggle': a particularist negotiation over the terms on which one gets inscribed in state records. Such administrative negotiation, because it focuses on the details of one's case, may end up becoming (although it does not necessarily have to be) inimical to social-movement organizing. Bill Sites deserves the credit for this insightful point.

One important feature of this sublime is its demobilizing effect. Speakers, whether staff or recipients, become overshadowed by the size of the nation. In the context of Bolsa Família, these speakers do not have rights that they could use to compel the nation to do something, as citizens would; instead, they have 'a great privilege', and the nation's enormousness makes it impossible to count on this privilege.

This particular version of the national sublime does create an identity-position for individual speakers, but that identity is not citizenship. It is the identity of the manager.

Management

It's always the woman who organizes things.	É sempre a mulher que organiza.

Speaking these words as she stared across her worn couch, just a few yards from the dry fields, Francesca tried to explain to me why the government preferentially paid Bolsa Família benefits to women. The specific feminine role that she invoked – always organizing – was one that I frequently heard farmers describe. They tended to refer to this role by using a particular word: management.

I often asked people to tell me the reasons behind Bolsa Família's gendered payout approach. In answering, interlocutors returned to a single theme, as can be seen in four responses from four different farmers:

(Woman, 34): Men don't know how to manage money, like this one here.	O homem não sabe administrar o dinheiro, tipo esse.
(Man, 39): Inside the home, women are better managers than men.	Dentro de casa, a mulher administra melhor que o homem.
(Woman, 33): He is the owner of the house. He's the head of the family. But he's not going to know how to manage money.	Ele é o dono da casa. É o chefe de família. Mas ele não vai saber administrar o dinheiro.
(Man, 38): A woman, sometimes she takes greater care. Buying clothes, sandals for the kids. She gets it, she holds on, she already knows where she's going to put that money to work.	A mulher, às vezes ela capricha mais. É comprar roupa, sandália para os meninos. Já pega, já segura, já sabe onde vai aplicar aquele dinheiro.

At The Havoc, a program supervisor agreed:

It's the woman who is going to manage the benefit.	É a mulher que vai administrar o benefício.

Management is the special task that Bolsa Família helps women accomplish, and this task comes with a special kind of knowledge, 'saber administrar' ('to know how to manage'). But if women are imagined to have a talent for management, it turns out that they are not the only ones. Strikingly, the rhetoric of management also helps Bolsa Família officials to understand *their* own identities. When I asked him to discuss the future of Bolsa Família policy, the municipal Bolsa Família coordinator explained simply that he and other staff 'só administra[m]' ('only manage'). Another municipal administrator offered a similar assessment of the staff's work:

Our only role is to do management.	O nosso único papel é de fazer gerenciamento.

The uncanny parallel here – Bolsa Família recipients and Bolsa Família staff resort to the same language when trying to understand their own roles – points towards an effect of the program. Bolsa Família, it seems, helps one to feel like a manager. And this occurs whether one receives the benefit or works in the office that passes it on.

To understand why, it is helpful to note that in the villages, 'to manage' ('administrar') does not only refer to activity related to Bolsa Família. It is a more general female task. Veronica, a young woman who was raising a baby at Maracujá while her husband worked in the city, described the woman's skill as follows:

It's knowing how to manage things inside the house and tell the man what's missing.	É saber administrar as coisas dentro de casa e dizer para o homem o que é que está faltando.

Clemencio, an older man with big dreams for expanding his farm, spoke specifically about the management of money.

[The man's role is] on the talking end of things. But on the financial end, no. The person who can organize the financial end inside a residence, I see that as the woman.	[The man's role is] do lado da fala. Mas do lado financeiro, não. Quem consegue organizar o lado financeiro dentro de uma residência, eu vejo a mulher.

The nature of management became even clearer in a series of conversations with Franco, a devout evangelical and one of the hardest-working cattle owners at Maracujá. Franco often spoke about management. In his view, to manage was not just a woman's job. Indeed, it was more like an existential position. From two different interviews with him:

A person has to know how to manage, even the little bit that he or she has.	A pessoa tem que saber administrar, mesmo o pouco que tem.
We're only the managers. God is the owner. […] 'I'm going to make man to manage what I made'.	Nós só somos administrador. Deus é o dono. […]
	'Vou fazer o homem para administrar o que eu fiz'.

To manage, in the dry valley, thus involved being in a relationship with an external party. For Clemencio and Veronica, this external party was a man; for Franco, it was God. What crucially mattered, though, was that the external party had some originary claim on the resource being managed. As Veronica put it, the end result of good management is telling the man what is missing. He, presumably, is the original source of objects and can resupply the household.

The implied presence of this external party helps explain why declaring oneself a manager is so often a self-*minimizing* claim. Using the word 'só' ('just'), Clemencio modestly declared himself to be 'just' the manager of his possessions ('nós só somos administrador' – 'we are just managers'); similarly, the Bolsa Família coordinator noted that he 'just' managed the program ('só administra' – 'just manage'). Here, we can see the point of connection between the words of Bolsa Família staff and the words of Bolsa Família recipients. Both use the discourse of management to make themselves seem small: they declare themselves to be 'just' managers, because someone much greater stands outside, whether that greater figure is God or the national political leadership. Bolsa Família, with center of gravity in Brasília, has been structured so that it makes everyone feel like a manager, recipients and staff alike. The resource always fundamentally belongs to someone else, and is never more than partially controlled by the various 'managers' – local power bosses, government functionaries, mothers – through whose hands it flows.

Inside this rhetoric, to say that you manage money is to say that it is not really yours. Management, moreover, produces a particular temporality. Managers cannot lay a permanent claim to the resource, and so their perspective on it, whether they are officials refusing to predict the program's future or recipients declining to count on the money, necessarily underlines its own temporary quality.

When one speaks of oneself as a manager, one is implying that there exists a real owner, a powerful external party. By referring to oneself as a manager, one places a distance between oneself and this Real, this foundational level: *We're only the managers. God is the owner.* And *by* speaking in this way, one helps to construct the sense that such a Real really exists and has a real claim, outside of oneself as speaker.

In each of the examples above, the speaker sets up a contrast between herself, as a merely provisional manager, and a particular kind of other, a Real other, who possesses the legitimate and permanent claim. Among Bolsa Família staff, the relevant contrast is between *politics* and *bureaucracy*. The Real other is the political decision-maker, as against the speaker, who casts herself as a functionary who only implements. For Bolsa Família recipients, the relevant contrast is between *household management* and *ownership*. The Real other is the man, whose ownership claim ('the owner of the house') stands in contraposition to the woman's managerial position (see Lavinas and Nicoll 2006; Suárez and Libardoni 2007; Mariana and Carloto 2009; Pires 2009, 2013; Gomes 2010; Rego and Pinzani 2013 on feminist reads of Bolsa Família; also see the journal special edition that features Pires 2013; for CCTs outside of Brazil, see Luccisano 2006; Molyneux 2009; Gruenberg 2010).

One further complexity marks the managerial position of the woman receiving Bolsa Família. When management is seen from a different angle, another Real other appears: not *the man*, but *the child*. To cite the words of a municipal program administrator I interviewed,

Generally, a woman manages better and is more committed to the children [....] She's more appropriate and has more sensitivity when it comes to managing the resources of the house.

Em geral, mulher administra melhor e tem mais compromisso com os filhos [....] Ela tem mais propriedade, mais sensibilidade, para administrar os recursos de casa.

Here, again, the woman 'only manages' the household – but now the money's real, legitimate targets are the children. The woman does not earn Bolsa Família as a payment, but rather *becomes appropriate* for it because of her alleged capacity to direct resources to children. In this discourse, then, the woman gets figured as household manager in a double sense: manager because the house is really owned by the man; manager because its resources are really owed to the child.[12]

Bolsa Família's managerial discourse evokes another buried dichotomy as well, namely the separation between *production* and *consumption*. Woman-as-manager does not produce; paradigmatically, she takes money and directs it towards consumption. She 'buys clothes, sandals [...] she already knows where she's going to apply that money'. She can consume because there is someone else, 'the man', to productively make objects come into existence.

When speakers conceive of the woman's activity as 'management' of money, they foreclose on an alternative view: that work inside the household is itself a kind of production,

---

[12]For more on questions of ownership of Bolsa Família, see Morton (2013a).

and that Bolsa Família is actually a wage or a payment for that production.[13] Rather than giving women money to manage, Bolsa Família would then be offering women an earned form of remuneration. Everything happens as if the rhetoric of management were designed to prevent such a logic from making sense, and instead provide a sort of covert support for the ideology of male labor-market productivity. For if women *manage* money – 'inside the house', as many interlocutors took care to emphasize – then their actions stand as the negative image of a different category, *making* money, or labor.

Bolsa Família rhetoric seems to bolster a political worldview in which there is a distinction between producing and managing. It may not be too much of a stretch to associate this worldview with the conception that left-wing political leaders have formed, in the past decades, about the state itself: the post-neoliberal state cannot produce, and hence leaves the creation of goods and services to the market (Ban 2013). The state's role, like that of the mother receiving Bolsa Família, is to manage the resources generated in the private sector. But whether or not Bolsa Família discourse connects to notions about a (female-gendered) state, the rhetoric of management, with its deliberate self-limitations and modesty, sets up 'manager' as a negative category that underscores the reality and dominance of the categories that stand in opposition to it. When people speak about management, they make it appear that there exists something strong and legitimate, something that is not management, called 'ownership', 'masculinity', 'national political leadership' or, above all, 'labor'.

Management is a midpoint, and it implies a class position; historically, managers have stood at the site of mediation between laboring and owning, between the working class and the bourgeoisie. Managers define themselves by distinguishing themselves both from the people above them and from the people below them. To excavate the specific class content of the rhetoric of management, then, is to note that managers always mark their position by opposing it to the Real, on one side or the other. On the one hand, managers do not really labor. On the other hand, they do not really own. They perform a role that gets cast as provisional, temporary or fake.

It is important to ask why this role has been so insistently reinvented, at a particular moment in history, to explain the action of women who receive Bolsa Família. We cannot content ourselves with arguing that 'management' refers to a traditional task of femininity in rural Brazil; in the villages, many interlocutors attest that women frequently did not handle money at all before Bolsa Família. To the extent that rhetoric of Bolsa Família management refers to earlier feminine duties, it also reconfigures and monetizes those duties. Clemencio specified that he saw women's financial involvement as a change of historical proportions:

Today [the woman] [...] is in first place, ahead of the man [...] both at work and in her own house [....] On the financial end she's getting to be first as well. The man is complementing her. It's not the woman who complements the [man] any more [...] because it's the woman who organizes the financial side [...] organizes the whole house [....] We're starting to get out from under the sexist law.	Hoje [a mulher] [...] tá em primeiro lugar do que o homem ... tanto no trabalho como na própria casa dela [...] O lado financeiro ela tá sendo primeiro também. O homem está complementando ela. Não é mais a mulher que complementa o [homem] [...] porque é a mulher que organiza o lado financeiro [...] organiza a casa toda [....] A gente começa a sair da lei machista.

---

[13]Note, here, the similarity to the Wages for Housework movement of the 1970s. This form of work has often been referred to as 'reproduction' (although, in my view, the distinction between production and reproduction risks enacting a subtle sort of segregation; the distinction is an unwarranted one, since all labor fundamentally 'reproduces' society).

Villagers, both men and women, increasingly 'get out from under the sexist law' as they avail themselves of the expanding opportunities for formal-sector employment in the cities and in agribusiness (Neri 2010, 15), opportunities that, even independently from Bolsa Família, are already placing cash in the hands of women. In this changing moment, why does the language of management come to settle on women? We can speculate, first, that management rhetoric allows for new female tasks to be assimilated inside a framework that still marks the difference between women and men. In describing women as 'managers' of money, one grants women the leeway to handle cash, without conceding ownership rights to them. The 'manager' category acknowledges women's newfound access to money, but acknowledges it by locating it as one pole inside an enduring gender binary.

There are other stakes to management rhetoric as well. The woman-as-manager motif has motivated the development theory that underpins conditional cash transfers, in Brazil and beyond (Thomas 1990, 660; Hoddinott and Haddad 1995, 77–78). This motif puts recipient women in the familiar mediator's position, between spouses and children on the one hand and the state on the other (Molyneux 2009; for an overview on women as 'resources in development', see Moeller 2013). Perhaps less obviously, the motif imports, for women in poverty, the class dilemma of an earlier generation of managers. Women are taught to understand themselves as agents working with a resource not fully theirs, on behalf of family members other than themselves. Neither the real beneficiaries of the money nor its real owners, they hold the responsibility of managing transience. With such a double distancing from the Real, this role is characterized by a permanent instability that makes it a difficult place from which to articulate a political position that can be heard as such. It is unsurprising that recipients of Bolsa Família, like other managers before them, do not launch social movements in their own names.

Conclusion

On 18 May 2013, 900,000 Bolsa Família recipients took to the streets. In 13 Brazilian states, they rushed the banks simultaneously on the same day, attempting to withdraw their benefits all at once and provoking cash shortages and chaos that lasted nearly a week. What motivated the incident was not an organized social movement but a rumor. Janúbia Silva Alves, a Bolsa Família recipient, described it to a reporter:

They're informing us in my community that the government is going to pay the next 3 months by the end of Sunday, and then cancel everything. My neighbor, who already got her money, said that the government wants to save money so that they can hold the parties for the Pope.	Estão avisando na minha comunidade que o governo vai pagar os próximos 3 meses até o final de domingo e cancelaria tudo. A minha vizinha, que já pegou o dinheiro dela, disse que o governo quer economizar dinheiro para conseguir fazer as festas para o papa (Rangel 2013).

The rumor, with its astonishing speed and reach, captivated media attention and led to official reassurances and worried investigations. Bolsa Família's own beneficiaries had managed eloquently to speak the truth about the program: that it operated by keeping them in an anxious state of permanent impermanence (Morton 2013b), always ready to withdraw their money lest it vanish.

Given the copious governmental efforts to mark Bolsa Família as a non-right, the news of its sudden demise seemed entirely credible to recipients. Different versions of the rumor offered different explanations. The money was needed, some said, to fund the Pope's visit; according to others, the cash would go towards the Confederations Cup soccer tournament. Indeed, when massive student protests broke out a mere 3 weeks later – Brazil's largest demonstrations in 20 years – the same theme of excessive spectacle would reappear. The students, represented by no coherent movement, excoriated the government for privileging the World Cup and the Olympics over fundamental rights to health and education. Despite the lack of a social movement to frame the issue, both the students and the Bolsa Família beneficiaries were challenging the post-Washington Consensus model of development without permanence. They were leaving open a space, a space in which it was possible to stand and call for more universal social guarantees.

Between bank rushes and student protests, Brazil's streets, in mid-2013, became an atelier for the refashioning of the notion of citizenship. Over the prior 30 years, 'citizenship' had served as a keyword for the construction of a post-dictatorial political imaginary, and the word had belonged inside the privileged vocabulary of organized social movements (Cardoso 1988; Holston 2008). Suddenly, chaotically, for no clear reason and with no clear movements, citizenship was again up for grabs.

In this paper I have argued that Bolsa Família, which has proven so vital in the livelihoods of one quarter of the Brazilian population, has also worked to confound social movements. Bolsa Família has opened up a new channel for linking millions of citizens to the nation. It has organized this channel according to the temporality of transience. And for the people who use this channel, Bolsa Família has sutured together a new subject-position: the manager's position.

But one can reject the role of the manager. Franciely, during the MST training session, had done just that. When Otilo asked her where Bolsa Família came from, she stated:

Franciely: The taxes that we pay on everything [....] The people who manage it are the government officials.
Otilo: Who manage it.
Franciely: But the money is mine.

Identifying the government officials as the true managers, Franciely finds space to think of herself as an owner. Her declaration, so unfamiliar in its logic, provoked nervous laughter from the crowd. If it had provoked something deeper, perhaps it could have marked the beginning of a way to dream of Bolsa Família in the terms of permanence.

Acknowledgements

I gratefully acknowledge the support of the Social Science Research Council and the Inter-American Foundation for this research. Thanks go to Rebecca Tarlau, Anthony Pahnke, William Sites, Julie Chu, E. Summerson Carr, Dain Borges and Jean Comaroff for their insightful comments. This paper is dedicated to the person whose wide-ranging lectures and far-reaching advice first made me love anthropology, Professor Enrique Mayer.

References

Ballard, R. 2013. Geographies of development II: cash transfers and the reinvention of development for the poor. *Progress in Human Geography* 37, no. 6: 811–21.
Ban, C. 2013. Brazil's liberal neo-developmentalism: new paradigm or edited orthodoxy? *Review of International Political Economy* 20, no. 2: 298–331.
Bauman, Zygmunt. 2000. *Liquid modernity*. Cambridge: Polity.

Bohn, S.R. 2011. Social policy and vote in Brazil: Bolsa Família and the shifts in Lula's electoral base. *Latin American Research Review* 46, no. 1: 54–79.

Branford, S., and J. Rocha. 2002. *Cutting the wire: the story of the landless movement in Brazil.* London: Latin America Bureau.

Canêdo-Pinheiro, M. 2009. Bolsa Família ou desempenho da economia? Determinantes da reeleição de Lula em 2006. *Encontro Nacional de Economia* 37: 1–20.

Cardoso, R.C.L. 1988. Os movimentos populares no contexto da consolidação da democracia. In *A democracia no Brasil: dilemas e perspectivas*, ed. F.W. Reis and G. O'Donnell. São Paulo, SP: Vértice, 368–81.

Casa Civil (Presidência da República, Brasil). 1993. Lei Orgânica de Assistência Social (Lei No 8.742, de 7 de Dezembro de 1993), http://www.planalto.gov.br/ccivil_03/leis/l8742.htm (accessed April 24, 2014).

Casa Civil (Presidência da República, Brasil). 2004. Decreto No 5.209 de 17 de Setembro de 2004, http://www.planalto.gov.br/ccivil_03/_ato2004-2006/2004/decreto/d5209.htm (accessed April 24, 2014).

De Brauw, A., D.O. Gilligan, J. Hoddinott, V. Moreira, and S. Roy. 2010. *Avaliação do Impacto do Bolsa Família 2: Implementation, attrition, operations results, and description of child, maternal and household welfare.* Washington DC: International Food Policy Research Institute, http://scholar.google.com/scholar?hl=en&q=Avalia%C3%A7%C3%A3o+do+Impacto+do+Bolsa+Fam%C3%ADlia+2%3A+Implementation%2C+attrition%2C+operations+results%2C+and+description+of+child%2C+maternal+and+household+welfare&btnG=&as_sdt=1%2C14&as_sdtp= (accessed August 28, 2010).

de la O, A.L. 2013. Do conditional cash transfers affect electoral behavior? Evidence from a randomized experiment in Mexico. *American Journal of Political Science* 57, no. 1: 1–14.

de Paula, R. 2009. Bolsa Família completa seis anos com investimentos de R$ 52,7 bilhões. 10/20/2009. Brasília: Presidência do Brasil, http://www.fomezero.gov.br/noticias/bolsa-Família-completa-seis-anos-com-investimentos-de-r-52-7-bilhoes (accessed August 1, 2010).

Fortes, A., and J. French. 2012. A 'Era Lula', as eleições presidenciais de 2010 e os desafios do pós-neoliberalismo. *Tempo Social* 24, no. 1: 201–28.

Gecas, V. 2000. Value identities, self-motives, and social movements. In *Self, identity, and social movements*, ed. S. Striker, T.J. Owens, and R.W. White, Minneapolis, MN: Minnesota. 93–109.

Gomes, S.S.R. 2010. Notas preliminares de uma crítica feminista aos programas de transferência direta de renda – o caso do Bolsa Família no Brasil. *Textos & Contextos (Porto Alegre)* 10, no. 1: 69–81.

Gruenberg, C. 2010. Pobreza, género y derechos en las políticas contra la pobreza. Conectando el género y el clientelismo en los programas de transferencias condicionadas. Paper presented at conference 'Clientelismo político, políticas sociales y la calidad de la democracia: Evidencia de América Latina, lecciones de otras regiones.' Nov. 5–10, Quito, Ecuador.

Hoddinott, J., and L. Haddad. 1995. Does Female Income Share Influence Household Expenditures? Evidence from Cote d'Ivoire. *Oxford Bulletin of Economics and Statistics* 57: 77–96.

Holston, J. 2008. *Insurgent citizenship: Disjunctions of democracy and modernity in Brazil.* Princeton: Princeton.

Kerbauy, M.T.M. 2011. Os programas de transferência de renda e o voto regional nas eleições presidenciais de 2010. *Opinião Pública* 17, no. 2: 477–92.

Lavinas, L. 2007. Gasto social no Brasil: programas de transferência de renda versus investimento social. *Ciência & Saúde Coletiva* 12, no. 6: 1463–76.

Lavinas, L., and M. Nicoll. 2006. Pobreza, transferências de renda, e desigualdades de gênero: conexões diversos. *Parcerias Estratégicas* 22: 39–76.

Lindert, K., A. Linder, J. Hobbs, and B. de la Brière. 2007. *The nuts and bolts of Brazil's Bolsa Família program.* Washington, DC: World Bank.

Luccisano, L. 2006. The Mexican Oportunidades program: Questioning the linking of security to conditional social investments for mothers and children. *Canadian Journal of Latin American and Caribbean Studies* 31, no. 62: 53–85.

Mansor de Mattos, F.A., and M.D. Carcanholo. 2012. Amenazas y oportunidades del comercio brasileño con China – lecciones para Brasil. *Revista Problemas del Desarrollo* 168, no. 43: 117–45.

Mariana, S.A., and C.M. Carloto. 2009. Gênero e Combate à Pobreza: Programa Bolsa Família. *Estudos Feministas (Florianópolis)* 17, no. 3: 901–8.

MDS (Ministério do Desenvolvimento Social e Comabte à Fome, Governo Federal, Brasil). 2013. Bolsa Família, http://www.mds.gov.br/bolsafamilia (accessed April 15, 2013).

Moeller, K. 2013. Searching for adolescent girls in Brazil: Corporate development and the transnational politics of poverty. *Feminist Studies*, forthcoming.

Molyneux, M. 2009. Conditional Cash Transfers: A 'Pathway to Women's Empowerment? *Pathways Brief* 5.

Morton, G.D. 2013a. Acesso à permanência: diferenças econômicas e práticas de gênero em domicílios que recebem Bolsa Família no sertão baiano. *Política e Trabalho* 1, no. 38: 43–67.

Morton, G.D. 2013b. O verdadeiro culpado do boato sobre o Bolsa Família. *Brasil de Fato*, 11 July.

Neri, M. 2010. *A nova classe média: o lado brilhante dos pobres.* Rio de Janeiro: Fundação Getulio Vargas.

Ondetti, G. 2008. *Land, protest, and politics: the landless movement and the struggle for agrarian reform in Brazil.* University Park, PA: Penn State Press.

Peixoto, V., and L. Rennó. 2011. Mobilidade social ascendente e voto: as eleições presidenciais de 2010 no Brasil. *Opinião Pública* 17, no. 2: 304–32.

Pires, F.F. 2009. A casa sertaneja e o Programa Bolsa-Família: Questões para pesquisa. *Política e Trabalho* 27: 1–15.

Pires, F.F. 2013. Comida de criança e o Programa Bolsa Família: Moralidade materna e consumo alimentar no semiárido. *Política e Trabalho* 38: 123–35.

Raney, A., and C. Heeter. 2005. Rough Cut- Brazil: Cutting the Wire. Witnessing a Land Occupation. PBS/ Frontline, December 13, http://www.pbs.org/frontlineworld/rough/2005/12/brazil_cutting.html (accessed May 16, 2011).

Rangel, S. 2013. Visita do papa seria uma das causas para suspensão do Bolsa Família, segundo boatos. Folha de São Paulo, 19 May.

Rasella, D., R. Aquino, C.A.T. Santos, R. Paes-Sousa, and M.L. Barreto. 2013. Effect of a conditional cash transfer programme on childhood mortality: A nationwide analysis of Brazilian municipalities. *Lancet* 382, no. 9886: 57–64.

Rego, W.L., and A. Pinzani. 2013. *Vozes do Bolsa Família: Autonomia, dinheiro, e cidadania.* São Paulo: Unesp.

Rennó, Lucio R., and Hoepers, Bruno. 2010. Voto estratégico punitivo: transferência de votos nas eleições presidenciais de 2006. *Novos Estudos – CEBRAP* 86: 141–61.

Shikida, C.D., L.M. Monasterio, A.F. de Araujo Jr, A. Carraro, and O.M. Damé. 2009. 'It is the economy, companheiro!': An empirical analysis of Lula's re-election based on municipal data. *Economics Bulletin* 29, no. 2: 976–91.

Snow, D.A. 2004. Framing processes, ideology, and discursive fields. In *The Blackwell companion to social movements*, ed. D.A. Snow, S.A. Soule, and H. Kriesi, 380–413. Oxford: Blackwell.

Suárez, M., and M. Libardoni. 2007. O impacto do Programa Bolsa Família: mudanças e continuidades na condição social das mulheres. In *Avaliação de políticas e programas do MDS – resultados volume 2 – Bolsa Família e assistência social*, ed. J. Vaitsman and R. Paes-Sousa. Brasília/ DF: Ministério do Desenvolvimento Social e Combate à Fome, 119–62.

Thomas, D. 1990. Intra-Household Resource Allocation: An Inferential Approach. *The Journal of Human Resources* 25: 635–44.

Wolford, W. 2010. *This land is ours now: social mobilization and the meanings of land in Brazil.* Durham, NC: Duke University Press.

Zucco, C. 2008. The President's New' Constituency: Lula and the Pragmatic Vote in Brazil's 2006 Presidential Elections. *Journal of Latin American Studies* 40, no. 1: 29–49.

Index

Note: Page numbers in *italics* represent tables
Page numbers in **bold** represent figures
Page numbers followed by 'n' refer to notes

ABRA 52n
accountability 72–4
activism 13, 77–81; autonomy 8; characteristics
7–9; dispossession 8–9; networks 7;
state-society collaborations 7–8; territorial
development 8
activists 122, 123; educational 93–4;
settlements 18–19
African Diaspora 167
African slavery 167, 204n
Africans 158; protection of rights 161
Afro-Brazilian communities 203n
agrarian struggles 42
agribusiness 103–5
agricultural credit 150–1
agricultural labor movement 41
agricultural production 20
Agricultural Production Cooperatives (CPA)
28, 28n, 31
agroecological programs: place-based education
112
agroecology 112
alliances 198–200; labor-peasant tensions 95–7;
networks 179; state-society 98–9
Almeida, A.W.B. 157, 164
alternative economies 9; development strategies
10–11
Amazon 122
Ansell, A. 195–214
Argentina: poor people's politics 198
Arruti, J.M.A. 203n
associations 143–4
autonomy: activism 8

Bahia: southern 133–56
Barcelos 180–8, **181**; indigenous mobilization
84–8; migration history 186; population 184,
184
Barcelos Indigenous Association (ASIBA) 185,
186, 187, 188

Barth, F. 166
Belo Monte dam: fisherman 81; groups related
to *76–7*; political voice 69–88; river changes
81; rivers 81
benefit amount: changes in 227–30
Bionatur 29
black: meaning of 204n
Black Experimental Theatre (TEN) 160
Bolsa Família 13
Brazil Quilombola (PBQ) 162
Brazilian Anthropological Association (ABA)
161–2
Brazilian Black Front (FNB) 160
Brazilian Institute of Geography and Statistics
(IBGE) 157
bureaucracy: policy 99–102

cacao lands: southern Bahia 133–56
Calha Norte (CN) 178
capital: accumulation 4; social 176
Cardoso, F.H. 25
Casca 168–70
centrality: persistence 21
civil society 75–81
civil society organizations (CSOs) 91
coalition building: framing 94–5
coercion 133–56
coherence 34
Cold War: and Cuban Revolution 46
collectivization: land 139
colonato (labor relations system) 42
Comerford, J.C. 206n
commodity agriculture 2
communism: strategy 44
Communist Party 43, 44, 45, 52
communities 73, 154; Afro-Brazilian 203n;
development projects 70; quilombola 157–72,
164; reform 137–8; self-identification 203
*Confederación de Nacionalidades Indígenas del
Ecuador, La* (CANAIE) 5

239

www.ingramcontent.com/pod-product-compliance
Ingram Content Group UK Ltd.
Pitfield, Milton Keynes, MK11 3LW, UK
UKHW010020280225
455677UK00023B/696